Food, Energy and Water Sustainability

Societies around the world face an increasingly uncertain future as social and ecological changes create pressure on resource governance, and this uncertainty calls for new models that illuminate the intersections of civil society, public sector and private sector resource management. This volume presents a diversity of collaborations between various governance actors in the management of the Food-Energy-Water (FEW) nexus. It analyses the ability of emergent governance structures to cope with the complexity of future challenges across FEW systems.

Divided into two sections, chapters in the first half of the book present a collection of case studies from around the world exemplifying how FEW nexus challenges are addressed in a multitude of ways and by a variety of actors,. Chapters in the second half offer broader perspectives on the management of FEW and underline the lessons that emerge from applying a FEW lens to the question of natural resource governance.

The varied examples in this book highlight that the management of FEW is often a question of reinventing, adapting, and building upon existing practices. Such practices are deeply embedded in unique socio-cultural, environmental, and political contexts as well as 'hard' infrastructures. Most of all, this edited volume seeks to communicate the wealth of ideas from committed individuals who continue to work to improve natural resource governance and our sustainable futures.

Laura M. Pereira is a post-doctoral research fellow at the University of Cape Town, South Africa and a researcher at the Centre for Complex Systems in Transition, Stellenbosch University, South Africa.

Caitlin A. McElroy is a Departmental Research Lecturer at the Smith School for Enterprise and the Environment and the School of Geography and the Environment, University of Oxford, UK.

Alexandra Littaye is a researcher at the School of Geography and the Environment, University of Oxford, UK.

Alexandra M. Girard is an independent researcher and consultant in environmental management and gender issues based in Sydney, Australia.

Earthscan Studies in Natural Resource Management

Global Resource Scarcity
Catalyst for conflict or cooperation?
Edited by Marcelle C. Dawson, Christopher Rosin and Nave Wald

Food, Energy and Water Sustainability
Emergent Governance Strategies
Edited by Laura M. Pereira, Caitlin A. McElroy, Alexandra Littaye and Alexandra M. Girard

Environmental Justice and Land Use Conflict
The governance of mineral and gas resource development
Amanda Kennedy

Soil and Soil Fertility Management Research in Sub-Saharan Africa
Fifty years of shifting visions and chequered achievements
Henk Mutsaers, Danny Coyne, Stefan Hauser, Jeroen Huising, Alpha Kamara, Generose Nziguheba, Pieter Pypers, Godfrey Taulya, Piet van Asten and Bernard Vanlauwe

Transformational Change in Environmental and Natural Resource Management
Guidelines for policy excellence
Edited by Michael D. Young and Christine Esau

Community Innovations in Sustainable Land Management
Lessons from the field in Africa
Edited by Maxwell Mudhara, Saa Dittoh, Mohamed Sessay, William Critchley and Sabina Di Prima

The Water, Food, Energy and Climate Nexus
Challenges and an agenda for action
Edited by Felix Dodds and Jamie Bartram

For more information on books in the Earthscan Studies in Natural Resource Management series, please visit the series page on the Routledge website: www.routledge.com/books/series/ECNRM/

Food, Energy and Water Sustainability

Emergent Governance Strategies

Edited by Laura M. Pereira,
Caitlin A. McElroy, Alexandra Littaye
and Alexandra M. Girard

Routledge
Taylor & Francis Group
LONDON AND NEW YORK

earthscan
from Routledge

First published 2018 by Routledge

2 Park Square, Milton Park, Abingdon, Oxfordshire OX14 4RN
52 Vanderbilt Avenue, New York, NY 10017

Routledge is an imprint of the Taylor & Francis Group, an informa business

First issued in paperback 2019

British Library Cataloguing-in-Publication Data
A catalogue record for this book is available from the British Library

Library of Congress Cataloging-in-Publication Data
A catalog record for this book has been requested

ISBN: 978-1-138-90409-5 (hbk)
ISBN: 978-0-367-35228-8 (pbk)

Typeset in Goudy
by Apex CoVantage, LLC

Contents

Foreword

First of all, I would like to congratulate Laura, Caitlin, Alex and Alexa on editing this volume. Their book is the culmination of dedicated work by four exceptional DPhil graduates of the University of Oxford's School of Geography and the Environment. It is always gratifying to see the accomplishments of early career-researchers; I commend the editors for the vision and hard work that has resulted in this timely volume. It is an important book for many reasons: not only does it provide insights into important questions about sustainability governance in the age of the Anthropocene, it also provides an opportunity to engage with the cutting-edge work of young researchers with their fingers on the pulse of what is new and innovative.

This book comes at a critical time. Questions of environmental and resource governance have been made acute with the shift in direction by the American administration. Declining trust in government is a global trend that further complicates issues of governance whilst threatening the very fabric of democracy. Despite the progress made since the Paris Accords, uncertainty about national and international governance as well as corporate environmental management have increased. Environmental and natural resource ministries have struggled to address these issues; made more problematic by the siloed nature of the resource portfolios they manage. The complex interdependencies of water, energy and food are challenging for governments, corporations, and societies. The sustainable management of these resources requires a fresh approach. This book provides examples of how to rethink governance.

We are heading towards an uncertain future where socio-ecological changes are driving the pressure for new models of resource governance. Each chapter in this book has an important lesson for how to improve resource governance, showing examples from around the world. Drawing from their interdisciplinary group of scholars, this book is a must-read for any practitioner or academic who is concerned with the governance of our global food, energy and water systems.

Prof Gordon L. Clark, July 2017

Acknowledgements

Thank you to all of our external review editors, without whose insightful inputs this book would never have been completed.

Prof. Sharmila Murthy, Suffolk University
Prof. Ralph Hamann, University of Cape Town
Dr Scott Moore, World Bank
Prof. Thomas Thornton, University of Oxford
Dr Sarah McGill, University of Oxford
Ms. Philile Mbatha, University of Cape Town
Prof. Gordon Clark, University of Oxford
Prof. Bill Freund, University of KwaZulu Natal
Dr Scott Drimie, Stellenbosch University
Dr Rafael Calderón-Calderas, Universidad Autónoma Metropolitana
Dr Timos Karpouzoglou, Wageningen University
Ms. Erin O'Donnell, Melbourne Law School
Dr Alvar Closas, International Water Management Institute

We also would like to extend a really big thank you to two interns in South Africa who helped with the formatting and editing of the chapters, Monique Classen and Andisiwe Mnkani. A very special thank you must also go to Jonathan Hall who slaved away getting the references in order and to Tomas Chaigneau for emergency sense-checking.

A very big thank you to you all for your contribution in making this book a reality!

Contributors

Dr Tomas Chaigneau
University of Exeter, Environment and Sustainability Institute

Tomas is an interdisciplinary social scientist interested in the relationship between people's wellbeing and the natural environment. This entails furthering our understanding of how coastal ecosystem services can contribute to material, social or relational wellbeing to identify opportunities for sustainable poverty alleviation or prevention. His research also focuses on interpreting and anticipating people's responses to natural resource management interventions along the coast. In particular, this has led him to investigate what predicts the support of individuals and communities towards the use of marine protected areas.

Jessica Cockburn
Department of Environmental Science, Rhodes University

Jessica Cockburn is a PhD candidate in the Department of Environmental Science at Rhodes University, Grahamstown, South Africa. Her current research focuses on understanding the enablers and barriers of stewardship for sustainable agriculture in multifunctional landscapes. The research takes a social–ecological system view of agriculture, and Jessica has a particular interest in the role of collaboration among diverse stakeholders across the landscape in enabling more systemic shifts towards stewardship of ecosystem services. She previously worked with WWF-South Africa in their Sustainable Agriculture Programme, and has also worked as a research fellow in Land Use Planning and Management at the University of KwaZulu-Natal.

Dr Samir Doshi
US Agency for International Development

Samir is a senior Scientist at the USAID Global Development Lab. He leads the initiative on Adaptive Programming Using Real-Time Data in how the Agency's programme delivery and implementation can be more adaptive, iterative and agile to promote a learning culture for improved decision making and outcomes. He also supported the Global Development Lab's work on health systems strengthening in the Ebola response in West Africa. He holds a PhD from the University of Vermont.

Professor Nick Eyre

Environmental Change Institute, University of Oxford, UK

Professor Nick Eyre is a Jackson Senior Research Fellow in Energy at the ECI and Oriel College. He also leads the ECI programme on Lower Carbon Futures. He is a Co-Director of the multi-university collaboration, the UK Energy Research Centre, leading its research work on decision-making. He is a Co-Director of the Oxford Martin Programme on Integrating Renewable Energy.

Dr Alexandra M. Girard

Independent Consultant in Gender and Natural Resources Management

Alexandra M. Girard is an independent consultant whose work focuses on the confluence of the socio-political and socio-cultural aspect of natural resources management. She specialises in questions of gender equity in the management of natural resources. Alexandra received her DPhil of Geography and MSc in Water Science, Policy, and Management from the University of Oxford as well as her MEng in Environmental Engineering from the University of Nottingham. Alexandra lived in Australia at the time of this book's publishing where she developed a keen interest in social and cultural influences on the Australian water markets.

Professor Jim W. Hall

Environmental Change Institute, University of Oxford, UK

Professor Jim Hall is Director of the Environmental Change Institute and is Professor of Climate and Environmental Risks in the University of Oxford. His research focuses upon management of climate-related risks in infrastructure systems, in particular relating to various dimensions of water security, including flooding and water scarcity. Professor Hall is a member of the UK independent Committee on Climate Change Adaptation. In 2010 he was elected as a Fellow of the Royal Academy of Engineering.

Dr Agostino Inguscio

Department of History, University of Cape Town

The author was a lecturer in Economic History at the University of Cape Town between May 2014 and October 2015, while this research was conducted. He joined UCT after having worked as a Post Doc in Economic History at Yale University. He holds a DPhil in Economic and Social History and a MSc in Economic and Social History from the University of Oxford.

Dr Timothy Karpouzoglou

Public Administration and Policy Group, Wageningen University

Dr Timothy Karpouzoglou is based at the Public Administration and Policy group at Wageningen University & Research, the Netherlands. He is currently working as a post-doctoral researcher for the ESPA project Mountain EVO "Adaptive governance of mountain ecosystem services for poverty alleviation enabled by environmental virtual observatories". Timothy has a strong interest in the social processes shaping ICTs as part of improving food, energy and water sustainability. Timothy has conducted research in diverse fields relating to sustainability including water resources management, environmental governance, resilience and political ecology with a particular emphasis on the Global South. Prior to his current position, he also worked at the Stockholm Resilience Centre and has obtained his PhD from the STEPS Centre, University of Sussex.

Vaughan Koopman
WWF South Africa

Vaughan Koopman has worked in the conservation NGO sector for 15 years and holds a BTech in Nature Conservation and a Masters Degree in Agriculture. As a social-ecologist he works with a range of stakeholders in the plantation timber, sugar, dairy, agri-business and government sectors on integrating water stewardship into business practice across multiple scales. Prior to this he worked for the Department of Water Affairs and the Department of Agriculture in South Africa on water quality and agricultural natural resource management.

Xiawei Liao
Environmental Change Institute, University of Oxford, UK

Xiawei Liao is a CSC (China Scholarship Council) – Oxford Scholar and a doctorate student working at the Environment Change Institute of the University of Oxford with Professor Jim W. Hall and Professor Nick Eyre. His work focuses on the future conflicts between water availabilities and water demands by China's thermoelectric power sector.

Dr Alexandra Littaye
Smith School of Enterprise and the Environment, University of Oxford

Alexandra is a Research Assistant at the Smith School of Enterprise and the Environment at the University of Oxford. She is also the founder of Azure, which supports Mexican indigenous farmers by selling pinole in Europe. She is particularly interested in the interplay between environmental and financial sustainability in business, and the narratives that underpin notions of food sovereignty around the world. Before joining Oxford's School of Geography and the Environment, Alexandra was a lecturer of Philosophy at the United Arab Emirates University in Abu Dhabi.

Dr Caitlin A. McElroy
Smith School of Enterprise and the Environment, University of Oxford

Dr Caitlin McElroy is a Departmental Research Lecturer in Enterprise and the Environment at the School of Geography and the Environment and the Smith School of Enterprise and the Environment at the University of Oxford. Caitlin is also the Director of the Smith School's Executive Education Programmes. Her research is concerned with the relationship between mining, the environment, and development from an economic geography perspective. She is particularly interested in the role of new institutional forms of mediating and managing financial investments related to the environment, development, and water infrastructure legacies.

Dr Alex Money
Smith School of Enterprise and the Environment, University of Oxford.

Alex Money is a research and teaching fellow at the University of Oxford. A former fund manager, he has over 15 years of practitioner experience in investment and industry. Alex is a programme director at Oxford's Smith School of Enterprise and the Environment, where his research interests include corporate water strategy, infrastructure, investment, renewables and development. He focuses on the opportunities for empirical research to bridge knowledge gaps between academia and industry. Alex also teaches on Oxford's undergraduate, postgraduate and MBA programmes, and is the director of a technology company that is backed by the University's seed fund. He received his master's and doctoral degrees from the University of Oxford.

Jaci van Niekerk
Department of Environmental and Geographical Science, University of Cape Town

Jaci van Niekerk is a Junior Research Fellow at the University of Cape Town. Much of her research is located within the social and environmental dimensions of economies based on biodiversity, with a particular focus on social justice. She has conducted research into the commercial use of indigenous South African plant species and the traditional knowledge associated with them, examining inequalities and blockages to benefit sharing in international trade chains. She has also investigated ways in which to protect, promote, and enhance the rights of small-scale farmers.

Dr Laura M. Pereira
Department of Environmental and Geographical Sciance, University of Cape Town, South Africa; Centre for Complex Systems in Transition, Stellenbosch University, South Africa

Originally from Johannesburg, South Africa, Laura's current position is as a researcher at the Centre for Complex Systems in Transition at Stellenbosch University working on sustainability transformations under the Guidance for Resilience in the Anthropocene: Investments of Development (GRAID) project. However, at the time of writing this book, she was a post-doctoral research fellow under the Bioeconomy chair at the University of Cape Town. Laura holds a BSc in Zoology, Ecology and Law from the University of the Witwatersrand and an MSc and DPhil from the University of Oxford. Laura has worked in South Africa, Brazil, Colombia, Mozambique, Nigeria and Kenya and her research focusses on the interface between development and environmental change. Her work is strongly embedded within a social–ecological systems approach and she is keen on exploring alternative methods in sustainability research. In particular, she looks at the interface between traditional knowledge systems and more formal systems of innovation with regards to food, nutrition and agriculture.

Brice Perombelon
School of Geography and the Environment, University of Oxford

Brice Perombelon is a Doctoral researcher at the School of Geography and the Environment, University of Oxford. His research seeks to find ways to understand and convey indigenous representations of geopolitics and geopower. It is based on a case-study of an Athapaskan community located in the Canadian High North.

Brice holds an MSc degree in Environment and Development from the London School of Economics and Political Science, a BA in Political Science and a double MA in International Relations and Geopolitics and Geostrategy from Sciences Po, Bordeaux France. He began his career working in the French Foreign Affairs Ministry.

He has taught international relations at University College London and postcolonial geopolitics at the University of Oxford.

Melissa M. Rohde
Department of Civil & Environmental Engineering, Stanford University, 473 Via Ortega, Stanford, California, USA

Melissa M. Rohde is a Groundwater Scientist for The Nature Conservancy and uses her expertise in biology, hydrology, and water policy to advance and advise sustainable groundwater management policies. Melissa has more than 10 years of international scientific research experience in Antarctica, Europe, South Asia, and the USA ranging across the fields of paleoclimatology, biogeochemistry, and hydrogeology. Her work has been featured in both peer-reviewed scientific journals and media outlets including the *San Francisco Chronicle*, *Los Angeles Times*, *Huffington Post*, *The Guardian*, *The New York Times*, and *USA Today*. Melissa has degrees in MSc Environmental Engineering & Science from Stanford University, MSc Water Science, Policy & Management from the University of Oxford, and BSc in Biology & Oceanography from the University of British Columbia. You can follow Melissa's musings on water, climate change, and the environment on her blog: www.reflectionsonwater.org.

Michael Valli
Aquasitions Group

Michael Valli is founder of Aquasitions Group, an investment firm focussed on global water scarcity. He previously managed a $100m+ portfolio of water and agriculture assets on behalf of institutional clients at an ASX-listed fund management company. Outside the financial world, Michael publishes academic work on food–energy–water issues using his background in economic geography. He holds a Master of Science (Water Science Policy and Management) from the University of Oxford and Bachelor of Science (Physics, Mathematics) from the University of Sydney.

Swathi Veeravalli
Geospatial Research Laboratory, Engineer Research and Development Center, US Army Corps of Engineers.

Swathi Veeravalli is an interdisciplinary research scientist with the US Army Corps of Engineers. Since 2010, she has been Principal Investigator on several basic and applied research projects where she focuses on developing the capability to better identify the impact of climate variability upon humans and the environment. Swathi is currently on detail to the US Agency for International Development's Office of Conflict Management and Mitigation. At USAID, she serves as a Climate and Environmental Security Advisor and focuses on maximising interagency learning and information between USAID and USACE while expanding her understanding of USAID's analysis tools and responses as well as policies and programmes in the context of complex emergencies involving violent conflict. Swathi also has several years' experience in international development, youth engagement and grassroots advocacy. She concentrates on understanding the dynamic role of governance and fragility as they affect and are affected by the human–water–climate nexus. She holds a Master's of Science in Water Science, Policy & Management from the University of Oxford. Swathi also holds a BA in International Affairs with concentrations in African Politics and Development from the George Washington University. She has worked and travelled in over 77 different countries.

1 An introduction to food–energy–water nexus thinking and sustainability governance

Laura M. Pereira and Caitlin A. McElroy

Introducing the book

This book explores emerging strategies for governing the natural resources of food, energy, and water. It is only since the 1960s that the governance of these resources has been formally addressed in many Western societies and although there have been great strides in environmental governance, our current approach is still insufficient. Globally we are struggling to provide food, energy, and water resources in ways that allow for equitable use, continuing development, and a sustainable future. Some of the challenges to achieving better environmental governance exist because the governance of food, energy, and water resources is inextricably tied up with the larger politics of understanding the sustainability and value of the environment, or, as Dryzek (1997) calls it, the politics of the Earth. However, the pressure to succeed in governing these resources has never been so urgent. Our food, energy, and water resources, which we often refer to as FEW from here onwards, are stressed to near breaking point, primarily from human impacts. Their failure would have profound consequences for the environment as well as for human wellbeing. It is in the context of this urgency that the chapters in this book seek to find new ways of thinking about how we can govern FEW resources.

In this introduction we present a short background to the science of sustainability and the many ways it is connected to governance. We have reduced this process to a rather straightforward hierarchy of theory and the practices operating from governance to sustainability. In the chapters that follow, we also provide illustrations of the politics of the Earth at work in myriad ways across different places and cultures. It is, however, the commonalities between case studies and the ways in which each author speaks to emerging governance strategies from a "nexus thinking" approach that links the focus of the chapters across the book. This is an approach that focuses on the complexity of connections between systems. Food, energy, and water systems are connected to each other in complex ways and thus we can call the interactions between these systems a nexus. In this book, we posit that the nexus of food, energy, and water systems is the central and fundamental feature of environmental sustainability.

This nexus approach and nexus thinking is simultaneously emerging in several fields of research as a response to the complexity of addressing global sustainability.

It appears to offer a perspective that appreciates the interconnectivity of complex systems and thus has the potential to help create ways to govern this complexity. As a result, we think it is important to explore its potential and to see what it might take to enable more nexus-based approaches to environmental governance. With the growth of the nexus-based approach there are also significant critiques that it is just another trend, and while time will tell if this is true for the language of the 'nexus', we believe that the underlying challenge of governing the relational features of interconnected systems will remain.

Further, the nexus approach can speak to the institutional complexities entangled within FEW resource systems. The institutions that have historically been responsible for managing human welfare are poorly positioned to take the necessary action. Governments around the world struggle to keep pace with environmental change and many lack the resources to employ even basic resource management, let alone innovative approaches. The private sector has developed corporate social and environmental responsibility activities that have worked in appeasement to environmental concerns. However, companies are only now recognising the fundamental changes in practice that will be necessary to conduct business in a world of growing environmental constraints. Communities, Non-Governmental Organisations (NGOs), and other stakeholders are becoming increasingly organised and influential in the areas in which they operate and over their issues of concern. Separately, these efforts are too little and perhaps too late. Through the nexus approach and its integrative governance options, there is an opportunity to integrate these groups and their efforts. We hope therefore to offer not only examples of best practice responses to complex sustainability challenges, but also to articulate some insights into the processes of implementing a multi-stakeholder approach to environmental governance.

With an ambition to influence and improve environmental governance, we have targeted this book to all those that might have a role in fostering and adopting improved environmental governance processes and especially for those concerned with the connections between food, energy, and water systems. We anticipate that other academics, researchers, policy makers, consultants, and environmental managers from many sectors will find this book and its case studies informative. Whilst the chapters are framed by scientific and academic literature, they also focus on the what, how, and why surrounding the actual practice of FEW governance. Thus, we hope that practitioners will also gain a greater conceptual understanding of the problems and solutions presented and that this understanding will improve their ability to adapt governance ideas to their specific needs. For those often removed from practice, we hope that the tangibility of practical examples of developing and deploying innovative governance approaches is useful.

The remainder of this introduction presents a primer to understand the nexus-based approach to resource governance used in this book. It starts with an overview of the state of environmental change and sustainability science. It follows this with a complementary section on the history and current approach to environmental governance. We then introduce the chapters to come and explain what they each add to the overall ambitions of this book.

Theoretical approaches

The discourse surrounding environmental sustainability has evolved over the past decades and provides an excellent foundation from which to conceptualise the nature of FEW governance challenges and the approach to potential solutions through a nexus-based approach. The planetary boundaries concept is a significant part of this discourse. It was developed by a group of scientists who suggested that the Earth has certain critical system thresholds such as climate, the nitrogen cycle, and freshwater, which, if transgressed, would be detrimental to the functioning of the planet. This would affect the wellbeing of the human species who have thrived under the relatively stable environmental conditions of the last 10,000 years (Rockström et al., 2009). In this book, we confront this tension of meeting the need for food, energy, and water within planetary boundaries that are already being breached. Society is facing a dilemma of increasing consumption dependent on a stable environment that is threatened by this very consumption (Leach, Raworth, & Rockström, 2013). It is now imperative that we find the resource governance relationships that can allow development to meet human wellbeing needs whilst not impacting the Earth irreparably. This can be described as a "wicked problem" (Rittel & Webber, 1973) or a "meta-problem" (Ramírez, Selsky, & van der Heijden, 2010). The efforts to solve this wicked problem drive the ideas behind the book chapters. Before we get to them however, the following theoretical framework provides some background into our existing understanding of the wicked problem of sustainable resource governance and why it is quite so wicked. First, we have tried to reduce a vast amount of research on environmental sustainability and governance into a rather simple hierarchy of concepts relevant to our approach to nexus-based governance. Then we go on a quick tour through the ways these concepts are integrated and deployed in governance practice. The chapters each draw from, critique, revise, and build on this existing knowledge with the examples they present.

The sustainable governance conceptual hierarchy

Figure 1.1 provides the map for the coming discussion. It reflects our conceptual perspective used throughout the book. It works to show the connections between broad sustainability, and its related discourses like planetary boundaries, to the decision making in complex systems, and then the governance approaches that hold it all together. It is represented as an inverted pyramid moving from a point of governance up to the breadth of sustainability. This is not to suggest that governance is less significant, but rather a matter of scale and impacts. A good strategic governance approach has the potential to broadly impact more widespread sustainability. In the next section we explain how each of these layers in the pyramid are conceptualised individually and, more importantly for the interests of this book, we explore how they are interconnected.

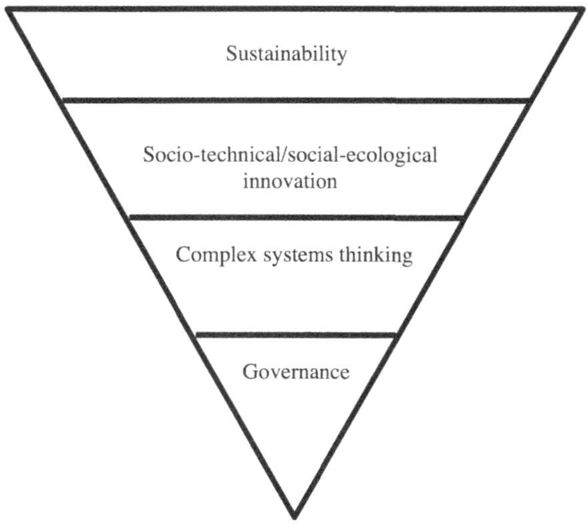

Figure 1.1 Hierarchy of theory: inverted pyramid

Sustainability: food, energy, and water (FEW) nexus thinking

We start exploring this pyramid at the widest level – that of sustainability. Within the discourse of planetary boundaries, food, energy, and water systems have been identified as critical due to their mutual dependence and projected scarcity, especially under the added pressure of climate change. The interdependencies of these systems and their associated risks were most eloquently elucidated by Beddington, who referred to this as "the perfect storm" facing humanity (2009). The concept of the food–energy–water nexus is rather self-evident. We are all aware of how these three systems are inter-dependent and how operating in one system often necessitates making a trade-off in another. For example, think of the trade-off between agriculture for use as food or for biofuels. These crops can be used as energy for our body or as energy for our cars. An example of a synergy is that of building dams for hydroelectric power generation that can also be a storage system for irrigation in agricultural systems – but this often leads to a a trade-off with biodiversity and ecosystem function. Further, there are some less direct, but equally important, resource inter-dependencies including how agriculture relies on the energy system, from powering tractors to synthesising fertiliser, and how important water is in energy generation, from hydroelectricity to burning coal where it is used as a cooling device. While these examples highlight the negative trade-offs between resource uses, there are also positive feedbacks that operate between these systems.

A positive feedback loop across systems means that an event in one resource system creates an amplified response in another resource system and that resource system then prompts an amplified response in the original system. This process leads to exponential changes in systems that are usually unforeseen. An example using the stressor of climate change illustrates the interconnectivities between the food, energy, and water systems and how changes in one system can result in non-linear feedbacks in the others.

In this example, the feedback loop starts with CO_2 emissions from the energy system leading to climate change impacts. These impacts are responsible for changes in rainfall and thus a decrease in agricultural production. A possible response to the increase in CO_2 emissions is the implementation of a biofuels policy, as was seen when the European Commission adopted their Renewable Energy Directive in April 2009. In this example, the intervention results in a market shift away from food crops like wheat, to energy crops like jatropha. This shift from food to energy crops can have a concomitant increase in food prices – as was seen in the 2008/2009 global food price crisis, (although this has been attributed to multiple factors, see Mittal, 2009). One possible response to rises in food prices is to increase production through agricultural intensification. This has the potential to lead to an increased uptake of water for irrigation as well as an increase in the use of fertilisers. As inorganic fertilisers are responsible for nitrous oxide emissions, this response creates a positive feedback loop that exacerbates the initial problem: the increase in greenhouse gas emissions that exacerbate the climate change stressor that the initial intervention was supposed to stop. The cycle starts again, but worse. The above example provides a hypothetical example of a simple positive feedback loop where climate change – the trigger that we tried to address – is amplified. It is a simple representation, yet it highlights the connections between food, energy, and water and how decisions made in one system can have often unforeseen long-term impacts in the other systems.

The growing risk associated with the inter-connectivity and inter-dependence of food, energy and water systems has caused organisations to reconsider their natural resource governance strategies; a shift highlighted in a 2014 report co-produced by South African Breweries (SAB) and the World Wide Fund for Nature (WWF) (SAB Miller & WWF, 2014). Two insights from this multi-country report are particularly relevant for setting up the theoretical underpinnings of this book. The first is that nexus issues play out very differently in contexts with different resource endowments: Brazil, with abundant natural resources, faces a different set of questions and challenges to a country like South Africa that has limited water and agricultural land, but vast reserves of coal and great potential for leveraging solar power (SAB Miller & WWF, 2014). The importance of a nexus understanding of natural resource governance is therefore not that it provides universal solutions, but that it enables a more holistic appreciation of the trade-offs and synergies that decision makers face when operating in these inter-connected systems. The second important insight is that, particularly at later stages of development, the inflexibility and archaic state of institutions can undermine countries' ability to respond to the complex interconnectivities of food, energy, and water systems as they face threats from shocks as diverse as climate change and demographic shifts. The lesson for developing countries is how to avoid institutional lock-in such that they cannot respond to the uncertainties of the 21st century (SAB Miller & WWF, 2014).

A fundamental shift that comes with nexus thinking is the imperative to break out of established siloes of how we think about different sectors. For example, at the national government level, there are usually separate ministries to deal with agriculture, water affairs, and energy. What the nexus demonstrates is that this segmentation of inter-dependent systems into different sectors is not helpful when it comes to long-term, strategic decision making. Furthermore, a lot about what makes the FEW nexus difficult to handle from a decision making perspective is that the food, energy, and water systems are all examples of social–ecological systems that contain characteristics of complex adaptive systems. What many of the chapters in this book show is that instigating a change in one system often has unintended and unanticipated consequences in a different system.

In the next section we move on from conceptualising sustainability through a framing of planetary boundaries and interconnected resource systems and responses. We go to the next level of the pyramid to understand social–ecological systems and socio-technical innovation. For this we outline some of the key reasons why strategically intervening in social–ecological systems can be so difficult, but also how under certain conditions, disruptive innovations can transition these systems onto more sustainable trajectories.

Social–ecological systems and socio-technical innovation

From our overview of how the broadest concepts of sustainability are related to the complexity of, and connections between, different resource systems, we move on to the next layer of the pyramid. Here we look to explain how we understand resource system complexity through the concept of social–ecological systems, as well as how socio-technical innovations are a useful way of thinking about how to improve the governance of social–ecological systems. Social–ecological systems (SES) are the complex articulation of social (e.g. economic, political, legal etc.) and ecological systems. Their complexity is created through multiple feedbacks present in the system at large. For example, the food system can be conceptualised as a social–ecological system: it consists of cross-scale, multi-level interactions between human and natural systems and has emergent properties arising from its dynamic, self-organising components (Ericksen, 2008; Folke, 2006). The complexities of social–ecological systems are part of what makes governing them for the already complex objective of sustainability so very challenging.

Carrying this food system example further, we see that there is a large technological component to the food system, particularly around the processing of agricultural products into food items. The intersection of social processes and technology is often truncated and referred to as something "socio-technical". In the case of a socio-technical innovation the innovation is not only a technological advancement, but also a social advancement in how that technology is used. The processing of food is often considered a socio-technical regime that has become established and embedded over time. Once embedded, transitioning a socio-technical regime towards a more sustainable alternative regime can be difficult and complicated – a characteristic that is demonstrated in some of our case studies (see Chapters 3, 4, and 5). Trying to understand socio-technical relationships in how we govern resources helps us to identify practices that are locked-in and path dependent, and perhaps also allows us to see where there is room for structural transformation in order to improve sustainability (Smith & Stirling, 2010).

To take our understanding of socio-technical innovation further, a conceptualisation of innovation over space and time is helpful. We understand that technology develops in an evolutionary manner over time as it builds on previous technologies through adjustments (Arthur, 2011). However, when we refer to socio-technical innovation we are not just considering how the technology itself is developed, but also the social and institutional spaces, such as the legal, political, economic, and environmental, in which the technology is developed and employed and how these structure its use. Socio-technical innovations, because of their combined social and technological considerations, can therefore be useful tools for policy and decision making related to the complexities inherent in social–ecological systems. These systems require more than just technological fixes for successful transformation. They require what Nelson and Nelson (2002) refer to as 'social technologies', or the institutional innovation that we see in socio-technical innovations.

In order to make our social–ecological systems move towards sustainability, we will need to harness the full range of human creativity and innovation potential (Westley et al., 2011). We turn now to a more specific discussion of how technology is currently developed in order to explain the significance of socio-technical innovations for social-ecological transformations. We know that much of the capacity for technological innovation resides in commercial enterprises, the private sector being a key actor in the creation of new pathways to sustainability (Allenby & Richards, 1994). However, there are warning signs that suggest that technological innovation, far from serving our needs, may indeed be driving development away from sustainability (van der Leeuw et al, 2011). Innovation has gradually become supply-driven rather than demand-driven. Today interesting new ideas or tools spawn companies that invest in marketing to stimulate product demand: "need" is created (Westley et al., 2011). Business has become "locked-in" to an unsustainable growth-oriented regime with a focus on optimizing shareholder value and externalising costs, which is at odds with a world made up of interconnected social–ecological systems. The evolutionary nature of innovation further means that new ideas or tools of change are in themselves driven by the existence of pre-existing technologies, which hampers our ability to innovate and risks that we will further reinforce unsustainable development pathways (Walker, Abel, Anderies, & Ryan, 2009). Overcoming this dynamic of unsustainable technological innovation requires finding the institutional frameworks to stimulate the kinds of innovation that solve rather than augment our environmental challenges (Steffen et al., 2011). These institutional frameworks will entail strategies from government and civil society that create conditions under which businesses that are creating broader social value are favoured over businesses that only focus on their own economic value creation (Loorbach, 2010).

This is a complex social and institutional process to manage technological development and innovation.

Unfortunately, some believe that systemic transformation only occurs once the current system has reached a "tipping point" whereby it can transform into a more desirable state (O'Riordan & Lenton, 2011). Crisis often has the effect of creating disruption at the institutional level. At such times, if there are innovative alternatives sufficiently well developed, this can tip the system into a more resilient state. This entails transformation of the institutions and governance structures that shape our cultural, political, and economic transactions (Westley et al., 2011). The transition towards sustainability requires radical, systemic shifts in value, beliefs, patterns of social behaviour and multi-level governance and management regimes (Clark 2001). However, as is the case with food, energy, and water systems, these are complex and adaptive systems that have certain characteristics that make such institutional interventions extremely difficult to get right (see Chapters 2, 8, and 10). We therefore need to explore the next layer of our conceptual pyramid and examine what we understand about the art of decision making in complex adaptive systems.

Decision making in complex adaptive systems

In presenting the next layer of our pyramid about complex systems thinking we introduce yet another way in which we situate "complexity". The social–ecological systems (ie. the food, energy, and water systems) described above have also been described as complex, adaptive systems (Berkes, Folke, & Colding, 2003; Ericksen, 2008). They are complex in that, as we have just seen, there is no single linear path to their development and change, nor does the behaviour of any one part explain what will happen across the whole system. They are also adaptive in that, despite the seemingly mixed relationships among the parts of the system, the whole system is able to and does change in response to disruptive events. Such systems are found not only in the natural environment, but also in economic and business systems. Understanding how decisions are made in these other complex adaptive systems is increasingly important to how we think about environmental governance. Furthermore, as environmental governance is more contingent on business and the financial sectors, integrating our decision making across these systems is critical.

Since the 1960s, businesses have come to recognise the increasing complexity of the environment in which they operate and have tried to develop tools to cope with the uncertainty of the future. Emergy and Trist (1965) have described this as "turbulence". Turbulent environments require different decision making tools for the future. Under turbulent conditions, traditional means of making decisions based on forecasting are no longer relevant because the future is no longer stable; rather it has become a moving target, and so there is no single correct projection to be deduced from how things have worked in the past (Wack, 1985, p. 73). Complex systems are inherently unpredictable and thus managers must learn to juggle shifting objectives (Holling, 2001). This is likely to result in a "messy" matrix of formal and informal institutions, which can only be adequately dealt with through a complex systems approach that heeds these particular characteristics and inter-dependencies (Scoones, 2005, p. 12).

Diversity is also a critical component of complex systems. Maintaining diversity (of individuals, world views, system function etc.) is important for organisational adaptability (Stark, 1999). A diversity of perspectives is critical for tackling wicked problems, but there are few management or research tools that allow for this diversity and multiplicity to be accounted for (see Chapter 10). Some developments in systems thinking have provided processes and methods for coping with the uncertainty that is generated from the diverse interconnectedness of a complex world (Lang & Allen, 2010).[1] Collaboration between actors is key to coping with uncertainty and developing tools for involving the perspectives of many different stakeholders is vital for collaborative governance (Ramírez et al., 2010).

Having common ground between diverse stakeholders when facing wicked problems like FEW governance creates a climate of cooperation and social adaptability that in turn should result in a positive response underpinned by shared values in understanding the problem (Alahi, 2010). Athanasopoulou (2010) discusses how these inclusionary mechanisms could be incorporated into the corporate social responsibility (CSR) framework. From the 1970s, interventionists were questioning how effective the "invisible hand" of competition was as an ethical regulator of large corporations that shape the environment in which they operate. There was a growing recognition that, despite the necessity of government regulation, it is often insufficiently knowledgeable, subtle or timely to reconcile the self-interest of corporate entrepreneurs with the greater good of society (Andrews, 1973, p. 58). In recognition of the social costs of economic activity, CSR has evolved from corporate philanthropy to contemporary strategic CSR, which is supposed to provide opportunities for corporations to focus corporate power on more socially desirable objectives and their own long-term operations (Andrews, 1973; Kramer & Porter, 2011).

Calling on a systems approach, and in particular the idea of needing a multiplicity of perspectives in order to understand a problem (or to generate useful ideas about the future), has implications for concepts of fairness and equality. Wicked problems reflect the coalescence of social, technical, and political concerns that cross boundaries of communities, organisations, or nations and so decisions on these issues cannot be made through a single lens without giving rise to unintended consequences and conflicts between stakeholders that could compound already turbulent situations (Alahi, 2010, p. 224). Such positive feedback and interconnectedness across traditional boundaries requires new analytical tools for decision making that take into account not only the characteristics of such dynamic and turbulent circumstances, but also the implications of the social dimension, such as deciding who gets to be included and how fluid power relations are constituted (see Chapters 10 and 11).

Stormer (2003 in Athanasopoulou, 2010, p. 246) argues that the current neoliberal economic paradigm, where profit maximisation is the primary obligation of corporations, inhibits CSR from moving beyond "enlightened self-interest".

Legal frameworks and the board's fiduciary duty to shareholders to maximise profit makes any other framing difficult to implement. A restructuring of the core of corporate governance structures is required in order to realise a shift towards principles of social and environmental responsibility. Stormer (2003) argues that the neoclassical justification of the stakeholder corporate model is irrelevant in an increasingly complex environment and so CSR needs to be understood within an ecosystem where businesses are only one actor within an interrelated system.

The recognition that a system of actors shape turbulent environments is a key element of Emery and Trist's (1965) argument. Increased participation of multiple perspectives creates shared values that can serve as a common ground thereby facilitating stability (not equilibrium) (Emery & Trist, 1965). The idea of involving multiple stakeholders can also be used to address issues relating to sustainability in corporate decision making. Elkington and Trisoglio (1996) argue that sustainable development is a strategic issue and it is therefore urgent to form win–win solutions between corporations and other stakeholders. The formation of international bodies like the World Business Council for Sustainable Development, the Global Reporting Initiative (GRI), and the Food Ethics Council, all dealing with corporate issues around sustainability, is evidence of the increasing relevance of a sustainability discourse across sectors. Despite this, we still lack understanding of how to respond to a call for sustainability in complex contexts like that of the food–energy–water nexus. In this book, we hope to make a solid contribution to addressing this dilemma by recognising the complexity inherent in social–ecological systems whilst providing examples of how this challenge has started to be addressed through some of the multi-stakeholder approaches alluded to above, in an innovative type of natural resource governance.

Governance in practice

While much of this hierarchy we pose is theoretically rich, it is not always obviously translated into practice. Governance in practice must reach through each of the layers described above, from decision making in complex adaptive systems to socio-technological innovations and right into the functions of sustainability. Indeed, the approaches to governance that might be able to do just this are the focus of each of the chapters of this book. Therefore, in this overview we conclude our conceptual introduction by focusing on what we mean by "governance" and how it has evolved in relation to sustainability and the environment. We have already introduced the necessity of that key feature of governance, decision making, occurring through inclusive participation of multiple stakeholders. Now we explore why "governance" is itself so variable and how it coordinates different institutions, systems, and mechanisms to facilitate decision making.

Lemos and Agrawal (2006, p. 298) define environmental governance as

> synonymous with interventions aiming at changes in environment-related incentives, knowledge, institutions, decision making, and behaviours. More specifically, we use 'environmental governance' to refer to the set of regulatory

> processes, mechanisms and organizations through which political actors influence environmental actions and outcomes. Governance is not the same as government. It includes the actions of the state and, in addition, encompasses actors such as communities, businesses, and NGOs.

Governance encompasses both the formal and informal processes of environmental management and is therefore adaptable to multiple perspectives and collaborations. In theory, environmental governance should provide the frameworks necessary for governing complex systems with a range of socio-technical approaches to sustainability. However, the realities of coordinating these collaborations and the mechanisms through which to do so require constant change and adaptation. Governance must be as adaptive as the processes and systems it is seeking to manage. The focus of the chapters in this book concerns this process of governance adaptation – to a changing climate, to new technologies, and to different social relationships. As discussed above, the chapters especially focus on the ways in which governance can and must adapt to our ever-improving understanding of the science of sustainability and our complex environments.

The next few paragraphs provide a brief introduction to the historical trajectory of environmental governance and how it has been adapted in the past few decades, and also suggests some key emerging themes that run through the chapters of the book. Western environmental governance as a clearly differentiated space of governance emerged in the 1960s with an increasing political discourse about the environment (Dryzek, 1997). During this period, the environment itself was conceptualised, popularized, and also politicised with events such as the first view of the planet from space; Rachel Carson's book *Silent Spring* (1962) and Garret Hardin's the *Tragedy of the Commons* (1968); and the instigation of formal government for the environment with then US president Richard Nixon's creation of the National Environmental Policy Act (NEPA) in 1970. For the first time on a global public scale the environment was presented as limited and threatened.

However, this viewpoint was not universally accepted and those wishing to change the way the environment was managed had to prove that their concerns were real and consequential, despite this being inconvenient to many powerful interests. As much as the issues of environmental governance have shifted from pollution, protected wilderness, and overpopulation, to contemporary concerns over climate change, biodiversity loss, and environmental justice (Dryzek, 1997), we still see the fight for the legitimacy of environmental governance in relation to other interests of "development" and business interests. Thus, different approaches to environmental governance over time have tried to meet environmental needs within the scope of politically expedient mechanisms and support by relevant stakeholders. Unfortunately, this governance strategy has been largely unsuccessful and we are reaching potentially irrevocable tipping points in the Earth's systems that require innovative strategies (Biermann, 2012).

Dryzek (1997) has documented environmental discourses of environmental survivalism, administrative rationalism, democratic pragmatism, and economic rationalism. These approaches have coalesced into a dominant Western conceptualisation of environmental governance as sustainable development and ecological modernisation. However, a green consciousness that imbues non-Western, non-capitalist conceptualisations of environmental harmony persists and is increasingly being legitimised. Some chapters in this book help illustrate how the current discourse of sustainability is tied to complex social–ecological systems and that consideration of environmental systems is open to methods of incorporating both Western (Chapters 7, 8 and 12) and non-Western (Chapter 11) conceptualisations of environmental governance.

The practical mechanisms of environmental governance are myriad and fast evolving. Lemos and Agrawal (2006) map out four different mechanical approaches to environmental governance tied to globalisation; decentralisation; market- and agent-focused instruments; and cross-scale approaches.

Globalisation approaches are tied to supra-state networks and agreements for environmental change. They are quickly able to involve many actors, work beyond legal commitments, and incorporate scientific and lay knowledge, but they are not the most effective means of implementing changes or responding to disruptions or unexpected changes in the systems they create (Lemos & Agrawal, 2006). The distribution of power among the states and actors involved has been criticised by those who see it as the perpetuation of existing ways of governing the environment that prioritises powerful states and the corporate sector (Falkner, 2003).

Next, the decentralised approaches to environmental governance are particularly aligned with those who view the management of natural resources as a problem of governing the commons at different scales (Ostrom, 2015; Ostrom, Burger, Field, Norgaard, & Policansky, 1999; Paavola, 2007). Political ecologists also point out that decentralised approaches allow positive changes in the relationships between power and governance, for there are some examples where local capacities, knowledge, and decision making are able to direct environmental governance (Bulkeley, 2005). However, like any system, the control and power dynamics nonetheless hinge on the effectiveness of its leadership (Dressler et al., 2010).

Third, market- and agent-focused instruments for environmental governance were developed increasingly in the 1990s. The loss of faith in governments to manage the economy and have the resources to manage the environment led to an increasingly capitalist driven, self-regulatory, neoliberal approach. Voluntary agreements about the environment, product certification schemes, and even carbon markets sought to regulate the environment in the absence of a regulator or formal government intervention. While government regulation is costly to deliver and prompts fears that it threatens international competitiveness of corporations, market and agent approaches often suffer from free-riders and/or can be less stringent to maintain the cooperation of actors antagonistic to any environmental governance.

Finally, cross-scale governance is often coordinated by non-state actors and makes use of the individual incentives driving the above mechanisms to create collaborative approaches. These can be hybrid partnerships between communities and the state materialising in co-management; or between the state and the market in public–private partnerships (PPPs); or even as various private–social partnerships that have developed between corporate actors and civil society (Lemos & Agrawal, 2006). It is these hybrid mechanisms, complete with their individual weaknesses, that the chapters in this book advance and further test; exploring the ways in which environmental governance is so variable in its process and, in particular, how governance must coordinate the interests of businesses, NGOs, governments, and society.

Governance themes for consideration whilst reading the book

Before one reads the empirical chapters in the book, we would like to highlight two themes to contemplate as they appear in the different contexts presented in the case studies: the tensions between the public and private sectors and the significance of scales of governance across space and time. Whilst the conclusion (Chapter 14) picks up on some cross-cutting commonalities that emerge across the chapters, these themes provide a broader critique that it is important to be aware of when reading each chapter.

A critical theme across the book is that of the role of business and the private sector in environmental governance. The relationship between private corporations and the public sector is not solely antagonistic (Chapter 6). Businesses are now operating in unknown territory when it comes to environmental sustainability and social justice issues. They are increasingly dependent on public sector governance and support structures, but also subject to its regulations. Similarly, governments at all levels are in need of the innovation, capital, and dynamism of private and market systems. The new alliances between civil society and business are also becoming increasingly more prevalent in the search for effective environmental governance strategies (Chapters 2 and 3) and tools, such as the SUSFARMS tool described in Chapter 8. Some chapters present solutions and compromises where private sector governance structures' ability to cope with the complexity of future challenges across food–energy–water systems are floundering or have failed and are thus requiring these new partnerships (Chapters 9 and 13).

As part of these examples, we see evidence that corporate social responsibility (CSR) was a short-term fix. In essence, it enabled corporations to continue externalising the deeper societal and environmental concerns embedded within their business practices. Increasingly, it is now being recognised that corporate strategy can no longer focus exclusively on profit maximisation without taking into account the impacts of these business practices on society and the environment. Companies must also respond and adapt to stricter governmental regulatory frameworks aimed at securing local economies and protecting human rights. This situation can at times be ironic. John Ruggie (2003) has pointed out that the private sector has been the lead actor separating global communities from the economy and now is being called upon by its investors, civil society, and its own risk management teams to act as a bridge to fix this relationship. Chapter 4 describes how the public sector needs the innovation of the private sector to overcome a locked-in fossil fuel energy system.

Another golden thread through the book is the contestation and compromise over the scales and levels at which governance occurs. There is considerable debate about the appropriate scale of governance for different challenges (Adger, Brown, & Tompkins, 2005). For example, there was recently support for the governance of water systems at the level of a watershed or a catchment area. While this may make sense in hydrological terms, there are few social and industrial systems that are aligned across an entire catchment and, even more problematically, many catchments lie across international boundaries, further complicating governance (Budds & Hinojosa, 2012). Similarly, global governance mechanisms dominated the early 2000s with sweeping supra-governmental initiatives around governance tools such as transparency and certification that were supposed to have direct impact at local levels (Cash & Moser, 2000). As a result, the social complexities of governance scales need to be acknowledged. Many chapters in this book present the challenges of scale mismatch in FEW governance, referencing context specific challenges and showing that there are some examples of how the scale problem can be met (see Chapters 2, 3, 5, 7, 9, and 12).

There are many supposed rules about "good" governance; that it must be equitable, transparent, just, collaborative, and accessible to all stakeholders (Graham, Amos, & Plumptre, 2003). We see these ambitions in the chapters that follow and we also see a commitment to approaches that do or could work in practice and not just in theory. Although governance is often far too abstract and conceptual, in this book we hope to bring it to life. We present a diverse spectrum of answers to the following questions: How can we enable complex systems to work together sustainably through different, real approaches to governance? What are the mechanisms of coordination and control? How can institutional governance and the governance of specific technical changes and different organisations fit together? How does good governance bring together the public and private sector towards common goals? We hope to present evidence that there is no simple or singular solution, but there are a number of examples of positive and compelling change, worthy of our further interest and investment.

Book overview

The book's unique contribution lies in its integration of recent academic research with innovative "knowledge from the ground" from practitioners affiliated to different types of organisations operating at different scales. Our book contributors have been deliberately chosen because they are a set of early career researchers spanning a diverse group of disciplinary backgrounds, but who also cross the spectrum of research and practice. The authors call home the UK, US, South Africa, Belgium, Australia, India, China, Greece, France, and Poland; and they have further conducted research in and write about Peru, the Philippines, Canada, and Mexico. Through the lens of these young, up-and-coming thinkers who have done cutting-edge research in this field and by using relevant and illustrative case studies, the book attempts not only to highlight problems, but also to posit workable solutions and new ideas for overcoming future sustainability challenges.

Some examples are taken from relevant fields like natural resource governance, whereas others illustrate key points through the use of a variety of examples rather than in-depth case studies. Cumulatively, these chapters aim to go beyond "business speak" or academic jargon to provide an updated conceptualisation of future resource governance. Several of these chapters present examples of corporations working on strategies that do not externalise social and environmental responsibility, but incorporate these as key outcomes of their business model. This is complicated by uncertainties about how far corporate responsibility extends with respect to resource use. Grappling with these issues, some chapters start to underline the ethical implications of power, inclusivity, and decision making in emergent resource governance structures. Many other organisations such as development agencies, governments, or NGOs are themselves trying to make sense of these changes. Not only are they seeking insights into new practices, they are also questioning the governance landscape in which new innovations, business models and practices will need to perform. Through the chapters, the book explores the emergence of these new governance strategies across a broad range of contexts.

The book is divided into two sections. We start the book with a deep dive into FEW governance led from the perspective of one system. Without a good understanding of what system complexity looks like in practice, it is difficult to see the relationships we refer to for nexus-based governance. The first half of section one (Chapters 2, 3, and 4) provides historical context for our approaches to resource governance. Across all three chapters we see the risks of path dependencies from previous forms of resource governance and the consequences of overlooked interdependencies between different resource systems. Chapter 2 by Melissa Rohde starts us thinking about resource interdependencies in India. She frames the current challenges to governing food, energy, and water in relation to the actions of the Green Revolution's ambitions to create food security and self-sufficiency for India. However, the policies created unanticipated socio-economic consequences tied to the interconnections and disruptive feedbacks between systems. We see a similar pattern of historical approaches serving as disruptive forces and complicating current governance in Chapter 3 by Alexandra Littaye, who carries us to Mexico to examine the governance of corn. She introduces the ways in which transnational entrepreneurship can address issues of the marginalisation of indigenous farmers in Mexico by exploring how pinole, a traditional food made of corn, is gaining ground as a heritage product and an "endangered" food in international markets. Then in in Chapter 4 Agostino Inguscio provides an historical account of the path-dependencies embedded in the South African story of energy production and the inequalities that are reinforced through this governance of the energy system. In each chapter we are shown how historical legacies effect how we understand the obstacles to future sustainability. Further, each chapter makes tangible the broader social, economic, and political intersections of resource governance.

The second half of section one (Chapters 5, 6 and 7) similarly consists of country-specific cases led primarily by an analysis of one resource system – in this case the water system and how it relates to the energy and food systems. These chapters carefully assess and analyse the current strengths and weaknesses of particular resource system governance approaches. In Chapter 5 Xiawei Liao, Nick Eyre, and Jim Hall discuss the challenges China faces because of its need for water to cool coal-fired power plants. This nexus of water-for-energy is further complicated by the geographical distribution of these resources across China and the inability of one policy successfully to address such a heterogeneous landscape. Swathi Veeravalli focuses on Peru in Chapter 6 and through the case of the asparagus industry, she explains the risks for Peru's long-term climate resilience and economic security that come with prioritising the short-term economic value of agricultural exports over water security. Then, in Chapter 7 Michael Valli and Alexandra Girard bring together the perspectives of a practitioner water-trader and an academic to analyse the nexus effects of the Australian Murray-Darling Water Basin water market. They unpick the consequences of this systems on agricultural production as well as other social, environmental, and economic outcomes with the aim of suggesting recommendations if it were to be adopted elsewhere.

In the second section of the book, the chapters investigate the broader implications of emerging FEW governance approaches. These chapters are more interested in our theoretical understanding of emergent governance strategies than the chapters in the first half of the book. They build on the practical understanding the first half presents and go further to press into the socio-technical and socio-economic structures needed for sustainable resource governance in complex systems. These cases are not always based within one country or a specific governance jurisdiction. They instead often focus on a particular governance issue or tool that may be place-based as with Chapter 8 (South Africa), Chapter 10 (The Philippines) and Chapter 11 (the Canadian Northwest Territories), but are most interested in governance lessons from these locations that could also be applicable elsewhere. For example in Chapter 8, Jessica Cockburn, Vaughan Koopman, Laura Pereira, and Jaci van Niekerk use the lens of institutional bricolage to explore critically how a new governance arrangement through the collaborative governance platform SUSFARMS was constructed. In their analysis, they draw on social learning theory and expansive learning methodology across difference institutions and scales. In Chapter 9, Alex Money uses transaction cost theory to propose a framework of five factors that guide a firm's decisions to manage corporate water risk and return. And, similarly, Chapter 10 by Tomas Chaigneau is primarily interested in the intersections of natural resource management and community support. He explores this in the context of three different marine protected areas in the Philippines in order to offer approaches to assessing the non-monetary costs and benefits to communities from these natural resource management programmes.

The final three chapters of the book also take a theoretical approach and particularly speak to novel governance approaches that incorporate indigenous knowledge systems (Chapter 11), emerging governance opportunities through ICT (Chapter 12), and the future for the application of such new technologies if applied to resource governance in the mining sector (Chapter 13). Brice Perombelon draws on his extensive research experience in the Canadian Northwest Territories and among communities faced with increasing commercial resource extraction in Chapter 11. In this chapter he "appraises how traditional indigenous land-use ontologies are used in and transformed by management practices that lead, from a Dene epistemological worldview, to the unsustainable pursuit of resource exploitation activities" and argues for the importance of culturally grounded sustainable development policies. In Chapter 12 Timothy Karpouzoglou, Laura Pereira, and Samir Doshi explain the potential of ICT (Internet and Communication Technologies) as a tool for resource governance able to be adapted at different scales and for different suggested purposes around the world. Finally, in Chapter 13, Caitlin McElroy looks to the future of resource governance using as an example of the challenges presented by the remote locations of mining activities and provides some suggestions of what it might entail to incorporate innovative governance approaches in this sector. She also emphasises the increasing effects of climate change that needs be addressed in new ways, which is also picked up as an important theme running through the book in Chapter 14. In our concluding chapter (Chapter 14), we review the key outcomes of each chapter and highlight some important themes that emerge time and again across the chapters of the book. This book can be read cover to cover or by individual chapters. It presents our larger argument about the progression, variability, and opportunity across the field of resource governance changes or you can pick the cases of most interest. Either way, we hope you see this book as a resource to broaden your perspectives of environmental and resource governance and just maybe find a few ideas to put to very good use yourself. We hope you enjoy it.

Note

1 Applying a systems approach also has implications for research because it legitimises theoretical and methodological pluralism (Midgley & Richardson, 2007). The multidisciplinary nature of this book emphasises the importance of diverse lenses to interrogate complex problems.

References

Adger, W. N., Brown, K., & Tompkins, E. (2005). The political economy of cross-scale networks in resource co-management. *Ecology and Society*, *10*(2). 9

Alahi, S. (2010). Conceptions of fairness and forming the common ground. R. Ramírez, J. W. Selsky, & K. ven der Heijden (Eds.), *Business Planning for Turbulent Times: New Methods for Applying Scenarios* (pp. 223–242). Oxford: Earthscan.

Allenby, B. R., & Richards, D. J. (1994). *The Greening of Industrial Ecosystems*. Washington, DC: National Academy Press.

Andrews, K. R. (1973). Can the best corporations be made moral? *Harvard Business Review*, 51: 57–64.

Arthur, W. B. (2011). *The Nature of Technology: What It Is and How It Evolves*. New York: Free Press.

Athanasopoulou, A. (2010). Turbulence and CSR: Is there a role for scenarios? In R. Ramírez, J. W. Selsky, & K. ven der Heijden (Eds.), *Business Planning for Turbulent Times: New Methods for Applying Scenarios* (pp. 243–260). Oxford: Earthscan.

Beddington, J. (2009). *World Faces "Perfect Storm" of Problems by 2030*. London: UK Government.

Berkes, F., Folke, C., & Colding, J. (2003). *Navigating Social-Ecological Systems: Building Resilience for Complexity and Change*. Cambridge, UK: Cambridge University Press.

Biermann, F. (2012). Planetary boundaries and earth system governance: Exploring the links. *Ecological Economics*, *81*, 4–9.

Budds, J., & Hinojosa, L. (2012). Restructuring and rescaling water governance in mining contexts: The co-production of waterscapes in Peru. *Water Alternatives*, 5(1), 119.

Bulkeley, H. (2005). Reconfiguring environmental governance: Towards a politics of scales and networks. *Political Geography*, *24*(8), 875–902.

Cash, D. W., & Moser, S. (2000). Linking global and local scales: Designing dynamic assessment and management processes. *Global Environmental Change*, *10*, 109–112.

Carson, R. (1962). *Silent Spring*. Boston, MA: Houghton Mifflin Harcourt.

Clark, W. C. (2001). A transition toward sustainability. *Ecology Law Quarterly*, *27*(4), 1021–1076.

Dressler, W., Büscher, B., Schoon, M., Brockington, D. A. N., Hayes, T., Kull, C. A., McCerthy, J., & Shrestha, K. (2010). From hope to crisis and back again? A critical history of the global CBNRM narrative. *Environmental Conservation*, *37*(1), 5–15.

Dryzek, J. S. (1997). *The Politics of the Earth: Environmental Discourses*. Oxford: Oxford University Press.

Elkington, J., & Trisoglio, A. (1996). Developing realistic scenarios for the environment: Lessons from Brent Spar. *Long Range Planning*, *29*(6), 762–769.

Emery, F., & Trist, E. (1965). The causal texture of organizational environment. *Human Relations*, *18*, 21–32.

Ericksen, P. J. (2008). Conceptualizing food systems for global environmental change research. *Global Environmental Change*, *18*, 234–245.

Falkner, R. (2003). Private environmental governance and international relations: Exploring the links. *Global Environmental Politics*, *3*, 72–87.

Folke, C. (2006). Resilience: The emergence of a perspective for social–ecological systems analyses. *Global Environmental Change*, *16*, 253–267.

Graham, J., Amos, B., & Plumptre, T. (2003). *Principles for Good Governance in the 21st Century*. Policy brief 15. Ottawa, Canada: Institute on Governance.

Hardin, G. (1968). The tragedy of the commons. *Science*, *162*(3859), 1243–1248.

Holling, C. S. (2001). Understanding the complexity of economic, ecological, and social systems. *Ecosystems*, *4*, 390–405. Retrieved from http://doi.org/10.1007/s10021-001-0101-5.

Kramer, M. R., & Porter, M. E. (2011). Creating shared value. *Harvard Business Review*, 89(1–2), 62–77.

Lang, T., & Allen, L. (2010). Reflecting on scenario practice: The contribution of a soft systems approach. In R. Ramírez, J. W. Selsky, & K. van der Heijden (Eds.), *Business Planning for Turbulent Times: New Methods for Applying Scenarios* (pp. 47–64). Oxford: Earthscan.

Leach, M., Raworth, K., & Rockström, J. (2013). Between social and planetary boundaries: Navigating pathways in the safe and just space for humanity. *World Social Science Report*. International Social Science Council.

Lemos, M. C., & Agrawal, A. (2006). Environmental governance. *Annual Review of Environmental Resources*, *31*, 297–325.

Loorbach, D. (2010). Transition management for sustainable development: A prescriptive, complexity-based governance framework. *Governance*, *23*(1), 161–183. Retrieved from http://doi.org/10.1111/j.1468-0491.2009.01471.x.

Midgley, G., & Richardson, K. (2007). Systems thinking for community involvement in policy analysis. *Complexity and Organization Emergence*, 9(1–2), 167–193.

Mittal, A. (2009). The 2008 food price crisis: Rethinking food security policies. G-24 *Discussion Paper no* 56. New York and Geneva: United Nations Conference on Trade and Development (UNCTAD).

Nelson, R.R., & Nelson, K. (2002). Technology, institutions, and innovation systems. Research Policy 31: 265–272

O'Riordan, T., & Lenton, T. (2011). Tackling tipping points. *British Academy Review, 18*, 21–27.

Ostrom, E. (2015). *Governing the Commons*. Cambridge, U.K.: Cambridge University Press.

Ostrom, E., Burger, J., Field, C. B., Norgaard, R. B., & Policansky, D. (1999). Revisiting the commons: Local lessons, global challenges. *Science*, *284*, 278–282.

Paavola, J. (2007). Institutions and environmental governance: A reconceptualization. *Ecological Economics*, 63(1), 93–103.

Ramírez, R., Selsky, J., & van der Heijden, K. (2010). Conceptual and historical overview. In R. Ramírez, J. W. Selsky, & K. van der Heiden (Eds.), *Business Planning for Turbulent Times: New Methods for Applying Scenarios* (pp. 17–30). Oxford: Earthscan.

Rittel, H.W.J, & Webber, M.M. (1973). Dimlemmas in a General Theory of Planning. Policy Sciences 4: 155–169.

Rockström, J., Steffen, W., Noone, K., Lambin, E., Lenton, T. M., Scheffer, M., Folke, C., Schellnhuber, H. J., Nykvist, B., de Wit, C., Hughes, T., van der Leeuw, S., Rodhe, H., Sörlin, S., Snyder, P. K., Costanza, R., Svedin, U., Falkenmark, M., Karlberg, L., Corell, R. W., Fabry, V. J., Hansen, J., Walker, B., Liverman, D., Richardson, K., Crutzen, P., & Foley, J. (2009, September). Planetary boundaries : Exploring the safe operating space for humanity. *Nature*, *461*, 472–475.

Ruggie, J. G. (2003). Taking embedded liberalism global: The corporate connection. In D. Held & M. Koenig-Archibugi (Eds.), *Taming Globalization* (pp. 93–129). Cambridge, UK: Polity.

SAB Miller, & WWF (2014). The water-food-energy nexus: Insights into resilient development. Retrieved from www.ab-inbev.com/content/dam/universaltemplate/ab-inbev/investors/sabmiller/reports/other-reports/water-food-energy-nexus-2014.pdf.

Scoones, I. (2005). Sustainable rural livelihoods: A framework for analysis. *IDS Working Paper 72*. Institute of Development Studies, University of Sussex, Brighton, UK.

Smith, A., & Stirling, A. (2010). The politics of social-ecological resilience and sustainable socio-technical transitions. *Ecology and Society*, *15*(1). Retrieved from http://doi.org/10.5751/ES-04565-170208.

Stark, D. (1999). Heterarchy: Distributing authority and organising diversity. In J. Clippinger (Ed.), *The Biology of Business: Decoding the Natural Laws of Enterprise* (pp. 153–180). San Francisco: Jossey-Bass.

Steffen, W., Persson, Å., Deutsch, L., Zalasiewicz, J., Williams, M., Richardson, K., Crum-Scheffer, M., Schellnhuber, H. J., & Svedin, U. (2011). The anthropocene: From global change to planetary stewardship. *Ambio*, *40*(7), 739–761. Retrieved from http://doi.org/10.1007/s13280-011-0185-x.

Stormer, F., (2003). Making the Shift: Moving from "Ethics Pays" to an Inter-Systems Model of Business, Journal of Business Ethics 44(4): 279–289.

Van der Leeuw, S., Constanza, R., Aulenbach, S. Brewer, S., Burek, M., Cornell, S., Crumley, C., Dearing, J.A., Downy, C., Graumlich, L.J., Heckbert, S., Hegmon, M., Hibbard, K., Jackson, S.T., Kubiszewski, I., Sinclair, P., Sörlin, S. & Steffen, W. (2011) Toward an Integrated History to Guide the Future, Ecology and Society, 16(4): 2–13. http://dx.doi.org/10.5751/ES-04341-160402

Wack, P. (1985). Scenarios: Unchartered waters ahead. *Harvard Business Review*, September: 73–89.

Walker, B. H., Abel, N., Anderies, J. M., & Ryan, P. (2009). Resilience, adaptability, and transformability in the Goulburn-Broken Catchment, Australia. *Ecology and Society*, *14*, 12. Retrieved from www.ecologyandsociety.org/vol14/iss1/art12/.

Westley, F., Olsson, P., Folke, C., Homer-Dixon, T., Vredenburg, H., Loorbach, D., Thompson, J., Nilsson, M., Lambin, E., Sendzimir, J., Banerjee, B., Galaz, V., & van der Leeuw, S. (2011). Tipping toward sustainability: Emerging pathways of transformation. *Ambio*, 40(7), 762–780. Retrieved from http://doi.org/10.1007/s13280-011-0186-9.

2 Securing Food, Energy, and Water in India

Shifting the governance landscape to tackle socio-economic challenges through integrated policies

Melissa M. Rohde

Introduction

The food, energy, and water sectors are inextricably linked. However, existing governance landscapes often make it difficult for governments to address and adapt to new and continuing challenges in securing food, energy, and water. This disconnect is often driven by a fragmented governance landscape with strong sectorial boundaries that segregate responsibilities, thereby hindering coordination and creating legal inconsistencies between sectors. These challenges exist in many countries around the world, but are most pronounced in India where more than one-tenth of the Earth's population resides. India's fragmented governance hinders its ability to sufficiently secure the food, energy, and water needs for their population. Endemic malnutrition, frequent power cuts, poor water quality, and rapid groundwater depletion are some examples of the ongoing challenges that India faces. To address these challenges and the policies that constrain them comprehensively, India would benefit from a more cohesive governance structure.

India is currently the world's largest agricultural groundwater user (Shah, 2009) with 35 percent of India's agricultural land dependent on groundwater for irrigation (World Bank, 2015). Intensification of agriculture and a monsoonal climate system (with distinct wet and dry periods) has increased groundwater dependency throughout India and contributed to unsustainable groundwater storage loss in the northwestern part of the country (Rodell, Velicogna, & Famiglietti, 2009). Increases in the demand for food, energy, and water resources to support India's large population are a major driver for the large decline in water availability being observed, but government policies are also largely to blame. These impacts are largely a consequence of multi-decadal policy making within a fragmented governance landscape, in addition to a diverse set of sociopolitical and environmental demands. India's trajectory to become the world's most populous nation by 2030 (United Nations, Department of Economic and Social Affairs, Population Division, 2015), leaves little repose for India's food, energy, and water woes unless the interdependencies between the food, energy, and water sectors are fully realized.

In less than a century, India has quickly evolved from autocratic colonial rule to the world's largest democracy. Early after Independence, a general distrust in foreign influence and markets drove India to adopt socialist-inspired policies that provided protection from world markets, state-ownership in the agriculture and power sectors, and the Green Revolution. The Green Revolution, dominated by isolated food, energy, and water policies in the mid-20th century resulted in subsidies for agricultural inputs (e.g. fertilizers, pesticides, and high-yield seed varieties), food grains, and electricity to pump groundwater for irrigation. These policies, although well intended to increase national self-sufficiency and food security, were successful in the beginning, but have resulted in poor management of water and energy resources in the 21st century. Distorted electricity and food grain prices have led to serious issues such as reduced market access for nutritious crops, high energy demands for agriculture, and groundwater depletion. Some progress has been made since economic liberalization began in the 1990s through privatization in the power sector and the integration of neo-liberal policies. However, unreliable transportation and electric grid infrastructure hinders the ability for businesses to fulfil the role where the government cannot. Despite progress being made since economic liberalization to open up India's markets, market distortions in the agricultural sector still persist, thereby driving a need for agricultural reform.

This chapter uses India as an example for how the confluence of the food, energy, and water sectors can complicate the management of any one of these individual resources and why addressing issues with the food, energy, and water sectors cannot be done in isolation of each other. India's governance landscape is slowly shifting away from its fragmented approach and sectorial divides based on a realization that the intrinsic interdependent nature of the food, energy, and water sectors must be reflected in policies and resource management to adequately address its challenges. The overall goal of this chapter is to highlight the emergence of India's new approach towards comprehensive policies to address food–energy–water interdependence, while bringing attention to the areas or policies that overlook an interrelated perspective for proper governance coordination.

The three main objectives are to describe: 1) the historical influences of India's colonial past on the policies that gave rise to the Green Revolution and dominance of unilateral policies; 2) how the confluence of economic liberalization and cross-sectorial issues promoted a shift in India's governance landscape; and 3) some examples of interdependent policies between the food, energy, and water sectors that are being used to resolve issues and meet growing demands. To illustrate the multi-dimensional and interdependent aspects of the food–energy–water nexus in India's new governance landscape, this chapter draws upon four examples. The first two examples focus on access to energy and the last two examples focus on access to food. The first example describes a water–energy policy scheme ("*Jyotigram Yojana*") that was enacted in northwest India as a way to combat groundwater over-exploitation and high demands on the energy grid. The second example describes the National Solar Mission, which takes a pro-market and pro-business approach to decentralize power generation and sustainably meet rising energy demands. The third example reviews how food pricing and trade laws are helping India promote crop diversification so that appropriate crops can be cultivated in more suitable hydro-climatic regions where groundwater exploitation can be minimized. The last example focuses on how the government's willingness to accept foreign direct investments from multinational corporations is helping India build sufficient cold chain and storage infrastructure to increase market access for high value crops (e.g. fruits, vegetables, dairy).

The emergence of unilateral policies within a fragmented governance landscape

India's colonial legacy and a quest for food security

Prior to colonial life in India, indigenous methods to manage water resources were widely practiced for a millennium, giving communities the necessary knowledge on rainwater harvesting so monsoon rains could be stored and water supplies secured during the dry season (Pandey, Gupta, & Anderson, 2003). Water supplies maintained with rainwater harvesting techniques throughout the year were sufficient to support food production with infrequent famines. However, after the British conquest of India in 1757, famines became a frequent feature of life in colonial India.

Soon after British rule began, India was plagued by the Great Bengal Famine of 1770, which resulted in the death of one-third of India's population at that time. This high death rate has largely been attributed to the dominance of extractive institutions under British control, such as the inflexible demands for land revenue by the British (Malone, 2011), which left cultivators impoverished with little food for their own sustenance. Likewise, the displacement of millions of acres of agricultural lands for export crops instead of domestic subsistence crops increased the vulnerability of Indians to food shortages (Davis, 2001). The introduction of large-scale irrigation systems and construction of canals under British control also eroded much of the traditional wisdom on rainwater harvesting that evolved specifically to cater India's monsoonal climate in pre-colonial India (Pacey & Cullis, 1986; Agarwal & Narain, 1997).[1]

After India's independence from British rule in 1947, food imports were required to curtail famines due to ongoing troubles with food production and storage. Haunted by their colonial past, India's dependence on food imports and international food aid triggered strong political motivations to achieve self-sufficiency in food production (Roth & Warner, 2007). Food policy began to take on a nationalistic flavour with the "Grow more food" campaign, launched in 1953 to reduce foreign dependence in the agriculture sector. These efforts were initially challenged by a lack of water and energy infrastructure and a loss of traditional knowledge in rainwater harvesting techniques that left most farmers dependent on rain-fed agriculture. After the occurrence of several drought-induced famines in Gujarat during the 1960s, it became apparent that rain-fed agriculture was insufficient (Bhatia, 1992; Shah, Bhatt, Shah, & Talati, 2008) and that India would remain food insecure if agricultural investments were not made. With the goal of achieving national food security in mind, direct investments in agriculture through government subsidies for agriculture inputs (e.g. fertilizers, pesticides, and high yield seed varieties) during the Green Revolution became a driving mechanism for improving agricultural productivity in India. However, the Indian agricultural economy quickly became entangled with the water and energy sectors through the formation of state-owned institutions and protectionist policy decisions that distorted the price of electricity and food grains.

The rise in energy access and its impact on groundwater

India's quest to achieve food security was powered by increasing rural access to energy so farmers could mechanize irrigation for their crops. In 1969, the Ministry of Irrigation and Power established the Rural Electrification Corporation, which helped government owned State Electricity Boards promote the use of energized irrigation pumps in rural areas. Initially, energy subsidies for agriculture achieved India's goal for self-sufficiency in food production (Birner, Gupta, & Sharma, 2011; Repetto, 1986). However, in the past couple of decades energy subsidies have come under scrutiny since they have created an unsustainable price structure that has resulted in annual losses for the State Electricity Boards. Political motives to win the support of rural voters have been a major determining factor in the provision of such high energy subsidies for farmers (Roth & Warner, 2007). The resulting sense of entitlement to these subsidies amongst farmers means that reforming the energy sector has become challenging (Dubash & Rajan, 2001). Energy subsidies have also undervalued the cost of water in the process, which has provided little incentive for farmers to conserve water and to grow crops that are suitable for India's predominantly arid climate.

Despite large subsidies in the energy sector, high levels of energy theft also contributed to unreliable energy supply on the grid. The combination of widespread theft from the electric grid, large energy demands from the agriculture sector, and high financial losses for the State Electricity Boards created enough political will to reform the energy sector in the early 1990s (Birner et al., 2011). The State Electricity Boards underwent privatization under the establishment of an independent power producer policy in 1991. The World Bank, who had played an important role in restructuring the policy in the state of Orissa, was key in ensuring that the restructured model was scaled up to the national level using a combination of reforms at national and state levels (Dubash & Rajan, 2001). However, this did not prevent the State Electricity Boards from annual fiscal losses due to their low pricing structure for agriculture customers. Although agriculture users consumed 32.5 percent of the electricity, they only contributed 3.36 percent of electricity revenues in 2000, which resulted in higher electricity prices for industrial and commercial users (Badiani & Jessoe, 2011).

Energy subsidies have been an important factor in the successful proliferation of the electric pump in India (see Figure 2.1), with more than 18 million wells and tubewells – although the exact statistics vary according to source (Rawat & Mukherji, 2013). Efforts to electrify rural areas are ongoing today, mostly in the Indian states of Bihar and Orissa, where a major dependence on diesel energy persists. The rise in electric pumps has not only been promoted through cheap

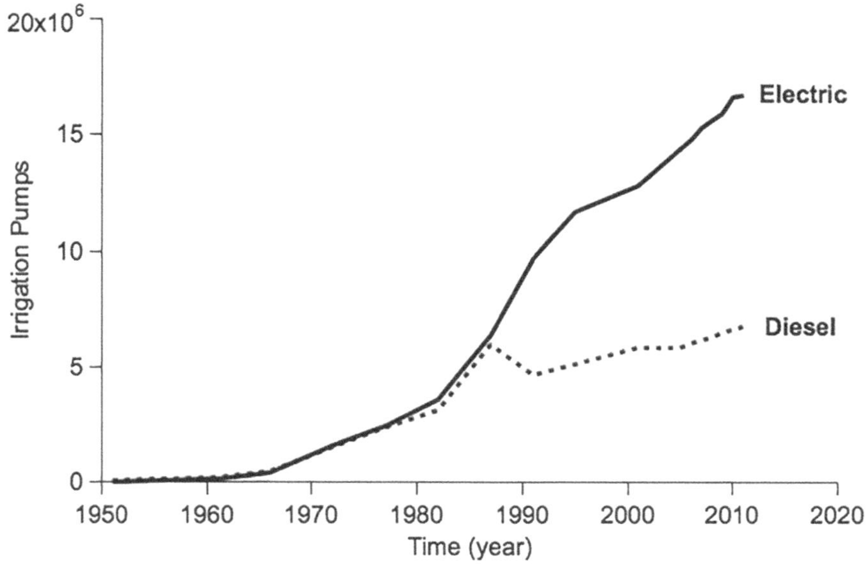

Figure 2.1 Number of irrigation pumps from 1950 to 2011 in India (Data Source: Ministry of Power 2011; India Energy Portal, 2005).

electricity, but also facilitated by groundwater rights law in India. Groundwater rights are directly linked to property rights, giving private landowners unrestrictive rights to groundwater underneath their land. The protection of groundwater law under property rights is a legacy of British colonial law, which favoured property rights and protected the state's right to control waterways for navigation and irrigation for economic gain (Padhy, 2008). The direct link between water and property rights has resulted in a largely informal water economy in India that has dissuaded the central government from imposing rigorous regulations on groundwater extraction. This has led to an absence of centralized institutions to regulate well construction, groundwater pumping, as well as monitoring water quantity and quality that have ultimately resulted in a classic case of a "tragedy of the commons". Although unintended, policy reforms made in the energy sector to address national food security have resulted in increasing water scarcity throughout India, which poses serious challenges for the water sector.

Food access through price setting and its impact on domestic markets

Efforts to enhance India's national food security have been largely administered by the Food Corporation of India (FCI), which was established under the Food Corporation Act in 1964. The FCI manages a Public Distribution System that centralizes food grain collection, transportation, storage, and the resale of food grains to families below the poverty line at a lower price. India spends 900 billion to 1 trillion rupees (INR) annually for the procurement of food grains from farmers and stockpiles it to flood the market when prices are high. In effect, this subsidy has a three-fold purpose: 1) to ensure that farmers receive an adequate return on upfront farming costs (e.g. pesticides, chemical fertilizers, irrigation, and high-yield variety seeds) for food grains; 2) to stabilize food prices in the domestic market; and 3) to stockpile food grains for the poor.

Food security was defined at the World Food Summit in 1996 as existing "when all people, at all times, have physical and economic access to sufficient, safe, and nutritious food to meet their dietary needs and food preferences for a healthy and active life" (Food and Agricultural Organization, 1996). India has prioritized increasing access to food grains through economics. This has been accomplished by the FCI setting Minimum Support Prices (MSP) for each food grain, which have a high caloric value and store well. As a result, the FCI has influenced crop choices by farmers, since MSPs give an economic incentive for the cultivation of certain crops over others. The MSP and other government policies (e.g. the provision of free electricity for irrigation and subsidies on chemical fertilizers) have ultimately supported the cultivation of water-intensive crops – rice and wheat – with the farmer incurring no additional costs for increases in water utility (Figure 2.2). By maintaining higher MSPs for rice and wheat production, the government of India has incentivized farmers to grow a surplus of food and subsequently created stagnation in the food grain market (Roth & Warner, 2007). In 2002, an excess of 48 million tonnes of grain was stored in grain silos, nearly double the 22 million tonnes of grain metric required to monitor national food security (Roth & Warner, 2007). The coexistence of malnutrition and poverty with such large food storage reserves sparked controversy as to whether agriculture policy reforms should be made to reduce excess grain production and water losses.

Food security efforts have also been managed by imposing strict international and trade laws. Import/export tariffs and non-tariff instruments have been used to prevent foreign food production from outcompeting domestic farmers and raising domestic food costs. India was considered to have one of the strictest borders for international trade during the mid-20th century until economic liberalization in 1991. In 1994, India signed the General Agreement on Tariffs and Trade (GATT) treaty, which was a commitment to meet World Trade Organization obligations for domestic and international trade policy. In the mid-1990s, trade laws relaxed for agricultural products, with the exception of cereals and pulses. Although trade liberalization has correlated with GDP growth and poverty reduction, an analysis by the International Food Policy Research Institute found these economic growth indicators not to improve caloric intake and other indicators of nutritional status in India (Panda & Ganesh-Kumar, 2009). India's trade laws were tightened again in response to global food prices increasing during the financial downturn in 2007/08 and continue to receive criticism that its strict trade laws are distorting its domestic food economy. After the release of India's 2013 Food Security Act, the World Trade Organization expressed concerns that the country's inefficient food grain

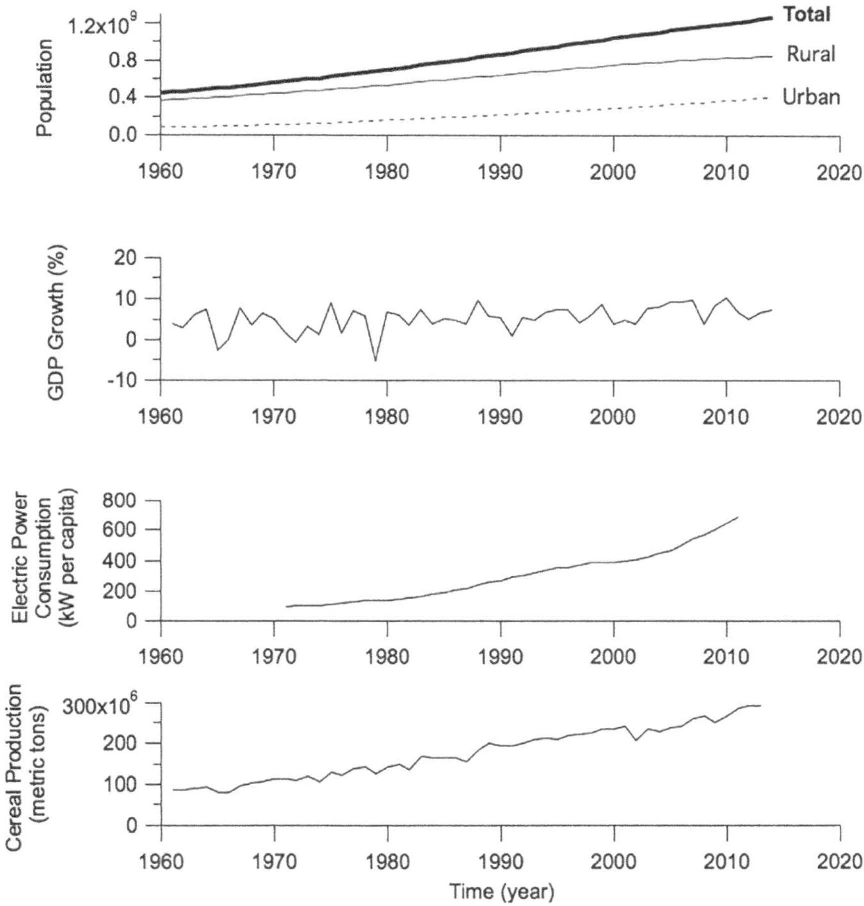

Figure 2.2 Trends in population, GDP growth, electric power consumption, and cereal production from 1960 to 2015 in India

(Data Source: World Bank, 2015).

subsidy program distorted global commodity and trade markets. Despite these claims, India has repeatedly defended its approach to tackling its national food security.

Economics aside, the above sections demonstrate how unilateral policy making in the quest for food security has comingled the food, energy, and water sectors resulting in endemic malnutrition, higher demands on the electric grid, and unprecedented groundwater depletion. Resolving any of these sectorial issues depends on cross-sectorial policies that take the interdependencies of the food, energy, and water sectors into consideration.

Economic liberalization and the path towards the confluence of food–energy–water policy strategies

The coincidence of cross-sectorial issues resulting from the dominance of centralized and sectorial policies in post-Independence India along with global forces of economic liberalization helped facilitate a shift in the governance landscape away from centralized policies. Economic liberalization emerged in India in 1991 after commerce opportunities began to arise from globalization in the 1980s. Globalization forces subsequently worked to change the policy landscape, motivating India to relax its regulatory regime and trade laws to become pro-market. The governance strategies that have evolved since the 1990s have better recognized the co-dependence of the food–energy–water sectors. As a result, India has begun to employ more integrated policy scheme across the water and energy sectors, such as the *Jyotigram Yojana* and the National Solar Mission, to tackle issues with meeting increasing demands on both resources. Additionally, changes in trade law and pro-market policies have together created a promising new policy landscape to encourage corporate and foreign direct investment to tackle its food and energy issues.

Rewiring the water–energy nexus: the Jyotirgram Yojana

India's growing demands in the water and energy sectors have been challenged by a legacy of unilateral energy policy decisions to subsidize heavily electricity and an absence of groundwater regulation that has resulted in both resources being undervalued. From the perspective of neoclassical economics, volumetric pricing by metering wells and tubewells would ensure that water and energy are valued during irrigation. It has been argued, however, that volumetric pricing of irrigation pumps within India's highly informal water economy would be politically infeasible due to the high transaction costs that metering would incur (Shah, Scott, Kishore, & Sharma, 2004; Shah, 2005a). Transaction costs would include the installation and maintenance of meters, reading meters, billing customers, and enforcing penalties for meter tampering and electricity theft. Given that more than 18 million tubewells exist in India, this would be a very costly and logistically difficult feat. These lessons were learned previously by the State Electricity Boards in the 1970s, when meters were used for volumetric pricing of electricity, but then later replaced with a flat-rate tariff payment structure based on the horse power rating of the electric pump set to minimize transaction costs (Shah, 2005b; Shah, Giordano, & Mukherji, 2012). Although a shift to flat-rate tariff pricing reduced transaction costs for the State Electricity Board, it also substantially decreased the marginal cost of electricity and the true cost of water to agricultural users.

A reliance on diesel and electric pumps across India depends on both the presence of an electric grid and the local hydrogeologic conditions. Diesel pump sets, which are less powerful than electric pumps, continue to dominate most of eastern India where alluvial aquifer conditions and a relatively wetter climate have both contributed to a groundwater table that is closer to the land surface. On the contrary, electric pumps are popular in much of northwestern and south-central India that are underlain by less productive hard-rock aquifers, which require farmers to pump deeper to access groundwater and diesel pumps do not suffice. The strong dependency on electric pumps has led some state political leaders, such as in Punjab, Tamil Nadu, and Andhra Pradesh, to refuse flat-rate tariffs all together and force the State Electricity Boards to provide free electricity for agricultural customers (Shah et al., 2012). This has largely contributed to the expansion of groundwater irrigation and done very little to signal the true value of water to its customers, resulting in groundwater overdraft and high electricity demands.

In Gujarat, flat-rate tariffs promoted excessive groundwater pumping and resulted in high demands on the electric grid. Insufficient revenue to support this demand gave the State Electricity Board no option but gradually to reduce the number of hours in a day that electricity was available for agriculture (Shah et al., 2012). In Gujarat, electricity was available only 10–12 hours a day in the early 2000s (Shah et al., 2008). Reducing the quantity of electricity was also combined with a reduction in the quality of electricity. Greater electricity demands, the rise in the use of capacitors (locally referred to as "totas") that increase voltage from single to triple phase, and an increase in electricity theft all greatly reduced the quality of power on the grid (Shah & Verma, 2008). This was not only problematic for the State Electricity Board, but also for farmers who were supplied low quality power for 8–10 hours a day with periodic outages. High demands on the electric grid due to illicit activity created inconveniences for farmers who rely on good quality power supply 30–40 days of the year (Shah, 2005b) to irrigate their crops during critical water stress periods in crop development. Unreliable and variable power voltage frequently damaged electric pump motors and created additional maintenance and replacement costs to farmers (Monari, 2002; Shah et al., 2008).

To circumvent all these issues and avoid pushback from farmers, the northwestern Indian state of Gujarat established a scheme called the *Jyotirgram* ("lighted village") *Yojana* under Prime Minister Narendra Modi when he was Chief Minister of Gujarat between 2003 and 2006. The *Jyotirgram Yojana* physically separated the domestic and agriculture electricity supply to better control pricing and accessibility. The scheme was based on three main policy recommendations given by the International Water Management Institute: 1) gradually raise flat-rate tariffs to meet the average cost of power consumed at a tubewell; 2) reduce agricultural electricity costs by using low-cost and off-peak power (at night); and 3) prioritize demand management of electricity through the provision of quality power at designated times (Shah et al., 2004; Shah et al., 2008). Although the infrastructure costs to rewire rural Gujarat under the *Jyotirgram Yojana* scheme were US$ 290 million, it gave the Gujarat Electricity Board the capacity to regulate groundwater pumping without negating access to electricity for domestic use (Shah et al., 2008).

The Gujarat experiment has proven to be successful in achieving a reduction in electricity use and groundwater over-exploitation. With 90 percent of groundwater withdrawal occurring through the use of an electrified tubewell in Gujarat, a 37 percent reduction in electricity was observed over the first five years (2003–08) of the *Jyotirgram* scheme (Shah et al., 2008). Other benefits included a reduction in power subsidies from the government, from US$ 788 million in 2001–02 to US$ 288 million in 2006–07 (Shah et al., 2008), all while achieving a 10 percent rise in agriculture GDP in contrast to the 2 percent rise for the rest of India (International Water Management Institute, 2011). The benefits of 24-hour domestic electric grid have been used to provide reliable and high quality energy to power schools and hospitals, and have helped broadly to improve rural life from uncertainties in the electric grid. Likewise, the agriculture power supply, although limited to 8 hours per day and pre-scheduled, provides high quality full-voltage power to irrigate crops and protect pump infrastructure, while rationing water consumption through power availability on the grid (Shah et al., 2008).

The *Jyotirgram* scheme is not without its drawbacks. Shah et al. (2008) have reported that the scheme has disproportionately placed a financial burden on marginal farmers who have larger bills to irrigate their crops in comparison to the lower flat-tariff pricing scheme. In addition, landless farmers who primarily work as labourers on larger farms that have a reduced need for irrigation are also disadvantaged with less work opportunities by the scheme. Other problems include illicit connections to the domestic power grid for agricultural activities in more remote villages, giving them 24 hours of triple phase electricity (Shah et al., 2012). Despite these drawbacks, the *Jyotirgram Yojana* is considered to be the second-best option to a volumetric pricing regime that would better reflect the true cost of water and one of the few model examples of how intelligent policies can address a wide range of issues that impact farmers.

Attempts to re-introduce a metered tariff are currently being experimented in northern Gujarat with researchers from Columbia University and the north Gujarat power utility (Uttar Gujarat Vij Company), but involve direct monetary incentives to farmers for installing meters to monitor power and water use (Shah et al., 2012). To successfully expand this into a policy, flat-rate tariffs will need to be higher than metered tariffs to make it economically appealing to farmers (Shah et al., 2012). This may be attainable in Gujarat where flat-rate tariffs are constrained by power allocation, but it is a much more difficult task in states with free and/or extremely low flat-rate tariffs for agricultural electricity use. Nevertheless, it was announced in 2014 by the Indian Ministry of Power that Gujarat's *Jyotirgram Yojana* is to be replicated in Rajasthan and Andhra Pradesh under the Integrated Power Development Scheme with a budget outlay of US$ 400 million. If the Integrated Power Development Scheme is successful as it has been in Gujarat, Rajasthan and Andhra Pradesh will benefit from a practical solution that reduces financial losses to electricity subsidies, while providing reliable power and curbing groundwater usage.

Redesigning the energy sector: The National Solar Mission

The Jawaharlal Nehru National Solar Mission is a major initiative of the Central and State Governments of India with the goal of deploying 20,000 MW of solar power by 2022. The National Solar Mission was launched in 2010 after discussions at India's National Action Plan on Climate Change in 2008. During the discussions, India's vast solar energy potential as a tropical country was realized, as was its potential to promote sustainable growth in the energy sector to address energy security issues and climate change challenges.

The 2022 goal is ambitious considering that India's current solar energy capacity is near 2.2 GW for both on and off grid units (Kapoor, Pandey, Jain, & Nandan, 2014). India's low solar capacity at the onset of the Solar Mission was largely due to the high capital costs of solar infrastructure in comparison to coal-based power generation. In order to make solar cost competitive with coal-based power generation, the primary role of the Solar Mission is to bundle policy and regulatory changes so that capital investment in solar energy and necessary technical innovations take place to lower costs. To achieve this, the Solar Mission is broken up into three phases: Phase I (2012–13), Phase II (2013–17), and Phase III (2017–22). Targets have been set for each Phase with plans to evaluate progress throughout. Solar power generation targets for each phase are 1000 MW, 10,000 MW, and 20,000 MW or more, respectively. The first two phases are designed to leverage the use of commercially available solar heaters and connect un-electrified villages to decentralized solar power, which are both market-ready and considered to be low-hanging fruit. Other objectives to accomplish during the first two phases are to construct a solar grid, increase domestic solar manufacturing capacities, and promote solar technology research and development. Each objective within the Solar Mission relies on a combination of policies and regulation changes to administer each target. These policies include: soft loan and refinancing; solar purchase obligations; by-law and certification regulations; government subsidies; and changes in the market environment.

Soft loans and refinancing options are targeted to support increases in efficiency and cost-reduction for solar heating technologies and their manufacturing, in addition to supporting an incentives package to increase manufacturing capacities for more advanced solar and concentrating collectors. Solar purchase obligations are targeted to support the construction of India's solar grid. A mandate for State Electricity Regulatory Commissions to increase the purchase of solar power generation from 0.25 percent in 2012 to 3 percent in 2022 was made on January 2011 in an amendment to the National Tariff Policy 2006. The solar power purchase obligation for states is likely to complement a Renewable Energy Certificate mechanism that would allow utilities and solar power generation companies to buy and sell certificates to meet the solar power purchase obligation. Regulatory efforts to incorporate solar heating units into the National Building Code and other building by-laws are being planned as a way to increase adoption. This, coupled with a certification and rating program, is being set to ensure quality standards. Government subsidies are being used to promote the use of solar energy where social benefits are provided. As India's rural electrification program is still working to expand the electric grid to off-grid villages, a 90 percent subsidy for solar lights will be provided through the market for these communities by enabling banks to offer low-cost credit. The government is also willing to consider

30 percent subsidies for the use of solar energy in projects that have social benefits such as in schools, dairy plants, self-help group efforts, and cold chain management for medicinal purposes.

In 2014, the Ministry of New and Renewable Energy announced that it would provide US$ 49 million in subsidies to install solar-powered irrigation pumps. Since the Indian government spends US$ 6 billion a year for power and diesel subsidies, it has a large incentive to increase the transition to solar pumps. Using revenue from carbon taxes on coal, this grant program is targeting the installation of 17,500 irrigation systems in partnership with a few state governments: Rajasthan, Tamil Nadu, and Maharashtra. Capital investment for solar pumps is often a prohibitive cost for farmers and economically infeasible due to electric subsidies in the grid. The Jawaharlal Nehru Solar Mission has attracted companies such as SunEdison and Jain Irrigation Systems Ltd. to expand into the solar irrigation market in India. In addition, the emergence of social entrepreneurship has increased a number of smaller companies in the solar market, particularly in the solar lantern market. Demonstration projects have been used by the Government and in partnership with corporate investors to cultivate an environment for solar research and development. The three-phase roll-out of the Solar Mission help provide a predictable incentive structure to better facilitate scalability and capital investment in the use of solar energy. However, more efforts are needed to understand how the expansion of solar-powered irrigation pumps will impact India's water scarcity issue.

Water–food nexus: food trade and virtual water

Food grain production has concentrated in northwest India, primarily Punjab, which has commonly been referred to as "the food bowl of India". Punjab became a major wheat producer for the country during the Green Revolution despite occupying only 1.5 percent of India's land area. Over the past several decades, Punjab has contributed some 40–50 percent of the rice production, 60–65 percent of wheat production and 20–25 percent of cotton production (Aggarwal, Kaushal, Kaur, & Farmaha, 2009). Punjab's productive agricultural industry consumes 85 percent of the water (Tiwana et al., 2007), and has largely been supported by government policies that provide free electricity to farmers resulting in the state having the highest levels of per capita electricity consumption in India (Tiwana et al., 2007). Food grains produced in Punjab have been traded domestically throughout India's Public Distribution System. Domestic food trade in India account for large transfers of "embedded" water, or virtual water trade from one region to another. Inter-state crop trade in India accounts for 106 km^3 of annual water use (13 percent of total water use), with Punjab being the largest exporter of virtual water and eastern India a net importer (Verma, Kampman, van der Zaag, & Hoekstra, 2009). Ironically, the major importers of inter-state virtual water trade are states, such as Bihar, Jharkhand, and Orissa, all located in the northeast of India where water is relatively plentiful (Verma et al., 2009). One explanation is that more favourable agricultural and energy policies in Punjab have brought down the costs associated with irrigation in comparison to the less economically developed, water-rich states in the northeast of India (Verma et al., 2009).

Understanding which regions in India are better suited to be virtual water exporters is necessary so that government policies can be directed towards regions where the hydro-climatic conditions are more appropriate to enhance agricultural growth. Despite the high MSPs for food grains in India, farmer debt is common (Tiwana et al., 2007). Economics are ultimately driving crop choice by farmers (International Food Policy Research Institute, 2007), and water is cheaply valued due to high energy subsidies. In 1985, a committee for the "Diversification of Punjab Agriculture" was established to increase the diversification of crops in Punjab; however, an interest in achieving national food security kept the MSP for food grains high and the dominance of rice and wheat cultivation in the state was preserved (Tiwana et al., 2007). Suggestions for the central government to provide funds to the state government for a "Crop Adjustment Programme" have been made to help provide an incentive for farmers to switch to alternative crops that consume less water (e.g. oil seeds, pulses) and have a greater demand in the market as opposed to wheat and rice (Roth & Warner, 2007). Without adequate economic incentives, farmers are likely to avoid the risk of diversifying crops and will continue the cultivation of wheat and rice as a consequence. Efforts to promote rice cultivation in water-rich northeast India are underway, but await rural electrification and supporting agricultural policies.

Food–energy nexus: pro-market approaches to improve cold chain and storage

The Food Corporation of India manages transportation and storage facilities for food grain production in India; however, India lacks adequate cold chain and storage infrastructure for the estimated 130 million tonnes of fruits and vegetables produced annually. It is estimated that 40 percent of fruit and vegetable production are wasted due to a lack of cold chain and storage infrastructure. This nutritious food waste translates into a financial loss of US$ 2.1 billion and contributes to the endemic malnutrition of India's growing population. There are approximately 6,300 cold storage facilities in India, and an investment of US$ 8.9 billion is required to meet the growing need in cold chain and storage infrastructure for fruit and vegetable production (Emerson Climate Technologies, 2013). Challenges faced by the cold chain and storage industry comprise rising real estate costs, high energy costs, an unreliable power grid, and an uneven network of temperature-controlled supply chains and cold storage facilities across India (Emerson Climate Technologies, 2013). The ongoing need for cold chain and storage facilities have only been recognized by the Indian government in the past couple of years. In 2012, the Ministry of Agriculture offered US$ 52 million to set up 451 cold storage facilities in India. It has also established in 2011 the National Centre for Cold Chain Development (NCCD), which is a multi-stakeholder task force comprised of government officials, industry, expert organizations, and grower associations that recommend technical standards for cold chain infrastructure to the government (National Center for Cold Chain Development, 2012). In addition, the government of India has approved foreign direct investment in multi-brand retail, which may facilitate infrastructure investment by foreign multinational corporations (Patibandla, 2014).

While half of the rural population in India (2011–12) depend on agriculture livelihoods, 25.7 percent of rural inhabitants reside below the poverty line (Indian Ministry of Agriculture, 2013). Government subsidies for food grain production and the storability of grain products post-harvest are appealing to smallholder farmers who have limited access to markets and cold storage infrastructure to meet market requirements. A market failure in the cold chain and storage infrastructure results from the absence of public–private partnerships and farmer cooperatives that can ensure that smallholder farmers are not excluded from horticultural sales in the market (Narrod et al., 2009). Smallholder farmer families often store grains as financial capital, so they can sell it back to the market when the prices are high again after harvest season. Despite the benefits of storing food grains as a risk management strategy, they offer marginal earnings and little nutritional value beyond caloric intake for households. Nutritional foods such as fruits, vegetables, and dairy are considered to be high value, but require cold storage and transportation infrastructure that are accessible to farmers. Inadequate and unreliable power supply and transportation routes (roads) to cold storage facilities that plague larger companies in India also limit adoption of these crops by farmers, amongst other factors such as water availability and high agriculture input costs.

A rural development program to help rural dairy producers connect to customers in towns and cities was launched by the National Dairy Development Board in 1970 and referred to as "Operation Flood". Operation Flood has been successful and resulted in the expansion of dairy cooperatives, increased milk production, increased rural income, and reasonable prices for customers. Investments in infrastructure are necessary to further enhance the production of high value and nutritious crops. This would benefit farmers and businesses alike, while increasing market access.

Conclusion

India is attempting to gradually shift away from insular centralized government policies towards developing a more integrated policy landscape across the food, energy, and water sectors. India's colonial legacy has played a large role in motivating much of the nationalistic policies during the 1950s–1970s that resulted in the Green Revolution. The expansion of business opportunities during the 1980s helped India shift its governance landscape towards integrating cross-sectorial policies in the 1990s. This new landscape has helped India begin to confront current socio-ecological issues within the food–energy–water nexus, which were inadvertently caused by policies established during the Green Revolution.

Water–energy policy schemes, such as the *Jyotigram Yogana* in the northwestern states of India, exemplify how tackling multi-sector issues simultaneously can increase water and energy supplies for the agriculture sector while saving government funds. On a national level, the Indian Solar Mission moves beyond traditional centric policies by creating partnerships with corporations, engaging entrepreneurs and using the market to roll out solar products. The need for cold chain and storage infrastructure has also recently increased government willingness to accept foreign direct investments from multinational corporations. However, India's willingness to support business needs to be supplemented with improvements to general infrastructure, such as roads and electricity, so that businesses can properly function. Growing demands in the food, energy, and water sectors will require more partnerships between the public and private sectors, as well as further changes in domestic and global trade laws and regulations. In this new governance landscape, it is hoped that improved market access and coordination between public and private institutions can better position India to target the interdependent challenges in the food, energy, and water sectors.

Note

1 Rainwater harvesting in India has recently been reintroduced by both Gandhian philosophy-based non-profit organizations and by the government as a water-management strategy.

References

Agarwal, A., & Narain, S. (Ed.). (1997). *Dying Wisdom: Rise, Fall and Potential of India's Traditional Water Harvesting Systems*. New Delhi: Centre for Science and Environment.

Aggarwal, R., Kaushal, M., Kaur, S., & Farmaha, B. (2009). Water resource management for sustainable agriculture in Punjab, India. *Water Science & Technology*, 60(11), 2905–2911.

Badiani, R., & Jessoe, K. K. (2011). Electricity subsidies for agriculture: Evaluating the impact and persistence of these subsidies in India. University of California Davis, Working Paper. Retrieved 6 August 2013, from http://econ.ucsd.edu/CEE/papers/Jessoe_4april.pdf.

Bhatia, B. (1992). Lush fields and parched throats: Political economy of groundwater in Gujarat. *Economic & Political Weekly*, *27*, A142–A170.

Birner, R., Gupta, S., & Sharma, N. (2011). *The Political Economy of Agricultural Policy Reform in India*. Washington, DC: International Food Policy Research Institute.

Davis, M. (2001). *Late Victorian Holocausts: El Nino Famines and the Making of the Third World*. London: Verso Books.

Dubash, N. K., & Rajan, S. C. (2001). Power politics: Process of power sector reform in India. *Economic & Political Weekly*, 3367–3390.

Emerson Climate Technologies (2013). The food wastage & cold storage infrastructure relationship in India. Retrieved 15 January 2015, from www.emerson.com/SiteCollectionDocuments/India%20Cold%20Storage%20Report%202013/Report_layout_Reduced.pdf.

Food and Agricultural Organization (FAO). (1996). Declaration on world food security. *Report of the World Food Summit*. Rome.

India Energy Portal. (2005). Number of irrigation pumps. Retrieved 15 July 2015, from www.iasri.res.in/agridata/05data/chapter%203/db2005tb3_38.htm.

Indian Ministry of Agriculture (2013). Pocket book on agricultural statistics 2013. Retrieved 2 February 2015, from http://eands.dacnet.nic.in/Publication12-12-2013/AgricultralStats%20inside_website%20book.pdf.

International Food Policy Research Institute (2007). *Withering Punjab Agriculture: Can It Regain Its Leadership*. New Delhi: International Food Policy Research Institute (IFPRI).

International Water Management Institute (2011). *Innovative electricity scheme sparks rural development in India's Gujarat State*. International Water Management Institute (IWMI).

Kapoor, K., Pandey, K. K., Jain, A. K., & Nandan, A. (2014). Evolution of solar energy in India: A review. *Renewable and Sustainable Energy Reviews*, 40, 475–487.

Malone, D. M. (2011). *Does the Elephant Dance?: Contemporary Indian Foreign Policy*. Oxford: Oxford University Press.

Monari, L. (2002). *Power Subsidies: A Reality Check on Subsidizing Power for Irrigation in India*. Public Policy for the private sector; Note no. 244. Washington, DC: The World Bank.

Narrod, A. C., Roy, D., Okello, J., Avendano, B., Rich, K., & Thorat, A. (2009). Public-private partnerships and collective action in high value fruit and vegetable supply chains. *Food Policy*, 34(1), 8–15.

National Center for Cold Chain Development (2012). National center for cold chain development. Retrieved 28 January 2015, from www.nccd.gov.in/AboutUs.html.

Pacey, A., & Cullis, A. (1986). *Rainwater Harvesting: The Collection of Rainfall and Runoff in Rural Areas*. London, UK: Intermediate Technology Publications.

Padhy, S. (2008). *Greening Law: A Socio-Legal Analysis of Environmental Human Rights in India*. ProQuest.

Panda, M., & Ganesh-Kumar, A. (2009). *Trade Liberalization, Poverty, and Food Security in India*. Washington, D.C: International Food Policy Research Institute.

Pandey, D. N., Gupta, A. K., & Anderson, D. M. (2003). Rainwater harvesting as an adaptation to climate change. *Current Science*, 85(1), 46–59.

Patibandla, M. (2014). Implications of foreign direct investment in India's retail sector. *IIMB Management Review*, 26(4), 214–221.

Rawat, S., & Mukherji, A. (2013). Poor state of irrigation statistics in India: The case of pumps, wells and tubewells. *International Journal of Water Resources Development*, 30(2), 262–281.

Repetto, R. C. (1986). Skimming the water: Rent-seeking and the performance of public irrigation systems. World Resources Institute.

Rodell, M., Velicogna, I., & Famiglietti, J. S. (2009). Satellite-based estimates of groundwater depletion in India. *Nature*, 460(7258), 999–1002.

Roth, D., & Warner, J. (2007). Virtual water: Virtuous impact? The unsteady state of virtual water. *Agriculture and Human Values*, 25(2), 257–270.

Shah, T. (2005a). Groundwater and human development: Challenges and opportunities in livelihoods and environment. *Water Science & Technology*, 51(8), 27–37.

Shah, T. (2005b). The new institutional economics of India's water policy. *African water laws plural legislative frameworks for rural water management in Africa*, January, 1–26.

Shah, T. (2009). Climate change and groundwater: India's opportunities for mitigation and adaptation. *Environmental Research Letters*, 4(3), 035005.

Shah, T., Bhatt, S., Shah, R. K., & Talati, J. (2008). Groundwater governance through electricity supply management: Assessing an innovative intervention in Gujarat, western India. *Agricultural Water Management*, 95(11), 1233–1242.

Shah, T., Giordano, M., & Mukherji, A. (2012). Political economy of the energy-groundwater nexus in India: Exploring issues and assessing policy options. *Hydrogeology Journal*, 20(5), 995–1006.

Shah, T., Scott, C., Kishore, A., & Sharma, A. (2004). *Energy-Irrigation Nexus in South Asia: Improving Groundwater Conservation and Power Sector Viability*. Colombo, Sri Lanka: International Water Management Institute.

Shah, T., & Verma, S. (2008). Co-management of electricity and groundwater: An assessment of Gujarat's Jyotirgram scheme. *Economic and Political Weekly*, 43(7), 59–66.

Tiwana, N. S., Jerath, N., Ladhar, S. S., Singh, G., Paul, R., Dua, D. K., & Parwana, H. K. (2007). *State of Environment: Punjab 2007*. Chandigarh: Punjab State Council for Science & Technology.

United Nations, Department of Economic and Social Affairs, Population Division (2015). *World population prospects*. New York. Retrieved from http://esa.un.org/unpd/wpp/Publications/Files/Key_Findings_WPP_2015.pdf.

Verma, S., Kampman, D. A., van der Zaag, P., & Hoekstra, A. Y. (2009). Going against the flow: A critical analysis of inter-state virtual water trade in the context of India's National River Linking Program. *Physics and Chemistry of the Earth, Parts A/B/C*, 34(4–5), 261–269.

World Bank (2015). India world development indicator dataset. Retrieved 15 July 2015, from http://data.worldbank.org/country/india.

3 The evolution of the narrative of corn in Mexico

From impediment to progress, to a commodity and as heritage

Alexandra Littaye

Introduction: how narratives affect food security

Food security is a topic widely argued in reference to food systems and the actors and institutions involved in bringing food from farm to fork. Food security is rarely discussed in relation to the way food is perceived, or the circumstances under which it is eaten or appreciated (Warde, 2016). This chapter traces the evolution of the narrative surrounding corn in Mexico as it refers to political, social, and cultural contexts. Different actors, from the government to revolutionaries and Non-Governmental Organizations (NGOs), have highlighted corn as a subsistence crop, inferior to its European counterpart wheat; a commodity crop and more recently, it has been seen as a heritage crop by international NGOs. Each of these narratives makes claims about corn as a means of governance that shapes the very identity of the Mexican people. Manipulation of these perceptions – and narratives of corn – has had profound effects on the social, economic and political fabric of Mexico and on its food security.

This chapter presents an argument in five parts. In the first, this chapter gives an overview of the narrative surrounding corn in Mexico over the past century. This section serves as an explanation of a contextual background to better understand the evolution of corn as an impediment to progress to corn as a global commodity. It explores the roots of its political dismissal as an emblem of indigeneity and of inferiority to European ideals. It then considers the reappraisal of corn as a symbol of Mexican identity during the Mexican Revolution. Corn is then articulated as a commodity by neo-liberal policies ushered in with the North American Free Trade Agreement (NAFTA)/TLC (Tratado de Libre Commercio de America Norte). As discussed, each of these narratives has had direct consequences on the food security and land rights of indigenous farming communities. The second part of this chapter explores how corn has been appropriated by various activist groups in Mexico as a symbol of Mexican heritage to fight against what is perceived as economic American imperialism. The third part delves into the case study of pinole, a traditional Mexican corn flour and the celebration of Mexican corn diversity by the NGO, Slow Food International. This case study, called the pinole project, underlines the current reality and poverty facing rural and indigenous communities in Mexico today as well as the climate change-related pressures that the food system faces. In the final section, this chapter makes an argument for a greater appreciation of 'transnational entrepreneurship' that rises from the struggle of local farmers in the face of an increasingly globalized food system; it also teases out how corn can be commodified as a heritage food for global elites.

This chapter addresses several key issues that are important to the governance of food, and particularly corn, in contemporary Mexico. These include: 1) filling gaps within the recent literature on food security and food narratives; and 2) understanding transnational entrepreneurship, as well as the ideas of 'immigrant transnationalism', an 'alternative form of economic adaptation of foreign minorities in advanced societies that is based on the mobilization of their cross-country social networks' (Portes, Guarnizo, & Haller, 2002, p. 278). Further, the presented pinole project typifies Fletcher's argument that global engagement results from the development of 'locally responsive cross-border activities' (2004, pp. 290–291) from which springs transnational entrepreneurship (Cordero-Guzmán, Smith, & Grosfoguel, 2001). This chapter concludes by arguing that claims about corn have had direct effects on the food security and health of Mexico's people. Though these effects have been mainly deleterious, more recently they are being harnessed and their governance reconceptualized by global grassroots movements to re-empower rural communities, address health issues linked to diets, as well as shed light on agro-biodiversity loss.

Methodology

Research for this chapter was supplemented by grey literature on the pinole project, constituted of two cooperatives called Amigos de Ozolco (in Mexico) and Blue Corn Alianza (in Philadelphia, USA). This grey literature included: the official websites of Blue Corn Alianza, the Facebook pages of Amigos de Ozolco, promotional brochures, journal articles (predominantly from Philadelphia Weekly), and reports released by Save the Children. This chapter uses discourse analysis encompassing both text and practice. This method of discourse analysis can elucidate the translocal nature of networks whilst addressing gaps identified within 'material geographies' (Goodman, 2015) and 'food studies' (MacDonald, 2013). This functional approach to discourse analysis focuses on the role that narratives play in the construction of reality and the meanings that emerge from these narratives, and that are shared (Bruner, 1991).

Data about the pinole project were collected during a four-year ethnographic research project between 2012 and 2016. During this period, three fieldtrips were made to Mexico, lasting in total six months and two visits were made to Slow Food's Terra Madre (an international biannual food festival) in 2012 and 2014 in Italy. In addition to months of participant observation, over 40 participants were interviewed. Three-quarters of the participants were recruited using the snowball method, starting from contacts in Terra Madre. Respondents were predominantly male, between 30 and 50 years old, due to the snowballing exercise, and the preponderance of male actors in alternative food networks. An attempt was made to rectify this discrepancy and as many women as possible were interviewed, although most members running various branches of Slow Food International (SFI) are men.

The central role of corn in political discourse

Tortilla discourse

There are several myths surrounding the creation of man in Mesoamerica, such as the Aztec creation myth, which claims man was created and perfected with corn. Corn has been at the centre of Mexican identity since the establishment of Mexico. Corn, according to Lind and Barham (2004, p. 52), is a 'thoroughly cultural artefact and its domestication and improvement is strongly correlated with the development of cultural complexity and the rise of stratified civilization of pre-colonial Mesoamerica'.[1]

The production of the Tortilla Discourse served the political and revolutionary aims of constructing a new society with an ideological hegemony that overlooked social inequalities. The Tortilla Discourse, a term invented by the scholar Pilcher (1998), represents the articulation of the attack launched against corn by Mexican elites. It had very practical applications for the way in which Mexicans created their realities, notably how they decided to envisage themselves within those realities. The first stirrings of the Tortilla Discourse in the 1880s saw that missionaries, grocers, chefs, social workers, agronomists, and politicians were enlisted in an attack on the tortilla, promoting its replacement by leavened wheat bread. Spaniards considered corn to be an inferior grain that produced coarse foodstuffs and tried to 'eradicate the existing culture of [corn]' (Pilcher, 1998, p. 3). This continued after Mexican independence and was reinvigorated in the later part of the 19th century by the newly emerging science of nutrition.

The Tortilla Discourse gained support with – and found justification in – the emerging, although still imprecise, science of nutrition. The birth of the Tortilla Discourse mirrored the shift amongst Mexico's elites away from tradition towards science, which held the promise for modernity since it represented 'objective truth'. Mexico's salvation was to be sought in science, a rhetoric, which still permeates its current political discourse. Corn was thus devalued from being a source of life throughout Mesoamerican history to being depicted as a root of the perceived debility of Mexico's mixed (mestizo) and indigenous populations.

Elites blamed corn for the failure of the national development campaign after Mexican independence in 1821, and considered it as the root for Mexico's backwardness, most notably amongst indigenous populations who were predominantly farmers. Wheat, associated with the Eucharist and European lifestyle, was perceived as superior. Bulnes, an influential intellectual during the regime of Mexican President Porfirio Diaz during the late 1800s, blamed the 'debility of Indians on their corn-based diet' (Pilcher, 1998, p. 82).

Post-Mexican Revolution

The Mexican Revolution in 1910 gave rise to conflict between competing elites and an opportunity for agrarian insurrection. The revolution undermined rights of farming communities and dispossessed many of their land as it led to the creation of policies that facilitated the expropriation and distribution of the rural population's land. Governmental control over the nation's resources by the revolutionaries was asserted through Article 27 of the Constitution in 1917. Through this policy-making, the revolutionaries sought to modernize Mexico and its rural population with agrarian and labour reforms embodied in the drafting of this Constitution.

The Tortilla Discourse thus achieved its most insidious effect by shifting debate away from the question of land reform, which could have benefitted *campesinos*[2] nutritionally, and directing it toward a dietary change that promised to make them more dependent on markets without offering corresponding nutritional gains (Pilcher, 1998). In 1940, 45 percent of Mexicans ate wheat bread occasionally. By 1950, 55 percent of Mexicans were eating it daily (Pilcher, 1998). Simultaneously, as DeWalt and DeWalt (1991, p. 194) show, towards the 1950s, Mexico was undergoing a transition that was effectively eliminating many small farmers: 'although employment in the commercial agriculture sector was growing, it was not growing at a fast-enough pace to absorb the farmers displaced from rural farm communities'. The large displacement of rural small farmers had several disastrous consequences on the political and social scale.

The Mexican food policy of 1953 sought to support prices for basic staples, which, despite supporting small-scale farmers, ended up benefiting large farmers inordinately (DeWalt & DeWalt, 1991). A programme called the National Company of Popular Subsistences (CONASUPO) was created in 1962 with the intention of guaranteeing the purchase and regulation of prices of corn and other products. This cheap food strategy kept the prices of staples down so that urban industrial wages could be kept low. Commercial farmers saw little incentive to grow staple crops whose prices were kept low by the government and instead, by the 1960s, under the doctrine of comparative advantage, they turned to producing export crops and crops to feed livestock. Small-scale farmers on the other hand, kept producing staple crops.

As a consequence of these policies, a fourth of all wheat, corn, and sorghum in Mexico was imported by 1980, which amounted to a substantial quantity of food. Small-scale farmers could not afford to cultivate the main staple crops as a result of the systematic subsidization of the government that went towards the development of commercial agricultural sector and that discriminated against small-farmers (DeWalt & DeWalt, 1991). Agriculture thus became a drain on the economy. By the 1980s, a third of Mexico's arable land lay idle (around 9 million hectares). Five years later, Reig (1985) estimated that between 40 percent and 50 percent of the population in Mexico were either undernourished or malnourished.

International influence

The Tortilla Discourse pushed policies with different effects (DeWalt and DeWalt, 1991). In the first instance, the implementation of a series of agrarian reform programmes sought the application of modern technology for an increase in agricultural productivity. These agricultural development initiatives were sponsored by agencies for international development, the World Bank, and the International Monetary Fund. This enabled the economic output needed for greater investment in research. Finally, the adoption of a comparative advantage approach enabled an 'outward orientation' and the adoption of politics that aimed at stimulating global competitiveness. These policies were aimed to promote global trade. This eventually led the Tortilla Discourse to embrace corn. Instead of trying to eliminate it, it sought to commodify it. This occurred on several occasions and at different levels of society. In a first such instance, dehydrated tortilla flour made an appearance on the market, bringing the CEO of Maseca, the organization that first produced it, into Forbes' list of billionaires in the 1990s. Again, corn's value was being captured by the very same elites that pulled it out of local production and consumption.

In a second instance, this industrial production mirrored the expansion of large agribusinesses, which relocated small-scale farmers and applied genetics and chemistry to high-yielding varieties of corn. Peasants were effectively displaced by the commercialized export-oriented agriculture. When Dr Norman Borlaug won the Nobel Peace Prize for his contribution to the Green Revolution, Mexico's *campesinos*' redemption seemed to lie not in the replacement of corn by wheat; instead, it relied on the transformation of corn through science and technology into a genetically modified commodity. Unfortunately, as Baker (2013) claimed, the requirements of the 'miracle seeds' were biased towards larger landholders because of government subsidies. Despite the Green Revolution, food production did not keep up with population growth between the 1960s and the 1980s and Mexico increasingly became dependent on large amounts of imported staples such as beans, corn, and wheat.

The Tortilla Discourse, the Mexican Revolution and the Green Revolution each in turn weaved narratives around corn as a means to establish and diffuse a particular political rhetoric and perpetuate particular practices of governance. These have spanned identifying corn as a symbol of inferiority and indigeneity, as impediment to cultural and economic advancement, as well as an obstruction to progress promised by science.

Neo-liberalism vs heritage preservation

Since the 1990s, opposing global narratives have been fighting for the authoritative articulation of corn's nature. The first is the neo-liberal discourse of free trade and corn as a global market commodity and the other is Slow Food's celebration of blue corn as a heritage crop by international networks of concerned consumers.

In an article exploring the relationship between the law and the cultural change in the production and consumption of corn in Mexico, Hernandez-Lopez (2008) details the successive deconstruction of the protection of Mexico's crop production by NAFTA and the 2008 treaty obligation, which effectively ended the tariff protection on corn. Hernandez-Lopez shows how NAFTA and the dismantling of tariffs on corn imports saw the price of corn rise by 206 percent in 1994 as

Mexican corn could not compete with its American counterpart. Between 1997 and 1999, the price of tortillas increased by 127 percent and between 2000 and 2002, by another 22 percent – due, in part, to the global fluctuations of corn market, which affected Mexico directly since it no longer had any tariff protection (Hernandez-Lopez, 2008).

The implementation of the NAFTA in 1994 and further legislation aimed at neo-liberal economic restructuring, enabled further privatization as well as sale of commonly held lands (Eakin & Lemos, 2006) whilst perpetuating the stigma of indigenous farmers as backwards and an impediment to progress – as first postulated in the Tortilla Discourse (Pilcher, 1998). The state withdrew participation in the basic grain market. Under the theory of comparative advantage, most of Mexico's indigenous farmers and their techniques were considered inefficient as average corn yields were lower than the US's (Pilcher, 1998; Baker, 2013). Neo-liberal policies increased corn imports and encouraged farmers to convert their land to crops that were supposedly at a comparative advantage, such as counter-seasonal and tropical fruits and vegetables (avocados and tomatoes).

NAFTA and other economic policies have favoured large-scale commercial agricultural holdings and intensive farming practices (Baker, 2013). Neo-liberal regulations dismantled the support and protection that the government gave to small-scale farmers, further marginalizing them and increasing the rate of urban migration. Mexican migration increased by 75 percent in the five years after NAFTA took effect (Washington Post, 2012). At the same time, consumer food prices rose, epitomized by the rising cost of tortillas, still the staple food in Mexico today.

Policy-makers deemed protectionism as backward – a thing of the past – and an impediment to modernity. The removal of a fixed price for basic grains, combined with the flood of subsidized corn imports from the United States, means that the price of corn is below the Mexican cost of production. Small- and medium-scale farmers are no longer able to sell their corn on the national market for profit. It is important to note that Mexican *campesinos*' subsistence production and corn production have remained stable since 1994 (Preibisch, Herrejón, & Wiggins, 2002). As Escobar Moreno points out, this is due partly to rural people becoming more economically insecure and hence growing corn for its contribution to household food security (Escobar Moreno, 2005). In 2010, 20 million Mexicans lived in 'food poverty' and another 25 percent of the population did not have access to basic food whilst one-fifth of Mexican children suffered from malnutrition (Carlsen, 2013).

The effects of NAFTA have diminished Mexico's food security: NAFTA triggered a decline in the real price of corn (Mexico's main agricultural import) on employment and on farmers' incomes. US corn exports to Mexico account for almost all Mexican corn imports (Prina, 2013). Mexico now has a high dependence on food imports from abroad that has adversely affected the Mexican economy since 2008, compared to 30 years ago (Sanchez & Moreno, 2013). Indeed, Mexico is no longer self-sufficient in corn production: it imports 30 percent of its corn (most of it transgenic) from the United States (Schepers, 2014).

NAFTA mirrors the Tortilla Discourse's emphasis on economic neo-liberalism as a route for modernity and progress. Hernandez-Lopez (2008) claims that ending the state's support for corn poses significant challenges for Mexican democracy. Corn, which was a source of life for Mesoamericans, is now a source of fuel as demands for ethanol rise, and a source of feed as the global demand for livestock increases.

This removal of tariff protection is mirrored by the penetration of fast food outlets in Mexico, which has caused devastating nutritional consequences. Today, as Ulijaszek, Mann, and Elton (2012) articulate, Mexico is an economically-emerging nation affected by globalization and beset with two types of nutritional health problems: undernutrition amongst its poorest people and obesity among its less-poor and middle-class citizens. They locate the origins of these health issues within social inequalities, colonialism, and early globalization. Ulijaszek et al. (2012, p. 262) claim that the industrialization of the food system, in which internationally branded fast foods have become indigenized by adapting to local tastes, is 'perhaps an unwitting accomplice of nutritional structural violence'. In this sense, the perception of corn as an impediment to progress was linked to the radical alteration of diets and to the dramatic obesity rates and malnutrition.

In 2001, researchers at the University of California, Berkeley, reported that genetic material from GM corn had already contaminated local populations of native corn cultivars (Quist & Chapela, 2001). Regardless, in 2009, Monsanto, Dow, and DuPont Pioneer were granted approval to grow GM corn for research in Mexico. On July 5th 2013, a collective lawsuit was filed by a coalition of activist groups to stop them. The coalition consisted of 53 individuals (scientists, journalists, chefs etc.) and 20 civil organizations advocating for the human right to a healthy environment for their development and wellbeing (Public Citizen, 2014). For the first time in court, the very narrative of corn was debated between proponents of conflicting ideologies: corn as commodity and corn as heritage.

The suit claimed that transgenic corn threatens the biodiversity of traditional varieties grown by subsistence farmers and smallholders throughout Mexico. Mexican farmers ensure the preservation of 59 unique landraces native to Mexico. This pool of genetic diversity is one of the cornerstones to facing the challenges of climate change, global food insecurity, chronic issues of hunger, obesity and malnutrition. *Milpa* farming – a traditional type of farming in Mesoamerica that produces, amongst other crops, maize, beans and squash from a relatively small section of land that is intensively farmed for two to three years and then left fallow for one year – is a technological pillar to maintaining agro-biodiversity as it involves the growth of a variety of crops and plants in symbiosis, directly preserving heirloom seeds and indirectly supporting the wildlife and biodiversity that form the ecosystem. Simultaneously, *milpa* farming is inherent to indigenous culture, a key to human development, social cohesion, and cultural practices in rural Mexico. More importantly, it is a pathway towards building resilience for the rural poor of Mexico by connecting them to global markets.

Pinole as endangered food

The pinole project is emblematic of this wider struggle facing Mexico and the fight of small-scale farmers for food sovereignty and large corporations. NAFTA has had a catastrophic effect on Ozolco San Mateo, a small village perched on an active volcano in Mexico that suffers one of the worst adolescent migration drains in the world (Amavizca-Ruiz et al., 2014), with many young men migrating to Philadelphia. Ozolco is home to an average of 2500 indigenous *milpa* farmers who grow blue, white, yellow and red corn. The development context of Ozolco is typified in the fact that they only acquired an antenna in 2012, which allowed them, for the first time, to use cell phones. In the case of Mexico's corn market, the US incursion of cheap corn has pushed many small-scale farmers such as Ozolco's farming community, to seek further business opportunities elsewhere, to be able to 'exploit competitive advantages beyond domestic markets' (Winch & Bianchi, 2006: 74).

To counter the effects of NAFTA and secure the preservation of blue corn, the grassroots movement Slow Food International (SFI) has been very active in attempting to revive the interest in pinole and the other 59 native Mexican corns. SFI believes that the industrial agro-food system is eroding the biological and cultural diversity associated with food, and has weakened local food traditions (Schneider, 2008, p. 389; Parkins, 2004; Leitch, 2003).

To support Mexico's agro-biodiversity, SFI recognized pinole as a product 'threatened by extinction' (Littaye, 2015). SFI uses biblical terminology to compare disappearing food traditions to species that were saved by Noah's ark. The movement aims to rekindle traditional food cultures seen to be 'at risk', promoting 'edible biodiversity' of 'exceptionally good' and 'forgotten flavours' (Petrini, 2007). Its founder, Carlo Petrini, conceived an 'Ark of Taste' – an online inventory or database of heritage food products created by the Slow Food International organization – to reconcile the priorities of both the gourmet and the environmentalist (Leitch, 2000, p. 111).

The organization's focus is centred on reactivating the local networks of farmers such as the producers of pinole not only to revive producers' confidence but also to encourage commercial interests in complementary ways. The organization's activities are mainly focused on food producers. *Campesinos* are not, as the Tortilla Discourse of NAFTA would have it, impeding progress. According to SFI's rhetoric, they are 'bearers of culture' (Petrini & Watson, 2001, p. 2) with invaluable tacit knowledge necessary for the preservation of traditional foods and recipes as well as the agro-biodiversity. Indeed,

> most of the world's hungry people are not urban consumers and purchasers of food, but peasant producers and sellers of agricultural products. Further, their high number is not a simple heritage from the past but *the result of an ongoing process* leading to extreme poverty for hundreds of millions of deprived peasants.
>
> (Mazoyer & Roudart, 2006, p. 10; emphasis added)

Though it offers a clear criticism of globalization, SFI is not an anti-globalization movement. On the contrary, SFI promotes itself as a 'virtuous globalization' organization. Unlike neo-liberal rhetoric, it believes in fostering global trade between small-scale, local producers with global networks of 'foodies'. The grassroots organization partners with academics, governments, and with 'food *corporations* who *wish* to *gain* access to the Slow Food membership base' (original emphasis, Chrzan, 2004, p. 121).

Amigos de Ozolco (Amigos), a Mexican organization based in Ozolco that produces pinole, took advantage of Slow Food's network. It partnered with the cooperative Blue Corn Alianza (Alianza), run by Mexican transmigrants in Philadelphia to distribute Amigos' pinole as a premium product throughout the United States. The partnership between these organizations is referred to throughout this chapter as the 'pinole project'. According to Alianza's website, the partnership was designed to preserve project members' hometown, Ozolco. Towards this end, the pinole project strives to revive Ozolco's failing local economy. The pinole project grew and sourced blue corn from Ozolco's farmers and sold it in Philadelphia through Mexican migrants who were born in Ozolco (Littaye, 2015, 2016). Pinole was then identified by SFI as 'endangered' and listed in 2012 on the Ark of Taste as a food threatened by the erosion of indigenous farming practices and devastating migration (Littaye, 2015, 2016). Mexican corn is now gaining increasing traction amongst global elites as a heritage food (Pilcher, 2012) whilst still being widely traded as a global commodity for feed, food, and as an alternative energy source.

Transnational entrepreneurship as solution?

In this vein, SFI leverages its reputation to help the creation of micro-markets (like village markets in Mexico City) and other promotional initiatives (Fonte, 2006) such as inviting pinole producers to the international gathering of Terra Madre in Italy in 2012. As SFI's expertise lies in marketing and publicity (Leitch, 2008), it can play a role in helping pinole producers by securing a consumer base internationally. The pinole project provides an example of 'cross-border co-ordinations' (Fletcher, 2004, p. 302), uniting local pinole producers to immigrant networks in the United States. Translocal ties between Ozolco and global markets and the United States have thus formed. They have also spurred economic activity as well as transnationalizing entrepreneurship, underlining how 'contemporary processes of entrepreneurship are no longer bounded to the local' (Yeung, 2009).

To support actors in peripheral economic spaces, internationalization endeavours need to create the transnational linkages that bring local and extra-local actors into play, whilst providing the financial security to turn subsistence farming into export farming in order to reach global markets. The case study of the *pinole* project typifies 'how economic globalisation is performed by small actors on the geographic margin' (Young, 2010, p. 838). SFI has since successfully applied for large grants from academic and governmental institutions as well as philanthropic foundations. In September 2014, SFI won a prize from the Universidad del Valle de Mexico, to develop the network of *milpa* producers and create the manual for the 'sustainable development of eco-gastronomy' in Mexico. In September 2015, after a successful application, SFI received a $250,000 grant by the Ford Foundation to identify, unite and support small-scale indigenous farmers in Mexico for two years. By being recognised and supported by international NGOs such as the Ford Foundation, the pinole project – and Mexican corn – are legitimized as fundamental to agro-biodiversity and to climate adaptation. In this sense, corn enters a new political and global debate. In addition, gathering these *milpa* farmers in a broader network facilitates SFI's aim to support them in gaining a foothold in the national or international markets, moving *milpa* producers from the margins of the economy of subsistence agriculture to the status of exporters and stewards of the environment (Petrini, 2007).

As part of its contract with the Ford Foundation, the SFI needs to link businesses in the United States with producers and ensure their partnership. SFI identified two enterprises with which to partner their budding network of *milpa* producers that will use their blue corn to develop a line of food products. Both companies are to buy produce at a fair price from the network of producers.

In the case of pinole producers, the de-localization of consumption is necessary to preserve local production and agro-biodiversity linked to blue corn. The pinole project can target the quickly growing demand of the gluten-free market by promoting pinole as a premium, healthy, gluten-free and heritage food from Mexico. Catering to a niche market of health foods and gluten-free consumers, the pinole project engages in a process similar to what Winch and Bianchi (2006) label a 'deep niche strategy'.

This process consists of narrowing what product an enterprise offers (in this case, the pinole project) in order to catch a segment of the market dedicated to a limited number of customers or geographical markets. Small companies or producers can maximize the effectiveness of their development capabilities and SFI can forge translocal links between farmers, consumers and markets. In other words, it could be a tripartite link between producers–businesses–retailers and, as at a global level, between international markets and Mexico. Engaging local (businesses) networks (Young, 2010) is a critical factor for stimulating product and market innovations (Huggins & Johnston, 2009). In this case, SFI helps market pinole as heritage food in its Ark of Taste, transforming it from an obscure local produce to a global commodity. The pinole project thus epitomises Fletcher's ideas that international entrepreneurship 'is a heavily contextualized and socially constructed activity that occurs through joint cross-border co-ordinations' (2004, p. 302). As this chapter underlines, these co-ordinations can be supported by SFI, highlighting 'how processes of internationalization come about and how owner-managers enact images or opportunities of international business as they construct locally responsive cross-border activities' (Fletcher, 2004, pp. 290–291). The narrative of corn as a heritage also has caveats. An important one is that SFI operates within a system that commodifies authentic, exotic, or health foods (Littaye, 2015) and takes advantage of the promotional and marketing strategies that have been linked to 'food fetishism and . . . consumer fetishism' (Paxson, 2005). Indeed, SFI is an actor that capitalizes on its links with small-scale farmers as well as large multi-national retailers. By postulating pinole as 'endangered', SFI is enshrining pinole in a sense of authenticity. Paradoxically though, 'authenticity necessitates that the ethnic groups essentialise and spectacularise themselves in order to attract customers' (Cook et al., 2008, p. 3). Slow Food's discourse reinforces a certain 'peasant essentialism' (Bernstein, 2006) whereby the *modern* and the *traditional* are opposed. Peasant essentialism is a global perspective that rests upon polarized concepts of the global market and local community (Hayden, 2003) whereby the culture of peasants is one untainted by a capitalist ethos. This trend has been reinforced by institutions and organizations that increasingly authenticate heritage foods by establishing qualification criteria that define, identify, and regulate which foods are registered as heritage. UNESCO's inclusion in 2010 of Mexican gastronomy in its List of the Intangible Cultural Heritage of Humanity exemplifies this trend. The narrative of a food or cuisine as heritage begs the question whether all peasant farmers are content to remain separate from or untainted by global or corporate agricultural systems, as Li (2015) points out.

Conclusions

This chapter has mapped out how the globalization of the food system can act as a challenge *as well as* an opportunity for the integration of small-scale farmers, producers, or their enterprises or projects within the global market. It aimed to demonstrate the role of narrative around corn in policy-making (as discussed in reference to Article 27 and the drafting of NAFTA) and the social and cultural impacts that discourse can have. As demonstrated, the Tortilla Discourse and neo-liberal rhetoric cast corn first as a sign of inferiority and then as a commodity to be traded globally. Each discourse in turn has had profound effects on the health of the Mexican people as well as the disposition of farmers' land. More importantly, this discourse has eroded the capacity of Mexico to feed itself. In its celebration of indigenous and local farmers, SFI has sought to redress this ill by valorizing corn as not just a national heritage but as a global and precious commodity. This chapter thus points to the role that national and international actors have in disseminating different narratives that directly affect the livelihood and health of local communities.

The pinole project reveals how actors perceive global markets as an opportunity to de-localize the consumption of pinole and commercialize it as a global commodity (Littaye, 2016). The narrative of corn as heritage, as promoted by SFI, helps form bridges across national borders or 'translocal ties' (Littaye, 2016) in stark contrast to the Tortilla Discourse that divided people according to their culture, race, and class or neo-liberal rhetoric and that sought to marginalize indigenous farmers. By postulating farmers as 'stewards of the environment', SFI underlines the intrinsic links tying local producers within a global context and 'the capacity of rural localities to engage with and shape globalization processes' (Woods, 2007, p. 492).

The meaning of food, and as described throughout this chapter, corn, is produced through the interaction of multiple narratives and discourses. Tracing differing interpretations of corn and its social, economic, and cultural relevance reveals the conflicting ideologies that have surrounded the crop, whether as a sign of inferiority, as commodity, or heritage. In this sense, the meaning of corn has evolved to become social and political. Each narrative is created by actors who construct meanings and identities for themselves (Silverman, 1998). As demonstrated throughout this chapter, the meaning of food has implications for people's food and nutritional security. This agency is widely underappreciated in the discussion of food security and policy-making.

International social movements, such as SFI, have a role to play in fostering transnational ties and support transnationalising entrepreneurship through their rhetoric celebrating local, 'endangered' foods. This narrative can be problematic as it imposes an understanding of corn as 'heritage' whilst simultaneously commodifying it for a global elite of foodies. There is a distinct lack of research on *how* foods are selected and identified as heritage for their potential commodifiability. Tellström, Gustafsson, and Mossberg (2006, p. 132), warn that 'studies skirt round an information "black hole" concerning the use of market research in choosing a food culture expression to ensure the best chance of economic success'. The consequences of the pinole narrative as heritage are still unclear on the production of corn or Mexico's food security; it begs to be further examined.

This chapter underlines the varying and changing role of the private sector, the government as well as international organisations in the everyday lives of farmers. It points to the changing dynamics of NGOs with relation to the networks they create or give access to. It highlights how the question of governance surrounding food security both nationally and internationally is a pressing and difficult one to investigate.

Notes

1 Mesoamerica represents the 'parts of Mexico and Guatemala where high civilizations flourished before the Spanish Conquest' (Brandes, 2009, p. 4).

2 Campesinos refers to rural peasants and farmers in Spanish.

References

Amavizca-Ruiz, J. R., Regalado-López, J., Álvarez-Gaxiola, J. F., Méndez-Espinoza, J. A., Mendoza-Robles, R., Galvanovskis-Kasparane, A., & Díaz-Puente, J. M. (2014). La migración de la población adolescente de San Mateo Ozolco, Puebla. *Agricultura, Sociedad Y Desarrollo, 11*(2), 181–199.

Baker, L. E. (2013). *Corn Meets Maize: Food Movements and Markets in Mexico*. London, U.K.: Rowman & Littlefield Publishers.

Bernstein, H. (2006). Once were/still are peasants? Farming in a globalising 'south'. *New Political Economy, 11*(3), 399–406.

Bruner, J. (1991). The narrative construction of reality. *Critical Inquiry, 18*(1), 1–21.

Carlsen, L. (2013, November 24). What we've learned from Nafta. *The New York Times*. 13–14.

Chrzan, J., 2004. Slow Food: What, Why, and to Where? Food, *Culture & Society* 7, 117–132.

Cook, I., et al., 2006. Geographies of food: Mixing. *Progress in Human Geography* 32(6), 821–833.

Cordero-Guzmán, H. R., Smith, R. C., & Grosfoguel, R. (2001). *Migration, Transnationalization, and Race in a Changing New York*. Philadephia, PA: Temple University Press.

DeWalt, B., & DeWalt, K. (1991). The results of Mexican agriculture and food policy: Debt, drugs, and illegal Aliens. In S. Whiteford & A. E. Ferguson (Eds.), *Harvest of Want: Hunger and Food Security in Central America and Mexico* (pp. 189–208). Boulder: Westview Press.

Eakin, H., & Lemos, M. C. (2006). Adaptation and the state: Latin America and the challenge of capacity-building under globalization. *Global Environmental Change, 16*(1), 7–18.

Escobar Moreno, D. A. E. (2005). Valoración de la agrobiodiversidad. Una aproximación desde la economía ecológica. *Revista de Geografía Agrícola, 35*, 7.

Fletcher, D. (2004). International entrepreneurship and the small business. *Entrepreneurship & Regional Development, 16*(4), 289–305.

Fonte, M. (2006). Slow food's Presidia: What do small producers do with big retailers? In *Between the Local and the Global* (Vol. 12, pp. 203–240). Bingley, UK: Emerald Group Publishing Limited.

Goodman, M. K. (2015). Food geographies I: relational foodscapes and the busy-ness of being more-than-food. *Progress in Human Geography*, 40(2), 1–10.

Hayden, C. (2003). From market to market: Bioprospecting's idioms of inclusion. *American Ethnologist, 30*(3), 359–371.

Hernández López, E. (2008). "Law, Food, and Culture: Mexican Corn's National Identity Cooked in 'Tortilla Discourses' Post-TLC/NAFTA" in *St Thomas Law Review*, 20(0): 573–593.

Huggins, R., & Johnston, A. (2009). The economic and innovation contribution of universities: A regional perspective. *Environment and Planning C: Government and Policy, 27*(6), 1088–1106.

Leitch, A. (2000). The social life of Lardo: Slow food in fast times. *The Asia Pacific Journal of Anthropology, 1*(1), 103–118.

Leitch, A. (2003). Slow food and the politics of pork fat: Italian food and European identity. *Ethnos*, 68(4), 437–462.

Leitch, A. (2008). Slow food and the politics of 'virtuous globalization'. In C. Counihan & J. D. Van der Ploeg (Eds.), *Food and Culture: A Reader* (pp. 45–64). New York: Routledge.

Li, T. M. (2015). Can there be food sovereignty here? *The Journal of Peasant Studies, 42*(1), 205–211.

Lind, D., & Barham, E. (2004). The social life of the tortilla: Food, cultural politics, and contested commodification. *Agriculture and Human Values, 21*(1), 47–60.

Littaye, A. Z. (2015). The role of the Ark of Taste in promoting pinole, a Mexican heritage food. *Journal of Rural Studies, 42*, 144–153.

Littaye, A. Z. (2016). The multifunctionality of heritage food: The example of pinole, a Mexican sweet. *Geoforum, 76*, 11–19.

MacDonald, K. I. (2013). The morality of cheese: A paradox of defensive localism in a transnational cultural economy. *Geoforum, 44*, 93–102.

Mazoyer, M., & Roudart, L. (2006). *A History of World Agriculture: From the Neolithic Age to the Current Crisis*. New York: NYU Press.

Parkins, W. (2004). Out of time fast subjects and slow living. *Time & Society*, *13*(2–3), 363–382.

Paxson, H. (2005). Slow food in a fat society: Satisfying ethical appetites. *Gastronomica*, 5(10), 14–18.

Petrini, C. (2007). *Slow Food Nation: Why Our Food Should Be Good, Clean, and Fair*. New York: Rizzoli International Publications.

Petrini, C., Watson, B., 2001. *Slow Food: Collected Thoughts on Taste, Tradition, and the Honest Pleasure of Food.* London: Chelsea Green Publishing.

Pilcher, J. M. (1998). *Que vivan los tamales!: Food and the Making of Mexican Identity*. Alberquerque: UNM Press.

Pilcher, J.M., 2012. Taco Bell, Maseca, and Slow Food: A Postmodern Apocalypse for Mexico's Peasant Cuisine?, in: Counihan, C., Van der Ploeg, J.D. (Eds.), *Food and Culture: A Reader*. New York: Routledge. 426–436.

Portes, A., Guarnizo, L. E., & Haller, W. J. (2002). Transnational entrepreneurs: An alternative form of immigrant economic adaptation. *American Sociological Review*, *67*(2), 278–298.

Preibisch, K. L., Herrejón, G. R., & Wiggins, S. L. (2002). Defending food security in a free-market economy: The gendered dimensions of restructuring in rural Mexico. *Human Organization*, *61*(1), 68–79.

Prina, S. (2013). Who benefited more from the North American free trade agreement: Small or large farmers? Evidence from Mexico. *Review of Development Economics*, *17*(3), 594–608.

Public Citizen (2014). *NAFTA's 20-Year Legacy and the Fate of the Trans-Pacific Partnership*. Available at: http://www.citizen.org/documents/NAFTA-at-20.pdf. [Accessed 20th January 2016].

Quist, D., & Chapela, I. H. (2001). Transgenic DNA introgressed into traditional maize landraces in Oaxaca, Mexico. *Nature*, *414*(6863), 541–543.

Reig, N. (1985). "Las tendencias alimentarias a largo plazo en México: 1950-1984." *Problemas del desarrollo* 16(61): 9–64.

The receding tide of illegal immigrants from Mexico. (2012, April 28). *The Washington Post*. Retrieved 5 February 2017, from www.washingtonpost.com/opinions/the-receding-tide-of-illegal-immigrants-from-mexico/2012/04/28/gIQACsu9nT_story.html?utm_term=.c395741523c2.

Sanchez J & Moreno M. (2013). The impact of agricultural policy in Mexico 1910-2012 and the effect of high food prices in the Mexican economy. *ECORFAN Journal-Mexico*, 4(9), 679–699.

Schepers, E. (2014). Mexico after 20 years of NAFTA: Poor and getting poorer. Peoplesworld. Peoples World, 6 Jan. 2014. [Accessed 20 April 2016]

Schneider, S. (2008). Good, clean, fair: The rhetoric of the slow food movement. *College English*, *70*(4), 384–402.

Silverman, D. (1998). *Harvey Sacks: Social Science and Conversation Analysis*. Oxford: Oxford University Press.

Tellström, R., Gustafsson, I.-B., & Mossberg, L. (2006). Consuming heritage: The use of local food culture in branding. *Place Branding*, *2*(2), 130–143.

Ulijaszek, S. J., Mann, N., & Elton, S. (2012). *Evolving Human Nutrition: Implications for Public Health*. Cambridge, U.K.: Cambridge University Press.

Warde, A. (2016). *The Practice of Eating*. London, UK: John Wiley & Sons.

Winch, G. W., & Bianchi, C. (2006). Drivers and dynamic processes for SMEs going global. *Journal of Small Business and Enterprise Development, 13*(1), 73–88.
Woods, M. (2007). Engaging the global countryside: Globalization, hybridity and the reconstitution of rural place. *Progress in Human Geography, 31*(4), 485–507.
Yeung, H. W. (2009). Transnationalizing entrepreneurship: A critical agenda for economic geography. *Progress in Human Geography*, 33(2), 210–235.
Young, N. (2010). Globalization from the edge: A framework for understanding how small and medium-sized firms in the periphery 'go global'. *Environment and Planning* A, *42*(4), 838–855.

4 Historical path dependencies and energy governance in post-Apartheid South Africa

Agostino Inguscio

Introduction

The South African energy sector is at risk of failing its citizens in two ways. First, it is too reliant on burning high carbon coal – 77 percent of South Africa's primary energy needs are provided by coal according to the South African Ministry of Energy – and this makes it difficult to enact a pathway for sustainable development (Jaglin & Dubresson, 2015). Second, it has not sufficiently dealt with its social inequalities, a crippling issue in a country that is affected by one of the highest levels of inequality in the world.[1] Although South Africa has gone from only 56 percent of the population having access to electricity in 1990 to 86 percent in 2014, still only 71.5 percent of the rural population have access to electricity (although this is up from a dire 26.4 percent in 1990) (World Bank, 2017). Considering that 95 percent of South Africa's energy is produced by the state-owned parastatal Eskom, the responsibility of changing this state of affairs rests largely on the shoulders of the South African government (Jaglin & Dubresson, 2015).

This chapter argues that the history of the energy sector in South Africa created institutional and technological problems, which need to be overcome to transform the energy sector towards less coal use and develop and encourage participation in sustainable practices. Further, the chapter argues that the current rigidities that impede the reform of the energy sector in South Africa are a product of the history of the sector even prior to apartheid.[2] The chapter uses secondary literature to explore this history, along with contemporary press material to investigate the present debate on, and perceptions of, the energy sector in South Africa.

The first section of the chapter reconstructs the history of South African electrification, outlining the establishment of path dependencies, the main obstacles to the successful diffusion of renewable energy in South Africa. This analysis begins with the foundational forces that led to the establishment of Eskom in 1922 (McDonald, 2009). The chapter then addresses the challenges faced by Eskom until the early 1990s. Understanding the historical challenges Eskom faced provides the framework to comprehend the reasoning behind the initial attempts to coherently develop renewable energies in South Africa. The chapter finally analyses the period between the mid-1990s and the early 2000s, which is crucial for our understanding of the initial investments in the renewable energy sector.

The beginning of South African electrification

The mining sector mainly drove the development of South Africa (Christie, 1984; Fine & Rustomjee, 1996). Before the mining boom, it would have been inconceivable for South Africa to become by the mid-1920s one of the world's highest producers and consumers of electricity per capita. The history of electrification in South Africa is intimately linked with the history of the mineral exploitation in the region. Indeed, while the first uses of electricity took place in urban areas, connected to the commercial life of port cities such as Cape Town, with the discovery of minerals the mining sector became the main driver behind the electrification of other areas. An early example of such interdependence is the electrification of Kimberley, the now capital of the Northern Cape province that borders Namibia and Botswana, – through coal-powered plants – and of its world famous diamond mines in 1882 (Jaglin & Dubresson, 2015).

Initially each mine was producing energy for its own needs, but with time, companies emerged that aimed to produce energy on a larger scale for the mining industry. Unsurprisingly, the owners of the mines were often the owners of the power stations. The Victoria Falls and Transvaal Power Company (VFTPC) was the first successful endeavour to produce electricity on a scale that went beyond the requirements of a single mine. As the name suggests, the initial plan of the shareholders was to channel the power of the Victoria Falls and of the Zambesi river to generate hydroelectricity to supply the growing energy needs of the South African mines. However, the existing technological constraints in the transmission of electricity over long distance made this proposition unrealistic. With the mining industry's growing demand for cheap electricity, the VFTPC set out to construct coal plants to produce the needed energy. This cheap fossil fuel, on which the VFTPC power stations relied, had the further advantage of being a by-product of the gold industry, as oftentimes coal has to be extracted first to reach the gold deposits. This reliance on coal burning firmly taking the lead over possible alternative sources of energy production, such as hydroelectric power, is one of the earliest critical junctures in the history of energy in South Africa (Jaglin & Dubresson, 2015).

In the period between the First and the Second World Wars, the South African state started playing a prominent role in shaping the energy sector. When the mining industry faced serious cost challenges, most prominently the rising cost of labour and energy, the state intervened. The state considered the energy sector – and its ability to supply cheap energy – as a strategic asset that could not be left to the private sector to manage independently (Clark, 1994). Eskom, among other parastatals, was the result of the government's decisions to inaugurate an era of state intervention in the economy. The Electricity Act of 1922 saw the establishment of the Electricity Control Board (ECB) and then of Eskom in 1923 as a government-owned power utility. Eskom was to be placed under government supervision and follow regulation from the ECB and its goal was to fuel the industrialization of South Africa by providing cheap electricity and by supplying the electricity necessary for the electrification of the railways (Clark, 1994).

To meet this goal Eskom was exempted from paying taxes and had the mandate to supply electricity as cheaply as possible 'neither at profit nor at loss' (Clark, 1994; Freund, 2011; Jaglin & Dubresson, 2015). In time, Eskom came to be seen as crucial to driving South African economic growth. While the interests of the mining sector were considered paramount, the establishment of Eskom was therefore an attempt of the South African state to intervene in the economy and develop its industrial sector more broadly.

The production of energy was very much to be shared with other operators, with Eskom remaining in 'a subordinate position to the privately owned VFTPC which served the lucrative mining contracts' (Fine & Rustomjee, 1996, p. 137); however this would quickly change through the action of the government (McDonald, 2009). The 1934 Klip Agreement, concerning the construction of the power plant of Klip, forbade private power-generating companies to invest further in the sector, paving the way to the acquisition of the VFTPC by Eskom. With the acquisition of the VFTPC in 1948, Eskom became a monopoly in the energy production sector in South Africa. Eskom's mandate was at this stage – having already achieved the electrification of railroads – to provide the mines with cheap electricity. Towards the end of the 1940s, the gold mines still consumed 59 percent of the electricity produced in South Africa. The powerful chamber of mines – one of the pillars of the Minerals-Energy Complex (MEC) – was the main advocate for the radical transformation of the energy sector (Fine & Rustomjee, 1996, Jaglin & Dubresson, 2015). Consequently, the cost of electricity was further lowered in 1952 (Clark, 1994).

With the acquisition of the VFTPC, Eskom was firmly on its path to create the energy sector of South Africa as we know it today. The expansion of Eskom following the achievement of its monopolist position confirmed the strategic objectives of the sector. Ben Fine and Zvareh Rustomjee define the ambitious 1950s power station construction programme unleashed by the nationalization of the electricity sector as 'one of the clearest state industrial strategies of South Africa' (Fine & Rustomjee, 1996, pp. 157–160). The outcome of this drive, which by 1960 more than doubled the capacity available in 1950, illustrates the strategic mandate of Eskom, which was not to reach a progressive electrification of the country, but rather to expand its capacity, since the mining industry was expanding and there were already shortages of power to the mines (McDonald, 2009).

The end of the 1960s did not alter this clear trajectory. The energy sector responded closely to the needs of the mining sector, and its main goal was to avoid any of the power cuts that had affected the mines in the 1950s and to continue providing cheap electricity to their ever-growing energy needs. At the same time, such '*gigantisme productiviste*' – as defined by Jaglin and Dubresson – was justified by the rapid pace of economic growth in the country – an average of 5.5 percent annually in the 1960s (Jaglin & Dubresson, 2015, pp. 26–27). No questions were raised about the environmental consequences of an electrification programme that depended solely on electricity generated by burning coal.

During this period, South Africa's attitude towards energy production aligned with that of the industrialized world during the post-war boom. The widely shared mantra was that, what economies and societies needed, was plentiful and cheap energy to sustain rapid growth, without concerns for the possible environmental consequences of such a model. In line with this credo, the achievements with regards to electricity production were such that, in 1962, the chairman of Eskom, H.J. van Eck, could state proudly that 'in South Africa, far from relying on so-called cheap labour we believe in the application of power' (Fine & Rustomjee, 1996, p. 183). What set South Africa apart was the fact that, thanks to apartheid, the majority of the population was excluded from consuming this energy.

This trajectory was all set to be maintained throughout the 1970s, with the planning of more power stations on the coal fields of the Eastern and Northern Natal, and then further in 1976, with the initial building of a nuclear power station (see Eberhard, Kolker, & Leigland, 2014). The plan was essentially to double South Africa's electricity capacity once more. Parallel to the power supply, the national grid that was to service the whole country, and that Eskom had started building in 1960, was concluded by 1973.

This ambitious building programme however was driven by an overestimation of the following decade's economic growth in South Africa, and consequently of the energy needs. The excessive investments in new productive capacity came at great financial costs, which forced Eskom to raise its tariffs by 166 percent in the period 1971–1977. This steep increase was seen as distorting the mandate of Eskom to provide electricity at the lowest possible price (Jaglin & Dubresson, 2015). For the first time, the 'giant' seemed to be heading for financial troubles because of unwise investment decisions that 'threatened to increase electricity costs unnecessarily' ('Eskom giant could be heading for trouble', 1977; 'Eskom's Expansion Programme: Power for the future', 1977). The contraction of South African economic growth, and consequently of the demand for electricity constituted a serious challenge to Eskom's capacity to generate revenue from the capital invested without further raising tariffs ('Fall in demand may cut Eskom's growth', 1977). Notwithstanding the financial consequences of the overestimation of demand for growth, the suggestions of scaling back the programme were ignored and the important investment programmes continued in the 1980s (Eberhard et al., 2014).

The initial crisis of the electricty sector

The concerns felt by South Africa's policy makers, as well as by the public, during the 1970s motivated a drastic change in the governance of the energy sector. It is essential to consider carefully along which lines this crisis developed, how it was perceived by South African society and, consequently, which paths were explored – and which ones remained closed – to propel the system forward. The 1970s saw the emergence of radically different challenges for the energy sector and the South African economy as a whole. This should not be considered in isolation, since the oil shocks of the first half of the 1970s had woken up the entire industrialized world to the reality of the limits of the planet's resources. However, South Africa and its energy sector had its own specific responses to the oils shocks, which can be best summarized by a mix of fear of a slowing economy, and hope for the role that a resource-rich and nuclear-hopeful South Africa could play in a world that craved energy resources.

The negative consequences of the 1970s oil shocks were compounded by the fact that South Africa was on the official boycott list of many oil-producing countries due to apartheid. The rising costs of oil imports were perceived as one of the main dangers for the economy. Emergency measures to save petrol were introduced and the price of petrol in the country steadily grew from October 1973 ('66-hour', 1976). The Minister of Economic Affairs stated, after having rejected a request to ease some of the restrictions put in place to save fuel, that the rising costs of oil imports were potentially disastrous for the South African balance of payments (The Argus Correspondent, 1975). He added on several other occasions that 'South Africa would be foolish to think that the price of gold could be used to compensate for the continuous rises in the prices of oil and other imported goods' (SAPA, 1977). With this statement, he was flagging for the first time the possible limitations of South African's energy silver bullet. These anxieties were further enhanced by the fact that South Africa was faced with an oil embargo enforced by the United Nations' security council ('Oil's Sanctions', 1977) because of its support towards Rhodesia (now Zimbabwe and Zambia).

From crisis to transformation

The 1970s are a watershed moment in which the rigidities of South Africa's dominant energy sector started to be exposed more clearly and could have sparked the transformation of the sector in a more sustainable and equitable direction. Instead business as usual in the mineral–energy complex renewed a system that would begin to crack again in the new, democratic South Africa.

During the oil crisis of the 1970s, South Africa first embarked on a series of explorations for off-shore oil, hence directly responding to the issue of rising worldwide oil prices. The Southern Oil Exploration Corporation, backed by the South African state, planned to invest R75 million over the period 1975–1978 in oil exploration (50 percent more than already spent – to no fruition – between the beginning of offshore probes in 1967 and 1975) (SAPA, 1975). It was reported at the time that 'oil fever has hit Mossel Bay and enthusiasm is high' (Own Correspondent, 1975). Explorations continued well into the early 1980s though they were economically unfeasible. An estimated R82 million had been spent on prospecting oil to no avail. Furthermore, the oil explorations transpired as shady 'deals' (Van Den Nieuwenhof, 1984). Queries in these regards posed to the Minister of Mineral and Energy Affairs (for example during the Ministerial interrogation on March 21st 1984) were nevertheless thwarted.

With the benefit of hindsight, this strategy could not have been a satisfactory one as even a lucky strike would have been insignificant for propelling the energy sector forward (Van Den Nieuwenhof, 1984). South Africa's second strategy was to invest heavily in further developing SASOL'S (South Africa Synthetic Oil Liquid) capabilities to produce oil from coal, through a costly procedure (Buys, 1975; 'The Sasol Story', 1975). The belief in the viability of the coal to oil process was based mostly on the reliance on the vast coal reserves of South Africa and on a gross underestimation of the costs involved in the building of the new SASOL II plant and operational costs. In less than a year since the start of construction, the costs of SASOL II had soared by 86 percent ('SASOL II', 1976).

Unlike the two previously mentioned strategies, coal promised an immediate solution to energy shortages and also the possibility of boosting South African exports by providing power for industry, as the development of nuclear energy remained a hope for the future. The world energy crisis created the basis for a South African coal export boom. The prospect of rapid profits involved several foreign companies, among them Shell, which stated that 'quite obviously' their investment strategy in South Africa was going to be focused on the country's coal reserves (The Argus Financial Editor, 1975; 'They're Shelling', 1975).

At the time, numerous experts had started to awaken to the fact that an increasing dependence on coal for energy would 'ultimately wreck the earth's environment' ('Don't burn coal', 1977). Worldwide, the peak of the energy crisis was accompanied by a feeling that the idea of harnessing solar power should no longer be relegated to the 'lunatic fringe of technology' (The Argus Bureau, 1975). The challenge that the energy crisis posed to the South African energy sector at the time could have fostered the development of renewable energies. A window of opportunity that could have shifted the system onto a more sustainable trajectory opened up, but was ignored. In South Africa, the possibility of using renewable energies as an alternative to coal was largely rejected. Instead, South Africa's optimistic views on coal power were guided by the conclusions of experts who convened in Istanbul for the World Energy Conference in 1977, who 'dismissed with deep scepticism such alternative power proposals as sun, wind and waves' (Stocks, 1977) and the general conviction that 'renewable energy sources [. . .] will not become significant until the next century' (Koenderman, 1977). As a result, the interest in adoption of such technological sustainability was low in a country with such a potential abundance of solar power. At the time, South Africa had invested only ZAR150,000 on solar energy research, which was negligible in comparison to investments in other energy sources. South Africa's conviction to ignore the opportunity to develop renewable energies, but rather to embrace the fossil fuel paradigm was further reinforced by the perceived lower environmental risks associated with the use of coal as an alternative to oil. The increase in exports in coal and, to a lesser extent, in iron ore, helped to overcome briefly the crisis of the 1970s (McDonald, 2009). The energy crunch was an unprecedented opportunity for South Africa's coal-based economy, regardless of the impacts on its environment.

Parallel to the coal export shaping the future energy sector, the socio-political characteristics of the country also played a part in the energy system. First, the energy sector was facing structural changes due in part to the crisis which led to accusation against Eskom's management of 'cloaking its affairs under a veil of secrecy' (Stocks, 1977). Furthermore, by the end of the 1970s, the vast majority of the South African black population in rural areas did not have any access to electricity. The most vulnerable part of the population had seen a marked decline in their capacity to consume any form of energy, given the rising prices of coal for the reasons outlined previously ('Domestic Consumption of Coal', 1975). This situation was a reflection of the socio-political context of South Africa, with policies that characterized what came to be defined as the second phase of apartheid, or Grand Apartheid, and which led to South Africa's growing isolation. This isolation due to the apartheid government's continuation with its plan of 'separate development' or racial segregation, emerged as one of the driving factors behind the changes in the energy strategy of the country.

As a consequence, on a country-wide level, the 1980s came to be known as the apartheid reform period for the attempts made to overcome the country's looming economic crisis that had arisen as a result of its politics (Clark, 1994, p. 65). In relation to the energy sector, the Ministry of Energy and Minerals established a commission. The commission's suggestions, released in May 1983, prompted a phase of radical transformation in the electricity sector. Following these recommendations, the Electricity Act (No. 41 of 1987) and the Eskom Act (No. 40 of 1987) were implemented. These Acts brought about changes to the way the electricity sector was structured. Contemporary commentators suggested that 'the alacrity with which government accepted the commission's recommendations "in toto" . . . suggests an uncommon sense of urgency' (The Financial Mail Editor, 1984).

Paving the way for structural reform

The 1986 White Paper on Energy Policy of the Republic of South Africa, the first such document produced by South Africa, reveals some of the concerns that were driving energy policy in the later years of the apartheid state (DME, 1986). The need for 'a structured and coordinated approach to solving the country's future energy problems' reinforced the disturbing realization that the 'bed' that the MEC had prepared for South Africa was starting to feel more and more uncomfortable and that 'future policy can still be honed and improved' ('Making a Plan', 1984).

The most pressing issue was to isolate South Africa from international energy shocks and from its dependence on oil imports. The white paper in particular stresses how the role of the state in energy affairs should be strong: '[E]ven in a market-oriented economic system', the white paper maintains, energy is a quintessential strategic sector that should be developed in the interest of the state's security and stability (DME, 1986). On the other hand, it marks a transition: for the first time in South African energy policy, the possibility of expanding the use of renewable energies in relation to the need of ensuring the strategic supply of energy received attention.

However, this occurred within a context of separate development. For instance, the drive towards 'Evaluating on a continuous basis all known indigenous energy resources, taking anticipated future needs thereof and development and applications thereof into account' (DME, 1986, p. 7), suggested the planting of 'about a million hectares of woodlot' to support the fuelwood consumption of rural communities ('No answer to energy crisis', 1984). This was a stark reminder of the unfeasible 1970s project of producing vast quantities of hydroelectric energy damming 'all the rivers in the Transkei' (see Box 4.1). It showed that the prospected

Box 4.1 The Transkei dams as energy strategy

An example of the ways in which the energy strategy was affected by the wider political economy is the considered solutions for the energy requirements of the political entities known as Bantustans – African homelands. In 1975, the Transkei was established within the borders of South Africa. The South African government believed that the newly independent state would embark on the construction of a 'huge hydro-electric scheme to provide electricity for the expected industrial development', which was anticipated to be 'immeasurable' (SAPA, 1975). The plan would have required – according to the Chief Minister of the Transkei, Paramount Chief Kaiser Matanzima – for 'all the rivers of the Transkei to be dammed' ('Power from the Transekei', 1976) and the South African government was going to pay for this. The project was expected to revolutionise the economy of the Transkei through the transformation of the 'state' as an exporter of electricity to South Africa, the expected consumer of the largest share of the electricity generated.

The proposed plan, which would have involved the construction of six major dams, 'some of which connected by tunnels cut under mountain ranges', did not come into existence ('Power from the Transekei', 1976). However, in matters of equitable access to electricity, the plan of Eskom and of the South African government for the most disenfranchised part of the society followed very much the strategy pursued by South Africa to turn an internal problem and responsibility into a formally foreign one – since the Transkei was granted independence – without any viable intervention.

expanding access to electricity by the South African state was still intended to happen within a framework of economic and political separation that saw the needs of the majority of the population as a lesser concern.

The 1980s and the growing opposition to the apartheid system brought about the perception that apartheid could not have lasted very long in its current form. Anticipating the broad political change on the horizon, a group of experts gathered to form the Energy for Development Research Centre in Cape Town. They aimed to explore the future possibilities of a South African energy sector that was to be based on a model of sustainable development, defined by its financial feasibility, social equality, and environmental sustainability. The research agenda pursued by this research group was to shape much of the 1990s break with the past (Jaglin & Dubresson, 2015).

Post-Apartheid energy policy

The beginnings of the 1990s were characterized by a period of intense negotiations between stakeholders involved in the democratic transition. The political agreement reached between the National Party and the African National Congress of Nelson Mandela framed a path for a post-apartheid South Africa that was characterized by the continuous reinforcement of a representative democracy and the liberalization of the economic order, with the aim of integrating South Africa into the global economy. The scope of the transitional agreements was to develop simultaneous economic policies of neoliberalism and redistribution to meet the two fundamental needs of economic growth: to compete in the global economy and redistribute resources to benefit the groups that were discriminated against by the previous regime.

For the first time, because of the changing political scenario, the energy needs of the majority of the South African population were taken into account while drawing a future energy strategy. Experts warned that while the normalization of oil prices in the early 1980s led South Africans to believe that the energy crisis was over, 'another "energy crisis" was deepening inexorably, irrespective of the short or long term effects of fluctuations in oil prices' (Eberhard, 1984).

Eberhard (1984) referred to an issue that frames two of the main objectives of the Energy for Development Research Centre (ERC). The energy system of a new democratic South Africa should urgently address the issues of inequality of access and the one of environmental sustainability. In a country that was – at the time – generating 60 percent of the total electricity supply of the African continent there was not an overall shortage of energy, 'only highly inequitable distribution of and access to energy resources' (De Villiers, 1984). One of the consequences of such an unequal distribution was that an estimate of 15 million blacks in rural areas depended on firewood for fuel, leading to increasing deforestation and the worsening of periodical droughts (De Villiers, 1984; Eberhard, 1984; 'Rural communities running out of fuel', 1984; 'No answer to energy crisis', 1984), and to women being forced to walk – in the case of one observed group in Kwa-Zulu – up to 19 km to collect bundles of fuelwood 'weighting about 40 kg each' (Eberhard, 1984).

The reform pushed forward by the ERC focused on the long forgotten issue of equitability of access to energy. As a consequence of the systemic discriminations inbuilt from an apartheid legacy, whites – 17 percent of the population at the time – consumed 67 percent of the electricity used in the domestic sector. They did so at significantly lower prices compared with impoverished black households that had to spend a much higher share of their income in acquiring energy (Labour reporter, 1984; De Villiers, 1984; 'The poor paying more for energy', 1984). In 1991, the new management of Eskom launched an extensive electrification campaign ('Electrification for all') in favour of the black population that had been disadvantaged, signalling that the organization was capable of adapting to a changing climate in the wider political scene ('The company that anticipated history', 2006; Jaglin & Dubresson, 2015; Bulbulia, 1990). The chief executive of Eskom – Ian McRae – hailed the fact that political constraints to reform were falling like 'leaves from a tree in autumn' (Drew, 1990). As a consequence, plans were put in place finally to address the structural inequalities of a system that had so far left 'two thirds of the population without electricity' ('A million electrified points', 1990a).

The actuation of this programme produced policy oscillations between what were sometimes perceived to be opposing aims. This balancing act 'from state to market and back again' was complicated by the disappointing trajectory of growth of the South African economy (see Eberhard et al., 2014). In 1990, the removal of the political ban on the parties that opposed the apartheid regime, allowed for the framing of the Reconstruction and Development Programme (RDP) relative to the issue of energy, which was quickly implemented after the elections of 1994. Let us consider how the programme fared in relation to the two main policy objectives formulated in the late 1980s: expanding access to electricity and the shifting of the South African energy sector towards a path of environmental sustainability.

The cardinal point of the envisaged energy policy is a National Electrification Programme, which, building upon the Eskom's Electricity for All campaign, promoted the rapid electrification of South Africa. The success of the programme was remarkable with 2.8 million households connected between 1991 and 2001, bringing the connection rate from 57.6 percent in 1996 to 70 percent in 2001 (Jaglin & Dubresson, 2015).

Expanding access to electricity was a clear priority for the governmental agenda and sidelined any serious attempt to find a solution to the environmental problems caused by the South African energy sector. It is important to remember that at this time in history, Eskom still ran a large surplus of energy supply. The challenge was not to produce more energy, but to get the energy to the people that had been excluded from the South African grid. As a consequence, renewable energy was seen mainly as a way of providing some level of electrification quickly to areas that were not yet connected to the grid, rather than to provide additional electricity.

Environmental implications of South Africa's energy structure

In 1995, South Africa had the unenviable record of being the third largest polluter per capita in the world (van Horen, 1996). It was evident at the time that these environmental costs subsidised the price of electricity, which was essentially borne not by its consumers, but by the poorest members of society (Levy, 2014, pp. 22–23). However, addressing this concern was not perceived as equivalent to the issue of electrification.

The political transformation in South Africa attracted foreign firms interested in investing in renewable energy. For instance, Shell and Eskom partnered to implement 6000 solar-powered systems for residential use (only 4700 were ever operational) (Winkler, 2005). Siemens, on the other hand, aimed to distribute its Sunsave product range in South Africa. These products would be a 'major factor in improving the quality of life in many communities' ahead of their connection to the grid ('Siemens is reaching out', 1990b). The company recognised that 'market awareness of solar photovoltaic' was low in South Africa and therefore it embarked on a parallel educational programme to build awareness among consumers ('Siemens is reaching out', 1990b).

These projects were not thought to be in competition with the grid electricity pursued by the Electricity for All strategy. They were aimed to quickly intervene to guarantee some level of access to communities that grid electricity could not reach (Clarke, 1991). All players in this field were keen to stress that 'the whole idea is not to compete with the grid of the public utility (Eskom), but to provide power where the grid doesn't reach', the main scope being 'enhancing rural family life' quickly (Staff reporter, 1995). Indeed, Eskom itself admitted that, at the time, one in five South Africans lived in areas that 'will never be economical to provide with conventional electrical power' (Herbert, 1995).

In 1998, the post-apartheid national policy for the energy sector was finally defined in the White Paper of the Department of Energy. This document provided a complete discontinuity with the past, reflecting the revolution that had taken place in the country. Its major objectives were to increase access to affordable electricity, improve energy governance, and stimulate economic development. Energy security – in a democratic South Africa finally open to the world – was not to be achieved through self-sufficiency, 'but through greater diversification and flexibility of supply' (DME, 1998; Levy, 2014).

Following the overarching policy lines of the GEAR (Growth, Employment and Redistribution) plan, which had supplanted the RDP, the government projects for improving energy governance consisted in opening it up to competition. To do so, Independent Power Producers (IPPs) could participate in the production of energy. The success of this strategy rested on the de-bundling of Eskom's functions – through the separation of the production and transport compartment and through the privatization of several power plants. The government-recognized IPPs should be allowed to enter the market. Therefore the Department of Energy was requested to produce a power plan to move in this direction. However, for this to happen the system had to show evidence of signs of its weakening first (See Eberhard et al., 2014).

While Eskom was still running a surplus of electricity production in 1995, initial warning signs started to emerge. Eskom warned that the gap between total capacity and demand started to become 'tight' for the first time in decades (Reuters, 1995; 'Eskom put under pressure', 1995). By 2003, power reserve margins were dropping and, without intervention on the supply and the demand sides, turned negative in a few years. This further crisis of Eskom was instrumental in the 2004 decision to announce that Eskom was to be unbundled. IPPs were invited to bid for 30 percent of new electricity produced. However, subsequent energy crises in 2008 and then again in 2014 where 'rolling blackouts' became the norm for all South Africa, have led to serious questions about the ability of fossil fuels to meet energy demand in South Africa.

An important step for the development towards renewable energy in South Africa emerged with the Renewable Energy Procurement Programme (Baker & Wlokas, 2014). Renewable energy has actually be touted as a game changer for South Africa – not only for meeting energy needs sustainability, but also because it has been one of the only sectors outside the mineral–energy complex that has created jobs in the country and it attracts foreign direct investment (Swilling, 2016). The challenges of a path towards an equitable and sustainable electricity production and distribution in South Africa are still many and the South African state is still very much committed to the development of new coal resources (McDonald, 2009). However, there are rising real costs of new coal-fired power stations (like Medupi) as renewables are now at grid parity per MW hour in South Africa (Swilling, 2016). South Africa's Renewable Energy IPP Procurement Program has now commenced (See Eberhard et al., 2014). Under this programme, the South African Treasury has introduced an innovative auction approach (rather than a feed-in tariff due to concerns about price-setting and rent-seeking), where renewable energy power producer bidders have to meet strict criteria including: lowest price, approved funding, a minimum of 40 percent South African entity participation and a minimum black ownership of the project company of 12 percent, and a proviso that local communities must also have a minimum 2.5 percent shareholding (Swilling, 2016). Finally, the shift towards a renewable energy solution for South Africa's crisis could indeed become a reality.

Conclusions

The South African energy sector has – from its beginnings until the fall of apartheid – followed a path that aimed to guarantee the profitability of the mining industry and the relative self-sufficiency of the country in terms of energy production. In doing so it became a tool that reinforced economic and social inequality and discarded the negative impacts that energy production has on the environment. Changing this requires a full understanding of the technological and institutional rigidities that the system carried from its past into its present. Among the lessons that can be learnt from the history of the energy sector, one is that the sector can be transformed radically along a sustainable path. Throughout history, the crisis of the sector led to its progressive transformation. Therefore, its current crisis could lead to a transformation that will reinforce the positive steps that have already been taken in the direction of making the sector more sustainable from an economic, social, and environmental perspective.

Notes

1 In 2014, South Africa's GINI coefficient based on expenditure data was 0.65 and 0.69 based on income data (World Bank, 2014).

2 Apartheid refers to the period in South Africa from 1948 to 1994 when under minority white rule, racial segregation was brutally enforced by the government.

References

The Argus bureau (1975, September 18). Last laugh for sun power men. *The Argus*.

The Argus correspondent (1975, August 27). Cost of oil worries SA says Heunis. *The Argus*.

The Argus Financial Editor (1975, July 25). As a result of the world energy crisis South Africa is heading for a coal export boom. *The Argus*.

Baker, L., & Wlokas, H. L. (2014). *South Africa's renewable rnergy procurement: A new frontier?* Energy Research Centre, University of Cape Town. Retrieved from www.erc.uct.ac.za/Research/publications/15-Baker-Wlokas-RE_frontier.pdf.

Berridge, G. (1977, December 25). Fall in demand may cut Eskom's growth. *Sunday Times*.

Bulbulia, S. (1990, July 3). Eskom develops an education programme. *Star*.

Buys, J. (1975, February 6). New Sasol will go up on Highveld farms. *Rand Daily Mail*.

Christie, R. (1984). *Electricity, Industry and Class in South Africa*. London: SUNY Press.

Clark, N. (1994). *Manufacturing Apartheid: State Corporations in South Africa*. New Haven: Yale University Press.

Clarke, J. (1991, May 29). The sun has enough power for Africa. *Star*.

De Villiers, R. (1984, April 16). SA's electricity denied to most. *Cape Times*.

Department of Minerals and Energy (1986), DME 1986 White Paper.

Department of Minerals and Energy (1998), DME 1998 White Paper.

Drew, R. (1990, November 13). Eskom chief hails fading of 'political constraints'. *Star*.

Eberhard, A. (1984, February 17). Real crisis: Fuel wood in underdeveloped areas. *The Argus*.

Eberhard, A., Kolker, J., & Leigland, J. (2014). *South Africa's Renewable Energy IPP: Procurement Program: Success Factors and Lessons*. Cape Town: World Bank Publications.

The Financial Area Editor (1975, June 22). They're shelling out R500 million with emphasis on coal. *Sunday Tribune*.

The Financial Mail Editor (1984, November 30). Dimming the lights. *Financial Mail*.

Fine, B., & Rustomjee, Z. Z. (1996). *The Political Economy of South Africa*. Johannesburg: Witwatersrand University Press.

Freund, B. (2011). *The union of South Africa: A developmental state?* Port Elizabeth: Economic Research Southern Africa. Retrieved from https://econrsa.org/wkshops/fresh/port_elizabethpercent20-percent20fruend.pdf.

Herbert, R. (1995, February 11). Solar power lights the way. *The Argus*.

Jaglin, S., & Dubresson, A. (2015). *Eskom. Électricité et pouvoir en Afrique du Sud*. Paris: Karthala.

Koenderman, T. (1977, July 31). Energy crunch is near, report warns. *Sunday Times*.

Labour reporter (1984, April 16). Whites enjoy cheap energy. *Cape Times*.

Lagerman, M. (1984, March 3). Rural communities running out of fuel. *Post*.

Levy, B., (2014) Working with the Grain: Integrating Governance and Growth in Development Strategies, Oxford: Oxford University Press.

McDonald, D. A. (Ed.). (2009). *Electric Capitalism: Recolonising Africa on the Power Grid*. Cape Town: HSRC Press.

Reuters (1995, January 26). SA growth: Eskom feels the pressure. *Cape Times*.

SAPA (1975, September 11). R75m for oil probe. *Dispatch*.

SAPA (1977, January 24). Gold rise will not offset jump in oil cost – Heunis. *Star*.

Staff reporter (1995, June 29). Solar power for rural development: American team leads the way. *The Argus*.

Star own correspondent (1975, November 1). New impetus on oil hunt. *Star*.

The Star's insight team (1977, September 16). Eskom giant could be heading for trouble. *Star*.

The Star's insight team (1977, September 27). Eskom's expansion programme: Power for the future. *National Mercury*.

Stocks, K. (1977, September 22). Eskom keeps public in the dark. *Star*.

Swilling, M. (2016, January 4). South Africa's renewable energy revolution. *Business Day*.

Unknown (1975, September 15). The Sasol story. *Star*.

Unknown (1976, February 13). SASOL II cost hits R1900m. *National Mercury*.

Unknown. (1976, May 1). Power from the Transekei. *National Mercury*.

Unknown (1976, October 13). 66-hour fuel blackout at weekends. *Star*.

Unknown (1977, July 27). Don't burn coal. *National Mercury*.

Unknown (1977, September 30). Oil's sanctions: This bogey is gaining strength at the U.N. *Financial Mail*.

Unknown,(1984, April 16). The poor paying more for energy. *The Sowetan*.

Unknown, (1984, April 16). Making a Plan to respond to the energy crises, Business Daily

Unknown (1984, April 17). No answer to energy crisis. *Day*.

Unknown (1990a, October 2). A million electrified points to improve the quality of life. *Business Daily*.

Unknown (1990b, October 2). Siemens is reaching out with the Sunsave range. *Business Daily*.

Unknown (1995, January 26). Eskom put under pressure to meet demand. *Star*.

Unknown (2006, November 30). The company that anticipated history. *Strategy Business Issue*.

Van Den Nieuwenhof, L. (1984, March 7). Search for oil: A world of costly secret deals. *Post*.

Van Horen, C. (1996). *The Cost of Power: Externalities in South Africa's Energy Sector*. Cape Town.

Winkler, H. (2005). Renewable energy policy in South Africa: Policy options for renewable electricity. *Energy Policy*, *31*(1), 27–38.

World Bank (2014). *South Africa Overview*. Retrieved from www.worldbank.org/en/country/southafrica/overview.

World Bank (2017). *Sustainable Energy for All (SE4ALL) Database*. Retrieved from http://data.worldbank.org/indicator/eg.Elc.Accs.Zs.

5 Water for energy in China

Xiawei Liao, Jim W. Hall and Nick Eyre

Introduction

An emerging global challenge

Water and energy are both critical and fundamental to human society's development and prosperity. They are also intrinsically and inextricably interlinked. Energy is required to produce, deliver, and distribute fresh water (e.g. seawater desalination), as well as treating wastewater (Siddiqi & Anadon, 2011). Water is needed in different processes of energy production, from upstream fuel extraction, operational phases of power generation – primarily as a cooling medium – to downstream power plant dismantling (Rio & Frei, 2009).[1] According to UN WWAP (United Nations World Water Assessment Program, 2014), about 90 percent of global power generation depends intensively on water. And thermoelectric cooling is estimated to be responsible for around 43 percent and 50 percent of total fresh water withdrawals within the EU and the US, respectively (UN WWAP, 2014). Such interdependencies, which are often called the 'water-energy nexus', have recently been brought to the forefront of scientific and policy debates. Indeed, while they have been traditionally governed as two independent issues with relevant policies being developed in isolation (Bazilian et al., 2011, Kahrl & Roland-Holst, 2008), the absence of integrated policies has increased water–energy trade-offs and related risks (Marsh, 2008).

Water has been identified as the biggest societal and economic risk for the next decade by the World Economic Forum (World Economic Forum, 2015). According to the IPCC report (2013), anticipated climate change will lead to a worrying trend of heightened drought risks, including increased drought periods and expansion of drought-affected areas, which will consequently amplify existing problems of water scarcity. UNESCO estimates that over half of the world's nations will face freshwater stress by 2025, and, by 2050, about three-fourths of the global population will face water scarcity (UN WWAP, 2012).

Water shortages and increased water temperature can put electricity generation at stake. This has already been witnessed during the 2007 and 2012 droughts in the US when many thermal generators were exposed to water-related shut downs and curtailments. The water demand by the power sector is expected to increase with the developing economy and population's growing demand for power production, while climate change is predicted to alter the spatiotemporal distributional pattern of water availability, e.g. low flows, thus reducing water security in some regions. This has led many researchers to raise red flags about the future of the energy and water sector in different parts of the world, e.g. the US (Sovacool & Sovacool, 2009) and the UK (Byers, Hall, Amezaga, O'Donnell, & Leathard, 2016).

The following section provides a quick overview of how water is used in the energy sector's life-cycle processes, e.g. fossil fuel extraction, as well as for power production. The rest of the chapter focuses on the case of China in particular and on water uses of power generation to provide an in-depth and comprehensive understanding of the water-for-energy issues in this region. A better understanding of the issues previously mentioned enables decision makers to come up with effective governance strategies to harness the potential syntheses and to avoid unintended conflicts. Finally, assessments of the coherence between water and energy policies at both national and regional levels demonstrate the rise of new governance challenges.

The impeding risks of climate change pose multi-dimensional governance challenges for development and resource management. The power sector contributes a significant proportion, about 40 percent, according to International Energy Agency, to the global greenhouse gas (GHG) emissions, considered as the primary cause of climate change (IEA, 2014). Water conservation in the power sector therefore may help climate change mitigation, a co-benefit that invites further exploration. Some potential synergies between energy policies and water conservation have been demonstrated (Bartos & Chester, 2014; Liao, Hall, & Eyre, 2016) and should be fully harnessed while potential conflicts between energy development and water policies (Qin, Curmi, Grant, Allwood, & Richards, 2015; Liao et al., 2016) should be avoided. Despite growing appreciation of the water–energy nexus, a better understanding of the whole system-of-systems as well as the governance integration of the two resources is needed (IEA, 2012; Schnoor, 2011; Lee, Ellinas, & House, 2010). The previous paradigm whereby each sector operates in its own silo – uninformed and unconcerned about its effect on the others – needs to be abandoned. This chapter explores a new paradigm to move these issues forward.

The significance of China

The aforementioned problem is especially pronounced in China, the world's biggest power consumer. China's economy is heavily based on coal, with coal-fired electricity production accounting for approximately 80 percent of the nation's total electricity generation (Zhang, 2012). Over the last decade, electricity consumption in China has almost quadrupled from 1,347 TWh in 2000 to 4,976 TWh in 2012. This rising trend is projected to continue with China's growing economy and population (Zhang, Zhong, Fu, Wang, & Wu, 2016). Even though China has pledged to reduce its share of coal-fired production, coal is still expected to dominate the country's energy portfolio.

China's water resources are limited. In 2012, it was calculated that even though the country's total surface water endowment ranked sixth in the world, the amount per capita stood at only one fourth of the global average (Ye & Zhang, 2013). In 2011, China issued its 'strictest' water management mechanism called the 'three red-line' policy, setting targets on total water use, industry and agriculture water efficiency, and water quality improvements. The first 'red-line' sets the water withdrawal cap at 635, 670, and 700 million m^3 in 2015, 2020, and 2030, respectively.

However, two distinctive features of China's water resources further exacerbate the potential water-energy challenge:

1 *Uneven spatial distribution and geographical mismatch with energy reserves:*

More than 80 percent of China's rivers are in the south while the northern areas, that contain 37 percent of China's population and 45 percent of its farmland, have only 12 percent of the country's water resources. As illustrated in Figure 5.1, a large amount of power generation capacities are installed in regions facing the most severe water shortages and over half of the country's proposed coal-fired power plants are placed in areas with high water stresses (China Water Risk, 2012; World Resource Institute, 2014). This geographical

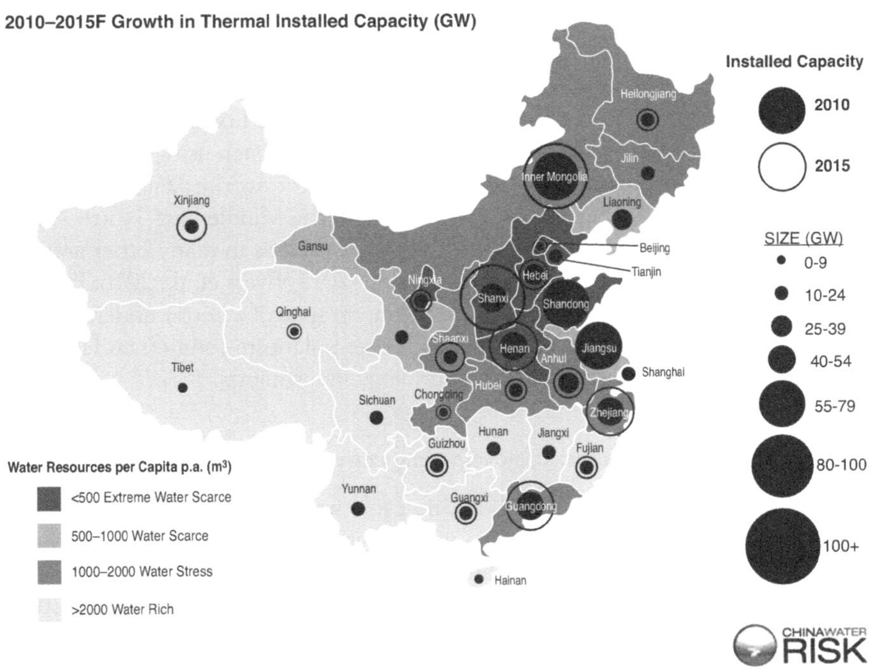

Figure 5.1 Denotes that albeit water resources are not evenly distributed among China's provinces, a large amount of power generation capacities are installed in regions facing the most severe water shortages.

mismatch is particularly significant with coal energy as coal extraction and coal-fired power production are done close to coal reserves to reduce production cost, despite using large quantities of scarce water. As a result, 11 Chinese provinces, mainly in the northeast, already face water scarcity (<1000 m^3 per capita per year) (Francis, Mehta, Cosgrove, Huang, & Chan, 2013). For instance, Shanxi, Shaanxi, and Inner Mongolia, the three provinces with the most coal reserves in China, contribute over half of China's coal output and a quarter of the thermal power generation with only 3 percent of the total water resources (Zhang & Anadon, 2013).

2 *Uneven intra-year temporal distribution*:

The seasonal distribution of China's water resources within a year is heavily impacted by its monsoon climate. South of the Yangtze river, precipitation between March and June makes up about 60 percent of the annual total, while precipitation between June and September accounts for around 80 percent in the north of the Yangtze river (Ye & Zhang, 2013). The least precipitation occurs in the winter, which means the hydropower capacity is also the lowest. However, China's national power demand also peaks in winter, as central heating is provided by Combined Heat and Power (CHP) plants in northern China and heating pumps is widely used in southern China.

Water use in the energy sector

Studies on water demand for the energy sector have increased markedly in the past decade. In 2005, Vassolo and Doll (2005) made the first attempt to present the water use by power generation via a bottom-up approach using the 1995 data of 63,590 thermal power stations located around the world. Water-for-energy nexus studies have also been carried out in Spain, the Middle East–North Africa (MENA) region, Jordan, the UK, and the US, as well as in many other areas in the world (Qin et al., 2015; Siddiqi & Anadon, 2011; Byers et al., 2016; Fthenakis & Kim, 2010). They all underlined the importance of a better understanding of water usage by the energy sector to avoid potential future conflicts and lock-ins with unsustainable infrastructure or technology investments.

Water for the energy sector: life-cycle perspective

Numerous studies have examined the life-cycle water use[2] of the energy sector, providing an understanding of the varying water usage across the sector's different life-cycle phases (e.g. Fthenakis & Kim, 2010; Mielke, Anadon, & Narayanamurti, 2010; Meldrum, Nettles-Anderson, Heath, & Macknick, 2013; Williams & Simmons, 2013). Despite such studies highlighting that power production takes up the lion's share of the energy sector's water use, which will be discussed in detail in the next section, water is also directly consumed in the extraction and process of raw energy resources, as well as in its conversion and transportation. Take coal for instance. During coal mining, water is mainly used to control dust and to cool down underground mining equipment. Depending on the quality, coal sometimes then needs to be washed to remove its impurities and sulphur before it can be burned to generate power, demanding more water consumption (Department of Energy, 2006). After extraction, transport can also be water intensive as it often involves pumping water (e.g. with slurry pipelines) (Meldrum et al., 2013).

Some characteristics of the life-cycle water uses of power production by various fuel types are listed in the Table 5.1:

Table 5.1 Life-cycle water uses of power production by various fuel types

Fuel type	*Characteristics*
Biofuel	By far the most water-intensive fuel. A substantial amount of water is consumed due to the evaporation when growing the stock, e.g. corn.
Natural Gas	Gas-To-Liquids of Natural gas and Coal-to-Liquids of coal involve a large amount of water input. China has thus abandoned its plans to build dozens of coal-to-liquid (CTL) plants in 2008 presumably due to the local water scarcity (IEA, 2012).
	Hydra fracturing for non-conventional fossil fuels, e.g. shale gas, raises an increasing concern. Even though the exploitation of shale gas is not anticipated in China in the near future, China has one of the largest shale gas reserves in the world, the development of it should be carried out with caution.
Hydropower	Water consumption of hydropower is very site-specific and varies considerably due to local climate, vegetation, and other factors (WRI, 2015).
	It is complicated by the various functions of reservoirs (e.g. water supply, recreation, and flood control) and the different ways of allocating evaporation to electricity generation (Macknick, Newmark, Heath, & Hallett, 2011). Hitherto, there is no commonly acknowledged way of allocating water consumptions in a multipurpose reservoir.
Renewables, e.g. wind and solar PV	It requires a considerably less amount of direct water use. Water is majorly used to wash wind turbines and solar panels.

In addition to the water uses mentioned previously, a certain amount of water is embodied in manufacturing infrastructure materials, e.g. wind turbines and solar panels, thus requiring indirect water input. Finally, the life span of power plants, normally ranging between 30 to 50 years, can also lead to further water consumption: building and dismantling a power plant after its retirement needs water input for various reasons, including dust control, domestic uses, and so forth.

Water for power production

Water is heavily used or consumed in various processes in the operational phase (within the power plants) of electricity production (Macknick, Newmark, Heath, &

Hallett, 2011). In coal-fired power plants: e.g., water is used to clean the coal, remove the lime-ash, for desulfurization, and for cooling the steam turbine as well as different auxiliary units. The water-consuming processes can be categorized as follows:

1) Industrial water use: which undergoes clarification and can then be used for different industrial purposes, such as lime-ash removal and wet desulfurization, as well as cooling of the steam that drives the turbine and other auxiliary engines (including induced air fan, forced draft fan, primary air fan, and so forth). Cooling water constitutes the biggest part of power production's water uses.
2) Domestic water use: it is consumed to meet the living needs of power plant workers.
3) Chemical water use: due to salinity enrichment, water boilers need to be regularly emptied and replesnished with demineralized water of the highest quality.

As seen from Tables 5.2 and 5.3, which detail the water consumption factors of different processes in coal plants, and Figure 5.2, which shows the water withdrawals and consumptions of different thermal-generating technologies and energy sources (e.g. gas-fired, nuclear, or concentrated solar power), water usage is primarily determined by the cooling system utilized. There are three types of

Table 5.2 Designed water consumption factor of power plants with open-loop cooling system of different capacities.

Capacity (MW)	*Wet desulfurization (without GGH, m^3/h)*	*Dry ash/lime-ash removal (m^3/h)*	*Coal transport (m^3/h)*	*Boiler make-up (m^3/h)*	*Domestic water use (m^3/h)*	*Total (m^3/h)*	*m^3/MWh*
2×300	100–110	109–195	46–83	80–150	5–10	340–548	0.567–0.913
2×600	190–230	155–247	60–107	90–170	5–10	500–764	0.416–0.637
2×1000	270–320	265–380	77–146	150–270	5–12	767–1128	0.384–0.564

Source: East China Electric Power Design Institute (2012)

Table 5.3 Designed water consumption factor of power plants with closed-loop cooling system of different capacities.

Capacity (MW)	*Wet desulfurization (without GGH, m^3/h)*	*Dry ash/ lime-ash removal (m^3/h)*	*Coal transport (m^3/h)*	*Boiler make up (m^3/h)*	*Domestic water use (m^3/h)*	*Cooling tower make-up (m^3/h)*	*Total (m^3/h)*	*m^3/MWh*
2×600	190–230	155–247	60–107	90–170	5–10	2092–2692	2592–3456	2.16–2.88

Source: East China Electric Power Design Institute (2012)

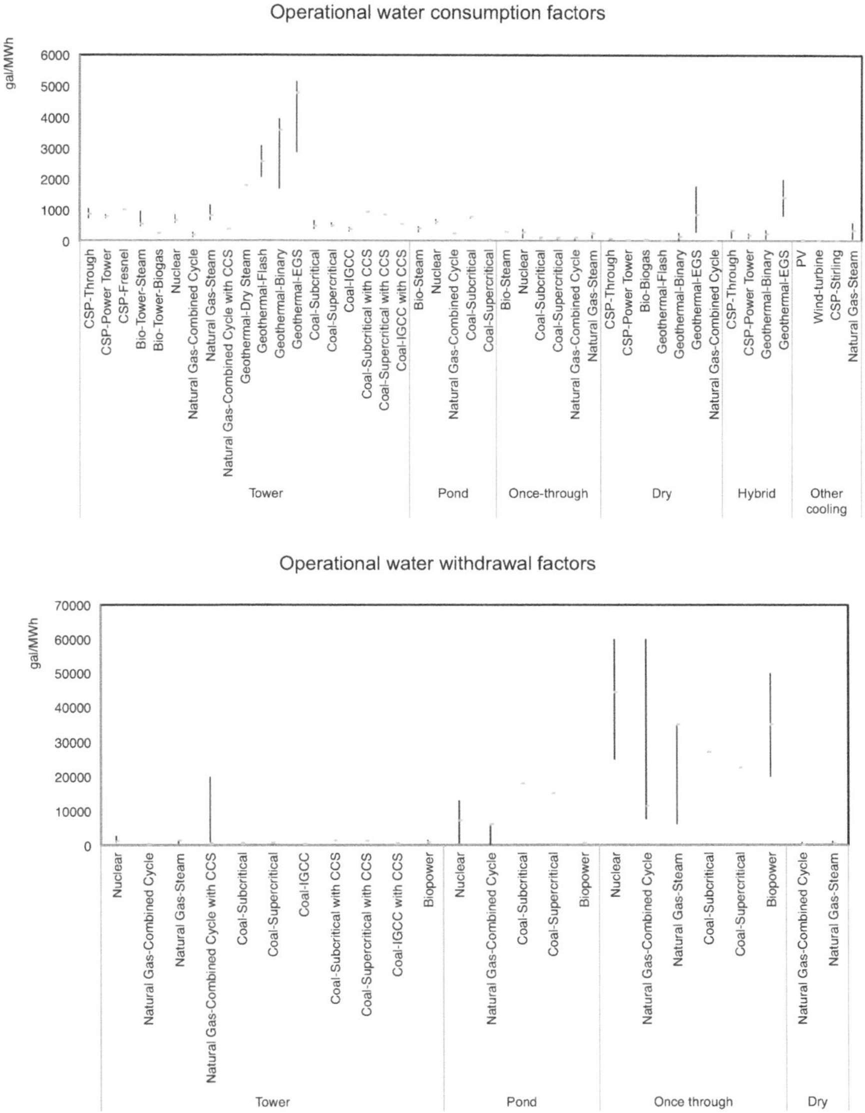

Figure 5.2 Operational water consumption (upper) and withdrawal factors for different energy sources and electricity generating technologies

(Data from Macknick et al., 2011)

IGCC: integrated gasification combined cycle. CCS: carbon capture and sequestration. CSP: concentrating solar power. Upper and lower ends represent maxima and minima, respectively. Grey dots represent medians.

cooling system being used: open-loop and closed-loop wet cooling, and air-cooling. Open-loop cooling systems require the largest amount of water withdrawal; closed-loop, or recirculated cooling, withdraws significantly less water; however, a larger proportion can be consumed due to evaporation, either in cooling ponds or cooling towers. Air-cooling, also called dry cooling, withdraws and consumes the least, almost negligible, amount of water.

Cooling technology affects the water consumption factors in two ways: first, nearly 100 percent of water abstracted by closed-loop cooling systems evaporates during the cooling circles while only 3 percent of water withdrawal of open-loop systems is consumed. Therefore, when compared to closed-loop cooling systems, consumptive water use in an open-loop system is 70–80 percent lower than in a closed-loop system, but water withdrawal is 30–60 times higher (World Resource Institute, 2015). Secondly, since cooling systems marginally impact the plant's efficiency, power plants with an open-loop cooling system consume less water in other processes due to its higher efficiency.

Apart from cooling technologies, energy types also bear a significant impact on power production's water use. To generate a certain amount of electricity, gas-fired power requires the least water while nuclear power needs the most. Moreover, water uses also differ by generating technology. For instance, application of Carbon Capture and Sequestration (CCS) results in increases of both water consumption and withdrawal (Byers et al., 2016).

Water for energy in China

In this section we review two key quantification studies on China's water-for-energy nexus on a regional level from a top-down and bottom-up perspective, respectively.

Multiple studies have examined the question of water and energy in China, although most are limited in terms of providing an overall assessment at the national scale. For instance, Pan, Liu, Ma, and Li (2012)'s study on water withdrawal and consumption of China's coal industry using average water intensity is limited to one type of technology. Similarly, Yuan, Lei, Xiong, Guo, and Zhao (2014) examined the water implications of coal-fired power generation in China with a bottom-up approach using data from power plants with capacity over 600 MW, which accounted only for around 10 percent of China's national thermal capacity. Other studies have accounted for a variety of technologies, for instance, Zhang and Anadon (2013) used a top-down input–output method and estimated China's energy sector's life-cycle water withdrawals, consumption, and wastewater discharges. However, although technologies and spatial variations were taken into account, no future assessment was provided. Feng, Hubacek, Siu, and Li (2014) on the other hand examined the emission and water consumption of eight electricity generation technologies (from coal to wind) using a hybrid life-cycle input–output analysis and demonstrated the long-term co-benefit of shifting energy source to low-carbon renewable ones for climate change mitigation and water conservation.

In terms of future assessments, different scenarios have been used by various authors. Liao, Hall, and Eyre (2016) have studied the future possibilities of China's water-for-energy nexus into 2050 based on WWF's scenarios while both Cai, Zhang, Bi, and Zhang (2014) and Qin et al. (2015) have utilized IEA's global energy scenarios to assess China's energy sector's future water uses. Although Cai et al. (2014) did not clearly differentiate non-consumptive and consumptive water uses and therefore underestimated the national total water demands by energy sector, they demonstrated the urgency of reforming China's energy structure to reduce the pressure of water scarcity. Qin et al. (2015) compared the projected future energy sector's water uses with the Industrial Water Allowances (IWA) by China's 'three red-line' policy and concluded that, under a Business As Usual Scenario, the IWA may be surpassed on a national level.

Regional results from a top-down perspective

To understand the various consumptions of water, Zhang and Anadon (2013) created a mix-unit Multi-regional Input–Output (MRIO) table. They extrapolated sectoral water uses by putting together (1) provincial monetary input–outut (IO) tables; (2) provincial and sectoral energy production, consumption, and trade data; and (3) provincial and sectoral freshwater withdrawal, water consumption, and wastewater discharge data. Surface water withdrawal and total water consumption, including surface water and groundwater, for electricity production are demonstrated in Figure 5.3.

The previous diagram underlines how consumptive water is generally one magnitude lower than water abstraction. A large amount of water abstracted is returned to the environment with or without treatment. Regionally, water is mostly consumed in coal-abundant areas, including Hebei, Shanxi, Inner Mongolia, and intensely populous areas, including Jiangsu, Shandong, and Guangdong. In Jiangsu and Guangdong, water consumption is directly correlated with water abstraction, while the correlation is reversed in other provinces, such as Shandong or Henan. The possible reasons will be discussed in detail in the following section.

Regional results from a bottom-up perspective

Water use intensity, measured as water use per unit of power produced, differs according to energy source and technology (Macknick, Newmark, Heath, & Hallett, 2011), therefore a bottom-up perspective reveals more details on these variations. As mentioned previously in section 2.2, water usage varies depending on the power units' capacity and the technology used, with water consumption decreasing as power plants' capacity increases (see Table 5.4).

For small-scale units, closed-loop cooling systems consume the most water while air-cooling systems consume the least. However, when the capacity exceeds 600 MW, air-cooling systems lose their advantage to open-loop systems in terms of water consumption, presumably because air-cooling systems are less efficient and require more electricity for the plant's own use.[3]

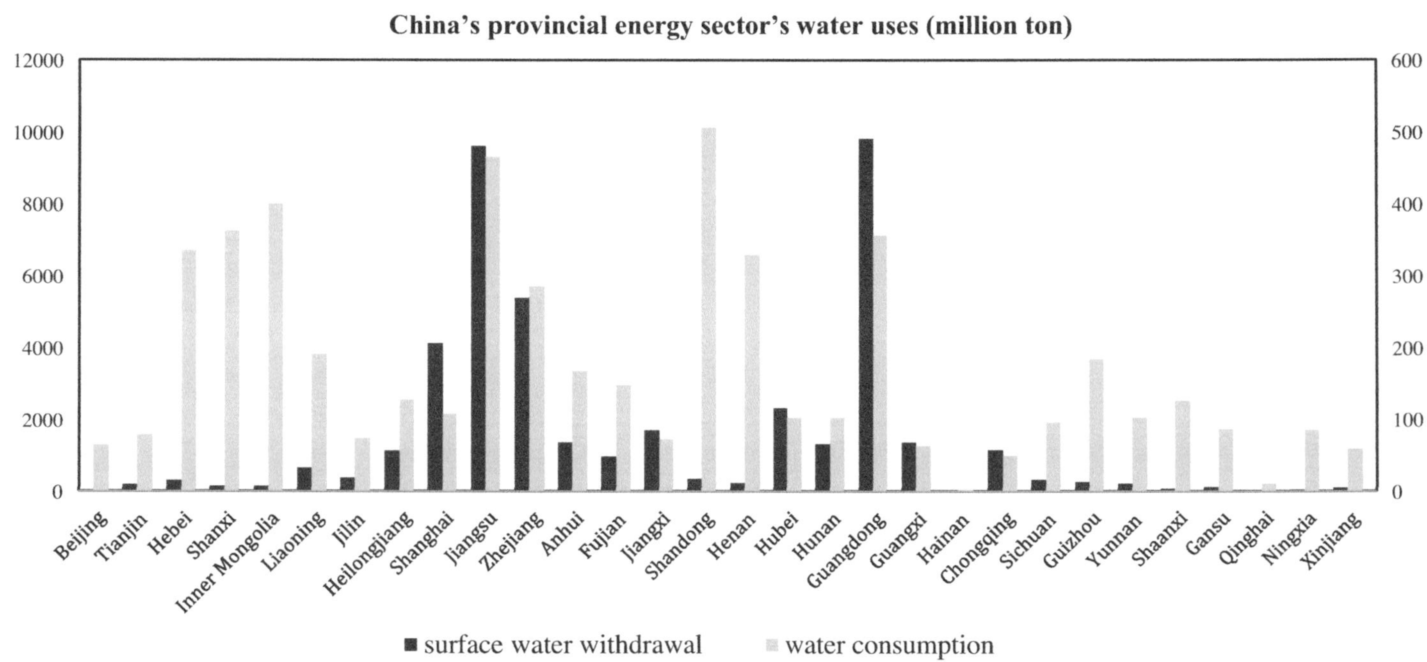

Figure 5.3 China's provincial energy sector's water uses
(Source: Zhang & Anadon, 2013)

Table 5.4 Average water consumption factors of units of different capacities with different cooling systems in China

Capacity (MW)	*Cooling system*	*Water consumption factor*
<300	Closed-loop	2.25
<300	Open-loop	1.23
<300	Air cooling	0.47
300–600	Closed-loop	1.93
300–600	Open-loop	0.52
300–600	Air cooling	0.41
>600	Closed-loop	1.79
>600	Open-loop	0.29
>600	Air cooling	0.31

Source: (China Electricity Council, 2012 and China Electricty Council, 2013).

The pattern of provincial water uses presented in Figure 5.3 can be understood using Figure 5.4, which present the capacities (MW) of different cooling technologies, including closed-loop, open-loop, and air cooling, in different provinces (Figure 5.4),[4] as well as the provincial water consumption by power sector (Figure 5.5). As we can see, closed-loop cooling systems are predominantly used throughout the country. Air cooling system is mainly utilized in the north, where water is lacking, while open-loop cooling technology is located mainly in coastal areas, where seawater is abundant. Indeed, the utilization of seawater and reclaimed water are effective options to reduce freshwater consumption, which is likely to be employed in coastal cities or cities with huge wastewater reclamation potentials. High water consumption in northwestern China (Hebei, Shanxi, Inner Mongolia), eastern China (Zhejiang, Jiangsu, Shandong) as well as southern China (Guangdong) can be explained by their high total installed capacity and corresponding power production. Water withdrawal is especially high in Jiangsu, Zhejiang, and Guangdong due to the abundance of open-loop cooling systems. With regards to the type of water used, surface water consumption prevails throughout the whole country, especially in intensively populated provinces with high power demands, such as Jiangsu, Shandong, or Henan, and provinces where closed-cooling technology is predominantly used, such as Guizhou, Anhui, or Xinjiang. To tackle surface water shortages in water-stressed areas, mainly in the north, (e.g. Hebei, Shanxi, Shaanxi, and Inner Mongolia), groundwater has been harnessed as a water resource. Reclaimed water utilization is also seen in the same areas.

The aforementioned studies, whether from a top-down or a bottom-up perspective, are consistent (except in Guangdong, where water consumption is overestimated with the top-down method) and underline the importance of looking at the water–energy nexus on a regional scale. Given that water and energy are both highly region-dependent issues in China, 'one-size-fit-all' policies normally lack

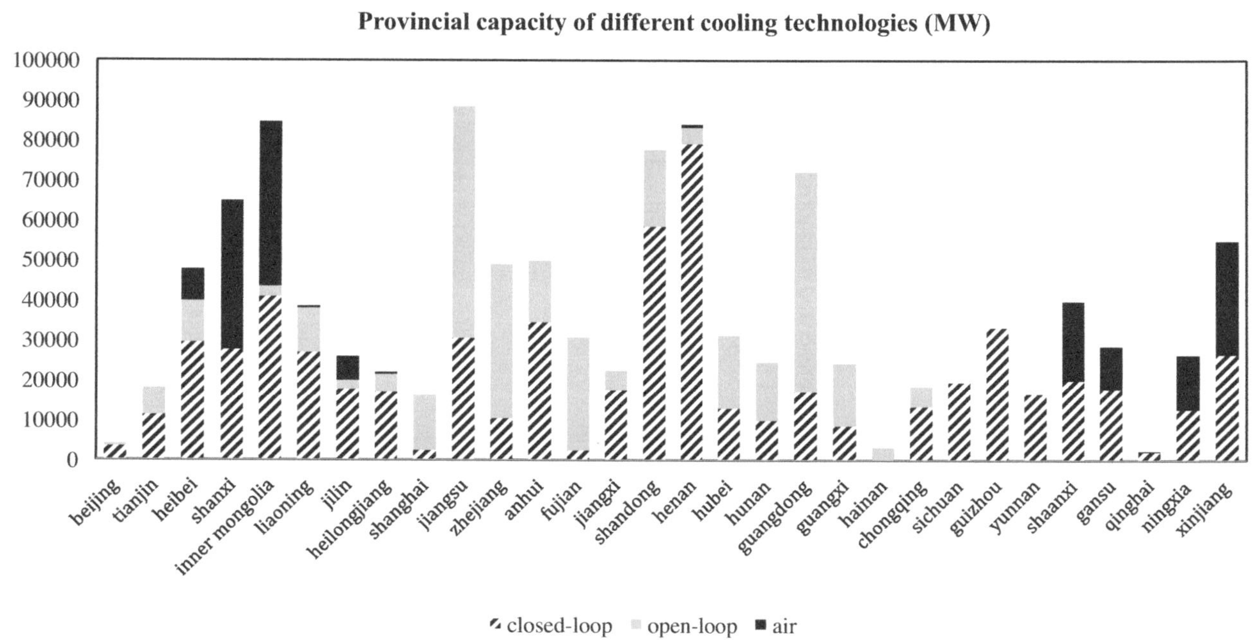

Figure 5.4 Provincial capacity of different cooling technologies

(Source: Liao et al., 2016)

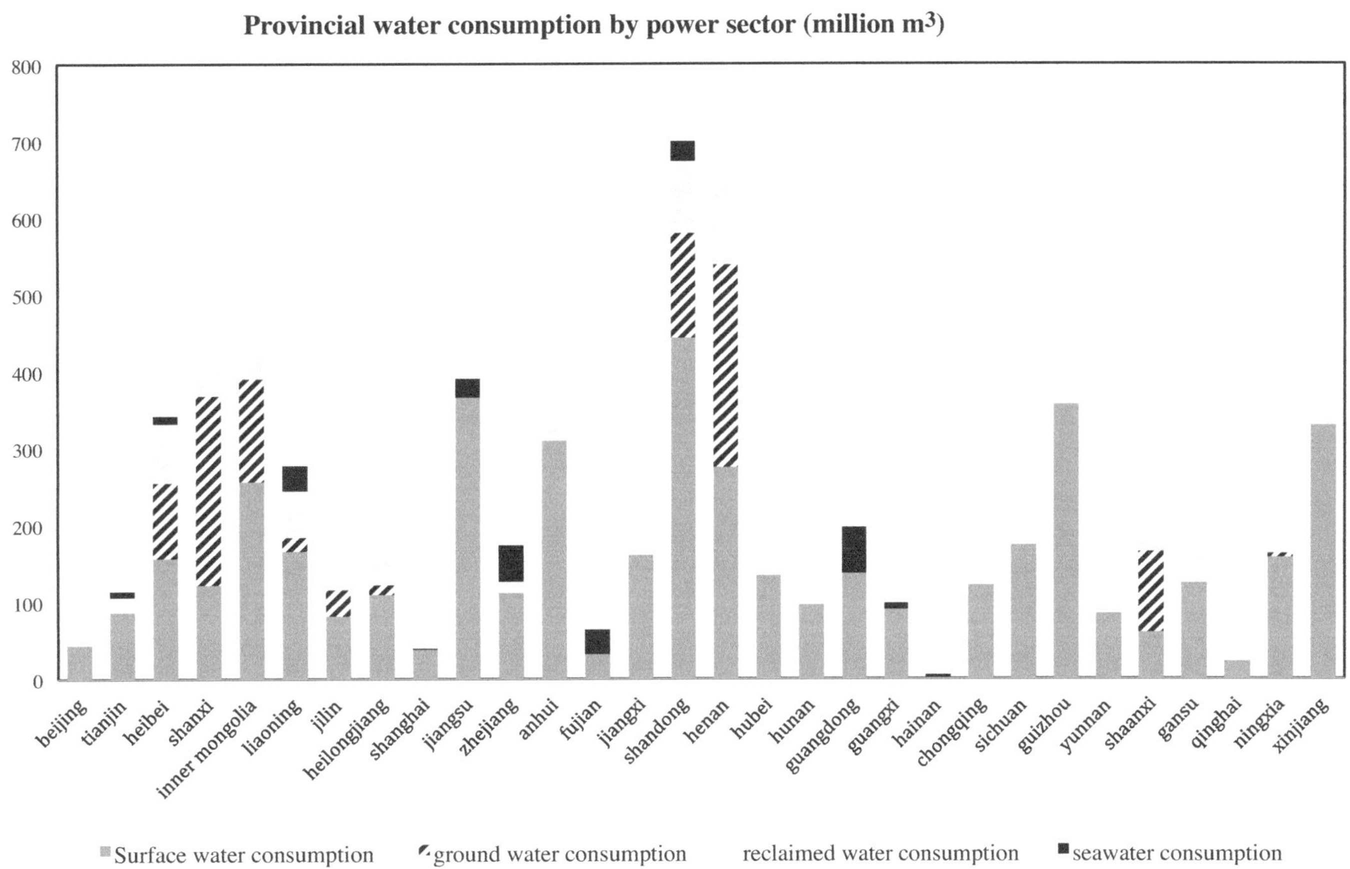

Figure 5.5 Provincial water consumption by power sector
(Source: Liao et al., 2016)

thorough considerations and may be ineffective. The following section examines the coherence between China's water and energy policies and discusses potential conflicts.

Challenges of water and energy governance

Coherence between water and energy policies

Some publications to date have pointed out the multiple benefits in setting energy policies aimed at cutting water uses in the power sector. Webster, Donohoo, and Palmintier (2013) found that, in Texas, a restriction on CO_2 emissions would also reduce water withdrawals by the power sector. Bartos and Chester (2014) demonstrated that the implementation of Arizona's Energy Efficiency Mandate and Renewable Portfolio Standard also helped to reduce its power sector's water usage.

Many researchers in China have intended to understand how China's energy structure transformation, complied with coal-fired generating technology advancement, and cooling system alteration, are likely to affect water usage (Liao et al., unpublished). Potential unwanted conflicts between water preservation and energy managements in China are forewarned by many scholars. For instance, in 2011, China issued its 'strictest' water management mechanism called the 'three red-line' policy, setting goals on total water use, water pollution, and water use efficiency. The first 'red-line' sets the cap of total water amount that can be withdrawn at 635, 670, and 700 million m^3 in 2015, 2020, and 2030, respectively. Using energy scenarios from the International Energy Agency (IEA) (2012), Qin et al. (2015) have assessed China's water demands by the energy sector in 2035 and compared it with the Industrial Water Allowed (IWA) by the 'three red-line' policy. They concluded that under a Business As Usual Scenario, water withdrawal is likely to exceed the IWA, unless water-conserving actions are taken, such as reconstructing power portfolio, implementing demand reduction policies, and expanding dry cooling technologies.

Considering the huge regional variations within China, Liao et al. (2016) further assessed China's power sector's compliance with IWA in 2030 with scenarios from WWF (2014) on a regional level. Five scenarios were examined: baseline scenario, where no policies other than the current ones are to be implemented; high-efficiency scenario, where China is devoted in improving its energy efficiency; high-renewable and low-carbon scenarios, where China meets its energy demand mainly by renewable or low-carbon energy sources respectively; under a technology change scenario, where open-loop cooling systems in southern regions (east, south, and central) and closed-loop cooling in northern regions are replaced by closed-loop and air cooling ones, respectively. The results show that there are tremendous discrepancies among different regions' energy sector's compliance with the national water policy (Figure 5.6). Specifically, water targets are relatively loose for northwestern and northeastern China. Water policy is too strict in the east and is likely to be violated under all scenarios – unless technology

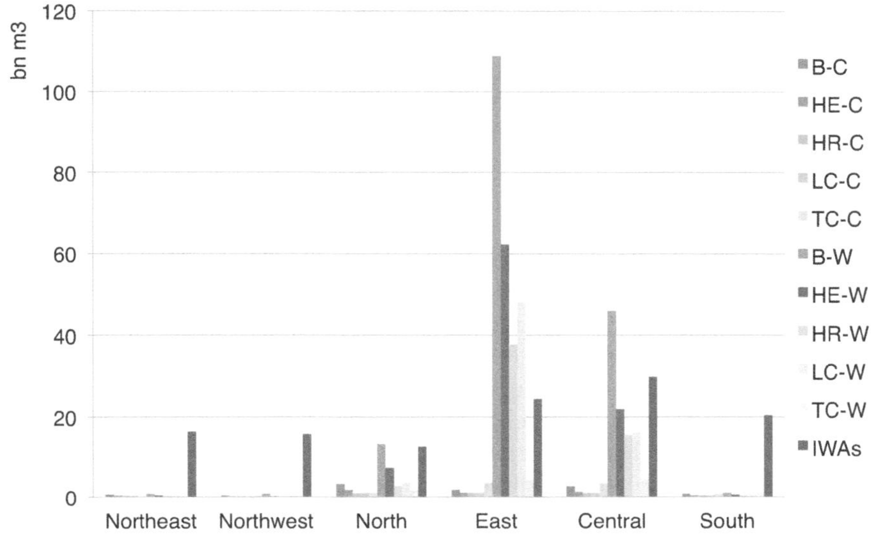

Figure 5.6 (a) China's energy sector's compliance with IWA set by the '3 Red-Line' policy in 2035 on a national level (Qin et al., 2015); (b) Comparison between regional water demands by the power sector with water targets set by the '3 Red-Line' Policy in 2030 in China

(W-Fresh water withdrawal in 2030; C-Fresh water consumption in 2030; B-Baseline Scenario; HE-High Efficiency Scenario; HR-High Renewable Scenario; LC-Low Carbon Scenario; TC-Technology Change Scenario; IWA-Industrial Water Allowed) (Liao et al., 2016).

change is facilitated – possibly due to the over-abundance of open-loop cooling systems and substantial water abstraction. For northern, central, and southern China, the water targets are considered stringent and are likely to be violated under a Baseline Scenario but could be abided with if energy efficiency is improved or energy structure is transformed, which is consistent with the national situations in Qin et al. (2015).

Hence it is foreseeable that uncoordinated or 'one-size-fits-all' policies may result in policy violations or governance conflicts, which could impair the credibility of respective institutions and even undermine further efforts. It is important to bring such complexities into future scientific studies as well as policy decisions.

Potential impacts by climate change

On top of the potential conflicts previously mentioned, climate change is expected to worsen the emergent problem. Van Vliet, Wiberg, Leduc, and Riahi (2016) have discussed these effects globally while other scholars have conducted some regional studies with greater details (e.g. in the UK with Byers et al., 2016).

Overall, studies have pointed to the three following effects of climate change: (1) alteration of the spatiotemporal pattern of river runoff and decrease of the low flows in some regions, thus reducing the hydropower capacities and increasing the demand for thermal power generation; (2) decline in low flows as well as reduction in the available water for thermal power's cooling use; (3) increase in water temperature, which will lower the cooling as well as generating efficiency of thermal power plants thus increasing the demand for cooling water intake.

Zheng, Wang, Cai, Kummu, and Varis (2016) conducted vulnerability assessment to water scarcity on China's thermoelectric power generation. They concluded that climate change will aggravate plants' vulnerability in the Yellow River Basin, but will have generally mild impacts on other regions in China. However, as low-flow situations and water temperature rises pose the biggest threats to reliable power production, the indicator adopted in Zheng et al. (2016)'s work (which only takes total water availability into account), is not able to fully address the potential impacts brought by climate change on China's power production. Indeed, a hydrological modelling reveals that eastern China is projected to experience a strong decrease in low flow and large increase in mean water temperature (larger than 2.0 °C) with climate change (Van Vliet et al., 2016). Thus, further assessments of the impacts brought by climate change on China's water-for-energy nexus need to be carried out.

Institutional arrangements governing water and energy in China

As Liao et al. (unpublished) have reviewed, the institutional arrangements for China's water and energy resources have gone through noticeable changes. In 1949, the Ministry of Water Resources and Ministry of Electric Power Industry were established in the People's Republic of China. In 1958, the Ministry of Water Resources and Ministry of Electric Power Industry were merged into the Ministry of Water and Electricity, which was divided again in 1979, combined again in 1982, and finally re-divided in 1988 until now. The repeated institutional transformation shows that governance of water and electricity has been an ongoing challenge in China.

Later, in 2002, the Bureau of Electricity was divided into five electricity production corporations and two grid companies. The five power production companies, namely, Guodian, Datang, Huaneng, Huadian, and State Power Investment, own about 50 percent of China's electricity-generation capacity and are responsible for power generation. The two grid companies, State Grid and China Southern Power Grid, are responsible for electricity dispatch, transmission, and delivery. The National Development and Reform Commission (NDRC) and local DRCs are responsible for short, medium, and long-term power construction and development plans while the Ministry of Industry and Information Technology and the local Commissions of Economy and Information Technology are responsible for electricity production coordination, demand management, and so forth.

Water resources are governed by the Ministry of Water Resources in China. The Ministry of Water Resources develops water utilization plans and provincial water quotas that are implemented by local water resource bureaus. Although there is a River Basin Committee set up for each river basin, water quotas set by River Basin Committees are usually used as a reference and their main responsibility is to coordinate if any conflicts over water use between different provinces within the same river basin, occur. The fragmented and contested authority of water resource management in China complicates and, sometimes, invalidates the enforcement of related polices and thus weakens the credibility of respective institutions.

After construction of a power plant is proposed by a power production corporation, it requires permission from the local Development and Reform Commission in order for power generation to fit into the overall power development plans and from the Water Resources Bureau so that its water consumption conforms to the regional water quota scheme.

Conclusion and future work

The nexus of energy and water is an emerging global issue, yet it is often overlooked. In recent years, an appreciation for the increasing interdependencies has grown considerably. On the one hand, accelerating climate change is projected to result in more frequent extreme climate events and higher water risks. On the other hand, the fast growing population and economy imply increasing power demands and ensuing water needs. The consequent water–energy conflict is especially pronounced in China, with the most prominent operational water use by power production and the governance conflicts caused by the current spatial pattern of China's power sector's water uses. The following points emerge from the previous sections:

- Water is needed throughout the life-cycle of the energy sector, from upstream fuel mining, to power production and downstream infrastructure dismantling. However, within the life-cycle of energy use, water is mostly consumed in the operational phase of electricity production. Cooling water dominates the water uses among various processes in the operational phase.
- The operational water use of power production is impacted by different factors. Cooling technology has the most influence. Plants with open-loop systems withdraw the most water, at about 30 to 60 times of that of closed-loop systems, and air-cooling systems withdraw the least. In terms of the water consumption, apart from closed-loop systems consuming the most, an air-cooling system consumes similar amounts of water as an open-loop system, possibly due to its efficiency disadvantage and consequent water increments in other processes.
- The power sector's water use is a region-dependent issue in China, which is manifested by studies from both a top-down and a bottom-up perspective. As freshwater resources are abundant in southern China, open-loop cooling systems are more commonly used and therefore water consumption is lower than in northern China where predominantly closed-loop and air-cooling

systems are employed. Moreover, using alternative water sources, e.g. seawater and reclaimed water, would reduce the power sector's freshwater usage.

- A governance dilemma appears regarding the aforementioned resources: there are possible conflicts between China's water demands from its energy development under different scenarios and the availability of water allowed by China's water policy's controlling the target on total water use, the IWA. On a national scale, the IWA is likely to be exceeded if the current status carries on. On a regional scale, the conflict is mostly pronounced in the east, where the IWA will be surpassed under all scenarios. On top of that, as some regions are projected to experience decrease in low flows and increase in water temperature under climate change – e.g. eastern China – prospective climate change may exacerbate the issue. Governing institutional transformation and improving policy coherence is necessary.

The aforementioned points revealed in this chapter point to the need for further research on the governance of the water-for-energy nexus. In particular, the following aspects could be further examined:

1 **Spatial analysis on future water demands by power sector.** Most power plants to be operated until 2050 will be built around 2020. In order to come up with potential effective policy implications and avoid 'lock-in' technologies, spatial analyses of future status need to be carried out.
2 **Temporal analysis on future water demands by power sector.** Extreme water–energy conflicts are most likely to happen during the winter when energy demand is especially high and water availability is low. In order to secure power supply all year round, a temporal analysis needs to be conducted.
3 **Hydropower's water implications.** While run-of-river hydropower is favoured due to its small water consumption and emission, its power generation can be disrupted if water is depleted upstream by droughts or anthropogenic abstractions (WRI, 2015). It is crucial to incorporate hydropower into the research framework.

Notes

1 Marsh (2008) has mapped out the comprehensive water uses in the upstream, transportation and downstream processes of electricity sectors.
2 Water use includes water withdrawal and consumption. Water consumption refers to water withdrawn from the environment but not discharged back to any water bodies (AQUASTAT, 1998).
3 As the life span of power plants stands at around 30 years (Word Bank, 2005) and almost all the power plants built after 2000 have a capacity of over 600 MW, it can be assumed that by 2030 all units with capacity below 600 MW will have been replaced by ones of 600 MW and beyond.
4 Liao et al. (2016) put together by far the most comprehensive dataset of China's coal-fired power plants with data from Carbon Monitoring for Action (CARMA), World Electric Power Plants Data Base (WEPP), Global Coal Plant Tracker (CPT),

and Enipedia. Over 1,000 coal-fired power plants are included. The cooling technology employed by a single power plant is identified individually via Google Earth. Operational water consumption factors and water source configurations are approximated by data from CEC (2012). Provincial average running hours of coal-fired power plants are used in order to calculate the electricity productions (National Statistic Bureau, 2014).

References

AQUASTAT (1998). AQUASTAT definitions. Food and Agricultural Organization of the United Nations (FAO). Retrieved from www.fao.org/nr/water/aquastat/data/glossary/search.html.

Bartos, M. D., & Chester, M. V. (2014). The conservation nexus: Valuing interdependent water and energy savings in Arizona. Envirnmental Science & Technology, *48*, 2139–2149.

Bazilian, M., Rogner, H., Howells, M., Hermann, S., Arent, D., Gielen, D., Steduto, P., Mueller, A., Komor, P., Tol, R. S. J., & Yumkella, K. K. (2011). Considering the energy, water and food nexus: Towards an integrated modelling approach. *Energy Policy*, 39, 7896–7906.

Byers, E. A., Hall, J. W., Amezaga, J. M., O'Donnell, G. M., & Leathard, A. (2016). Water and climate risks to power generation with carbon capture and storage. Environmental Research Letters, *11*, 024011.

Cai, B. M., Zhang, B., Bi, J., & Zhang, W. J. (2014). Energy's thirst for water in China. Envirnmental Science & Technology, 48(20), 11760–11768.

China Electricity Council (CEC) (2012). Notification of energy efficiency benchmarking and competition data of 2012 national 600MW thermal power units (in Chinese). Retrieved from http://kjfw.cec.org.cn/kejifuwu/2013-04-07/99877.html.

China Electricity Council (2013). Notification of energy efficiency benchmarking and competition data of 2013 national 300MW thermal power units (in Chinese). Retrieved from http://kjfw.cec.org.cn/kejifuwu/2014-08-14/126214.html.

China Water Risk (2012). China: No water, no power. Retrieved from http://chinawaterrisk.org/resources/analysis-reviews/china-no-water-no-power/.

Department of Energy (DOE) (2006). Energy demands on water resources: Report to congress on the interdependency of energy and water. Washington, DC.

East China Electric Power Design Institute (2012). Investigation on designed water withdrawal and water consumption statistics of Shanghai's big thermal power plants. Shanghai: East China Electric Power Design Institute. (Internal materials).

Feng, K. S., Hubacek, K., Siu, Y. L., & Li, X. (2014). The energy and water nexus in Chinese electricity production: A hybrid life cycle analysis. Renewable & Sustainable Energy Reviews, 39, 342–355.

Francis, S., Mehta, K., Cosgrove, J., Huang, S., & Chan, W. S. (2013). China coal and power, the water-related challenges of China's coal and power industries. Hongkong: HSBC Global Research. Available: http://www.longfinance.net/images/reports/pdf/hsbc_chinacoalpower_2013.pdf

Fthenakis, V., & Kim, H. C. (2010). Life-cycle uses of water in U.S. electricity generation, Renewable & Sustainable Energy Reviews, *14*, 2039–2048.

Intergovernmental Panel on Climate Change (IPCC) (2013). *Contribution of Working Group I to the Fifth Assessment Report of the IPCC*. Cambridge, UK: Cambridge University Press.

International Energy Agency (IEA) (2012). *World energy outlook: Water-energy nexus*. Retrieved from www.worldenergyoutlook.org/resources/water- energynexus/.

International Energy Agency (IEA) (2014). *CO2 Emissions from Fuel Combustion*. Paris: OECD/IEA.

Kahrl, F., & Roland-Holst, D. (2008). China's water-energy nexus. *Water Policy*, *10*(S1), 51–65.

Lee, B., Ellinas, L., & House, C. (2010). Water and energy security. In J. Osikena & D. Tickner (Eds.), *Tackling the World Water Crisis: Reshaping the Future of Foreign Policy*. London: The Foreign Policy Centre, P. 29–34. Available at: http://fpc.org.uk/fsblob/1228.pdf.

Liao, X. W., Hall, J. W., & Eyre, N. (unpublished). *Empirical Study on China's Energy Policies' Impacts on Water Conservation.*

Liao, X. W., Hall, J. W., & Eyre, N. (2016). Water uses of China's thermoelectric power sector. Global Environmental Change, *41*, 142–152.

Macknick, J., Newmark, R., Heath, G., & Hallett, K. C. (2011). Review of Operational Water Consumption and Withdrawal Factors for Electricity Generating Technologies. Golden, CO: National Renewable Energy Laboratory. NREL/TP-6A20-50900. doi: 10.1088/1748-9326/7/4/045802

Marsh, D. M. (2008). *The Water-Energy Nexus: A Comprehensive Analysis in the Context of New South Wales*. Sydney, Australia: University of Technology.

Meldrum, J., Nettles-Anderson, S., Heath, G., & Macknick, J. (2013). Life cycle water use for electricity generation: A review and harmonization of literature estimates. Environmental Research Letters, 8, 015031.

Mielke, E., Anadon, L. D., & Narayanamurti, V. (2010). *Water Consumption of Energy Resource Extraction, Processing, and Conversion*. Cambridge, US: Harvard Kennedy School.

National Bureau of Statistics of China (2014). *Chinese Energy Statistics Yearbook*. Beijing, PRC: China Statistics Press. (in Chinese).

Pan, L. Y., Liu, P., Ma, L. W., & Li, Z. (2012). A supply chain based assessment of water issues in the coal industry in China. *Energy Policy*, *48*, 93–102.

Qin, Y., Curmi, E., Kopec, G. M., Allwood, J. M., & Richards, K. S. (2015). China's energy-water nexus assessment of the energy sector's compliance with the '3 Red Lines' industrial water policy. *Energy Policy*, *82*, 131–143.

Rio Carrillo, A. M., & Frei, C. (2009). Water: A key resource in energy production. *Energy Policy*, *37*, 4303–4312.

Schnoor, J. L. (2011). Water-energy nexus. Environmental Science & Technology, *45*(12), 5065.

Siddiqi, A., & Anadon, L. D. (2011). The water energy nexus in Middle East and North Africa. *Energy Policy*, *39*(8), 4529–4540.

Sovacool, B. K. and Sovacool, K.E. (2009) Identifying future electricity-water tradeoffs in the United States. Energy Policy, 37(7), 2763–2773.

United Nations World Water Assessment Programme (UN WWAP) (2012). The United Nations *World Water Development Report 4: Managing Water under Uncertainty and Risk*. Paris, UNESCO.

United Nations World Water Assessment Programme (UN WWAP) (2014). *The United Nations World Water Development Report 2014: Water and Energy*. Paris, UNESCO.

Van Vliet, M. T. H., Wiberg, D., Leduc, S., & Riahi, K. (2016). Power-generation system vulnerability and adaptation to changes in climate and water resources. Nature Climate Change, 6, 375–380.

Vassolo, S., & Doll, P. (2005). Global-scale gridded estimates of thermoelectric power and manufacturing water use. Water Resources Research, *41*(4).

Webster, M., Donohoo, P., & Palmintier, B. (2013). Water-CO_2 trade-offs in electricity generation planning. Nature Climate Change, 3, 1029–1032.

Williams, E. D., & Simmons, J. E. (2013). Water in the energy industry, an introduction. BP. Retrieved from www.bp.com/energysustainabilitychallenge.

World Bank (2005). Annex 1: Detailed technology descriptions and cost assumptions. In technical and economic assessment of off-grid, mini-grid and grid electrification technologies (Vol. 2). *Energy Sector Management Assistance Programme (ESMAP) Technical Paper Series: ESM 121/07*. World Bank, Washington, DC. Retrieved from http://documents.worldbank.org/curated/en/2005/01/9488629/technical-economic-assessment-off-grid-mini-grid-grid-electrification-technologies-vol-2-2-annex-1-detailed-technology-description-cost-assumptions.

World Economic Forum (2015). *Global Risks 2015*. Geneva, World Economic Forum.

World Resource Institute (2014). *Identifying the global coal industry's water risks*. Retrieved from www.wri.org/blog/2014/04/identifying-global-coal-industry%E2%80%99s-water-risks.

World Resource Institute (2015). *Opportunities to Reduce Water Use and Greenhouse Gas Emissions in the Chinese Power Sector*. Beijing, World Resource Institute.

World Wide Fund for Nature (2014). *China's Future Power Generation*. Beijing, World Wide Fund for Nature.

Ye, Wenhu, & Zhang, Yong (2013). *Environment Management* (3rd Version). Shanghai: Higher Education Press.

Yuan, J. H., Lei, Q., Xiong, M. P., Guo. J. S., & Zhao, C. H. (2014). Scenario-based analysis on water resources implication of coal power in Western China. *Sustainability*, 6, 7155–7180.

Zhang, C., & Anadon, L. D. (2013). Life cycle water use of energy production and its environmental impacts in China. Environmental Science & Technology, *47*(24), 14459–14467.

Zhang, C., Zhong, L., Fu, X., Wang, J., & Wu, Z. (2016). Revealing water stress by the thermal power industry in China based on a high spatial resolution water withdrawal and consumption inventory. Environmental Science & Technology, 50, 1642–1652.

Zhang, W. (2012). *Greenhouse Gas Emissions Reduction Capacity and Cost Analysis of Chinese Power Sector*. Peking University: Beijing, China.

Zheng, X. Z., Wang, C., Cai, W. J., Kummu, M., & Varis, O. (2016). The vulnerability of thermoelectric power generation to water scarcity in China: Current status and future scenarios for power planning and climate change. Applied Energy, *171*, 444–455.

6 The case of Peruvian asparagus

Water governance trade-offs under climate change

Swathi Veeravalli

Introduction

This chapter examines how Peru's focus on economic growth prioritizes water resource decision making for the benefit of agricultural productivity above the longer-term impacts of climate change (Hepworth, Guemes Delgado and Kjell, 2010). In reviewing Peru's economic success with its agro-exports, we discover incompatibilities between its climate policies and the future impacts of climate upon its hydrological resources. Consequently, there is a trade-off between accounting for climate induced hydrological transformations and accounting for sufficient water quantity to develop its agricultural sector. Further climate and meteorological information needs to be incorporated into water governance decision making to understand how current water usage can affect long-term availability of water. Peru is a case study that offers valuable insights into how increased exposure to climate risk occurs despite the country's existing climate policies.

Climate change impacts manifest primarily through the hydrological cycle (Vorosmarty, Green, Salisbury, & Lammers, 2000). Observed warming over several decades has been linked to changes in the large-scale hydrological cycle such as: increasing atmospheric water vapour content, changing precipitation patterns, intensity and extremes, reduced snow cover and widespread melting of ice, and changes in soil moisture and runoff (Bates, Kundzewicz, Wu, & Palutikof, 2008; IPCC, 2014). As the character and severity of impacts from climate change depend on both exposure and vulnerability, it is vital to examine how water resource governance decision making accounts for these changing impacts. To illustrate the challenges in water resource governance, new approaches are necessary in order to provide decision makers with a better understanding of the temporal implications of their decisions. The necessity of taking such an approach is highlighted in this chapter by focusing on two main tenets. First, we discuss how, despite Peru's climate adaptation policies, there is a mismatch with its decision making concerning its water resources that prioritize agro-exports. Second, we unravel how decisions regarding one resource (in this case, agricultural commodities) can have detrimental effects upon the country's vulnerability to climate change by affecting its water resources.

Analyzing the tensions and security trade-offs between climate security and short-term economic security is useful to highlight the principle challenges associated with water resource decision making. Generally, decisions about water governance occur on a daily to annual time-scale (Liu et.al., 2007); decisions about climate security can occur on an annual to decadal time-scale. As a result, water and climate change are usually addressed in different policy and institutional contexts. Thus, in this chapter, we argue that water governance decision makers need to consider how their decisions could result in negative tradeoffs. Such tradeoffs could simultaneously compromise existing adaptation to climate change and render their country vulnerable to the impacts of climate change.

Governance, path dependency, risk and trade-offs

Water governance is the range of social, political, economic, and administrative systems in place to develop and manage water resources, and the delivery of water services, at different levels of society (Rogers & Hall, 2003). Water governance is complicated by the fact that the social, political, economic, and administrative systems it seeks to govern are themselves characterized by uncertainty, dynamic complexity, variation, and are scale dependent (Bodin & Crona, 2009; Liu et al., 2007; Berkes, Colding, & Folke, 2003; Allan, 1999; Levin, 1998). Water governance can thus be characterized as seeking to manage competing priorities while minimizing risk. If risk is not evidenced, governance decision making rarely takes such unanticipated perturbations into account. Pierson (2000) argues that often institutional and political arrangements are hard to change and historical path dependencies are relied upon in order to maintain and consolidate power. Peters, Pierre, & King (2005) argue that path dependency adopts historical institutional perspectives that reduce an organization's capacity to cope with change while obscuring the ability to understand complex processes or realize alternatives. In this manner, path dependency creates a situation that cannot deal with unaccounted perturbations such as climate change.

To account for the impacts of climate change, it is important to examine the principles of risk (Beck, 2006; Matten, 2004; Allan, 1999). Considering risk within water governance is difficult because there are always opportunity costs, or trade-offs, for each decision. Trade-offs occur when a change in one entity is linked to a change in another (Stearns, 1989). Such change doesn't have to be detrimental, but can be if not carefully considered. Trade-offs can vary between stakeholder objectives, risk and productivity, short- and long-term goals (Janssen & Anderies, 2007).

Managing trade-offs successfully is a balancing act composed of understanding and responding to both expected and unexpected change. Without this flexibility, a system may become so adjusted to a particular type of change that it grows vulnerable to other unknown changes and shocks since it may not have the coping capacity to deal with such varied change (Nelson et al., 2009). In these cases, the opportunity cost of decision making turns out to be a trade-off between short-term costs/gains against long-term costs/gains.

Water resource governance is a complex challenge and, with climate change, it becomes even more challenging due to temporal trade-offs. Climate change is seemingly easier to ignore because it is believed that its effects are mainly long term. However, we know that short-term climate-induced disasters, such as floods, contradict these perceptions. Understanding the temporal trade-offs associated with climate change requires understanding complexity. We turn to the case study of Peru to understand why it is critical to consider long-term water governance trade-offs in the context of climate change. We examine how water resource governance is caught between prioritizing economic development utilizing its agro-exports and its climate adaptation policies. A trade-off between short-term economic gains against increasing longer-term climate vulnerability is consequently created.

Agrarian reform and water governance in Peru

In this section, we briefly review Peru's history of its agrarian reform's contribution to a highly sectoralized approach to water governance. A country heavily dependent upon its natural resources for export, the transformation of Peru's landscapes has been driven by land reform, economic liberalization, and armed insurrection (McClintock, 1984). Landscape change in Peru began as early as the 16th century with the invasion of the Inca Empire by Spanish conquistadores. From moving the political capital from Andean Cuzco to coastal Lima and diverting water flow from the Andes to the coast, new agricultural holdings known as haciendas emerged. Haciendas were land grants given to Spaniards to encourage their entry into agriculture and to provide much needed revenue to the Spanish (Keith, 1976). Haciendas, in this manner, became a land tenure system with wealth and power concentrated in the hands of a few conquistadores at the expense of millions of poor peasants forced to give labour, goods, and money to their landlords or *hacendados* (Klaren, 2000).

It was not until the mid-20th century that haciendas became increasingly untenable due to unrest and a great sense of injustice among the peasantry (Fajardo, 2013). The system provided the context for a successful coup by an armed rebellion to take over the Peruvian government (Fajardo, 2013). Within days, the new government began taking control of haciendas in the name of its people, redistributed it to landless peasantry and created new agricultural cooperatives.[1] Scholars argue that this agrarian reform ultimately failed by simply transferring problems from the hacienda system to the newly redistributed properties and continuing the unsustainable water management policies of the previous regime (Mayer, 2009).

Water and land property rights transitioned towards centralized state control with large agrarian cooperatives having massive access to water (Boelens, 2008). Water use rights were then awarded based on land size, with larger allotments receiving larger quantities in order to encourage large-scale production of cash crops. Agrarian reform in effect made irrigation water and water management, state responsibilities (Boelens & Seemann, 2014). During the same period, the

Ley General de Aguas, or Water Law, passed in 1969 with a high emphasis on water for agricultural productivity. This law stipulated that water, with no exception, is the property of the State, and its ownership is inalienable and imprescriptible, with no private ownership of water or acquired rights over water (Government of Peru, 1969). The law did devolve management to water user associations and irrigation commissions, but the main responsibility still lay with the State. With water governance effectively centralized, the emphasis on water for irrigation remained ever present "with an ever-increasing paternalistic and protectionist tendency, and an acceptance of that situation by both agricultural producers and representative organizations" (Huamanchumo, Pena, Silva, & Hendriks, 2008, p. 303).

Peru's national government began issuing a series of laws relating to decentralization in 2002. These laws required regional and municipal government to develop plans and budgets in a participatory manner. Developed through dialogue between civil society and local governments, these efforts moved away from the previous regime's highly centralized tendencies (Hordjik, 2005). Utilizing these principles of decentralized governance, Peru's decision making is divided into three levels: national government, 25 regional governments and 1,834 local municipalities. The dimensions of decentralization include transferring centralized government responsibilities to regional and local governments. Its main objectives for decentralizing power were to strengthen and protect the rights of all Peruvian citizens. Interestingly, one of the other tenets for decentralization include promoting regional and local economic development, wherein the State and regional governments are required to establish favourable conditions for business and private industry (Reilly, Alvites, Tello, Velasco, & Villaverde, 2010).

In the 2000s, water policy reform based on integrated water resource management principles took place. It was based primarily on two factors: 1) the shift away from the State-centered hierarchy; and 2) increased competition between a growing number of water users.[2] Decentralization of water management from the national ministries to local regions and districts took place over a decade, but bias towards agriculture remained – even after the reforms, the Ministry of Agriculture retains the most authority and power (Ore, 2005).

In 2002, Peru released the National Strategic Exports Plan, of which the agriculture and agro-industry were critical components. Contained within this plan was the Agricultural Operative Export Plan that sought to turn the country into "a big player in large scale agricultural farming and commerce" (International Trade Center, 2004). The creation of large reservoirs and other large-scale public irrigation schemes were heavily subsidized by the government (Oxford Business Group, 2016). As local water user associations, representing community led small-scale agriculture as well as domestic consumption, were institutionally and organizationally weak at best (Huamanchumo et al., 2008), the export-oriented agriculture and extractive industries were able to maintain their control over access to water resources.

In 2008, the Water Resources Law 29338 created the National Water Authority (ANA), attached to the Ministry of Agriculture. Focusing on decentralization, the ANA created 14 administrative water authorities, 72 local water authorities, and 17 water resource basin councils (Government of Peru, 2016). According to Peru's National Water Resources Policy and Strategy of 2009, it delineates 80 percent of consumptive water use for agriculture with 18 percent for population and industrial use and the remaining 2 percent for the mining sector (Government of Peru, 2009). Since the passing of the Water Law and other reforms, large-scale farming of asparagus, rice, and sugar production have increased significantly. These large-scale export-oriented agricultural projects are responsible for hydrological demand increases (Rios, 2007).

Budds and Hinojosa (2012) contend that the scale and nature of water use in Peru has significantly changed since the 1990s with successive government reforms focusing on decentralization as well as economic liberalization. Although the water reforms were intended to address inequities and disparities of water access, on the contrary, forms of access and exploitation of water within Peru have become more contested than ever (Lynch, 2012; Pastor, 2013). These reforms assisted the growth in the export-oriented agriculture and extractive industries (Pastor, 2013; Lynch, 2012; Ore, 2005). Both industries were heavily reliant upon water usage and placed heavy demands upon already stressed water resources. For example, highland communities have reported significant water flow decrease in their customary irrigation schemes (Budds & Hinojosa, 2012). Hepworth et al. (2010) argues that poorly designed and weak institutional capacity of water laws have allowed a few large agribusinesses to monopolize preferential access to water.

For vulnerable groups in Peru, water insecurity and deficient water availability for livelihood food production commonly reflect unequal distribution of water volumes, quality, and services within unequal power structures (Boelens & Seemann, 2014). The increases upon water demand coupled with the development of new irrigation infrastructure presented challenges for water governance decision makers who were not at all accommodated within existing water policy (del Castillo (2006); in Budds & Hinojosa, 2012). Although existing water agreements at local levels appeared to take local community needs into consideration, these local governance structures "have been challenged by the Peruvian state's modernization policies" (Boelens & Seemann, 2014, p. 5).

Water governance structures aggravate vulnerabilities to water scarcity by stimulating water demand and encouraging polluting land uses (Lynch, 2012). Competing interests for water by agricultural and mining industries mean that water access for communities has been ignored (Alegria, 2014). Decentralized water management combined with centralized economic and political interests create a context that is vulnerable to conflict over water access (Pastor, 2013). The country's National Water Authority warned that without action to slow aquifer decline, severe water shortages could lead to plunging farm output and job losses within years (Comisión Técnica Multisectorial, 2009).

It becomes apparent how a water governance system that prioritized water for productive uses – especially irrigation – at the cost of all other uses became institutionalized. Bury et al. (2013) suggest that additional pressures on water resources, including demographic change as well as other competing complex political and economic factors, transformed the agricultural sector within Peru. Thus, the evolution of Peru's water governance has been driven by tensions between the country's economic, social and political systems.

Perhaps of most importance was the country's penchant for international trade agreements to develop its extraction and export agriculture (Lynch, 2012; Bury, 2005). Water governance decision making has effectively been held captive by the country's national economic strategy that prioritized export-led growth. Peru's Ministry for Agriculture and Irrigation as well as the Ministry for Production continue to develop policies to benefit and increase the added value of the agricultural sector. Declines in revenues from Peru's traditional exports, such as coffee, has vindicated the country's decision to continue diversifying production through their irrigation schemes (Oxford Business Group, 2016). To highlight Peru's focus on the short-term economic priorities of the country at the expense of the longer-term implications of climate change vulnerability, we turn to an analysis of the impacts of climate change upon Peruvian water resources.

Water and agriculture in Peru

Peru as a case study offers critical insight into the human–water relationship. Peruvian livelihoods are intimately linked to natural resources as the country's GDP is highly dependent on income from various crops. Since 2012, Peru has exported an average of 7.4 billion USD in agricultural produce (Duff & Padilla, 2015). With a total land area of 128,522,000 ha (FAO, 2016), Peru is a mountainous country. From 1961 to 2014, it is estimated that agricultural land expanded from approximately 13 percent (16,707,860 ha) to 19 percent (24,419,180 ha) (FAO, 2017; World Bank, 2016). From 1994 to 2012, irrigated land increased from 792,394 ha to 970,536 ha between 1994, and 2012 (Salazar, 2013). Table 6.1 summarizes Peru's major water sources.

Table 6.1 Peru water sources

Region	*Total*	%	*Irrigated*	%	*Rainfed*	%
Coast	1,686,777.58	24%	1,469,422.55	57%	217,355.03	5%
Mountains	3,296,008.11	46%	989,481.65	38%	2,306,526.45	51%
Rainforest	2,142,222.09	30%	120,995.68	5%	2,021,226.40	44%
Total	7,125,007.78	100%	2,579,899.88	100%	4,545,107.88	100%

(adapted from Salazar, 2013)

Table 6.2 Peru's renewable water resource

Renewable Freshwater Resources	*Total*	*mm/year*
Precipitation	1,738	mm/year
Surface water produced internally (e.g. glacial runoff)	1,641,000	million m^3/year
Total renewable water resources (surface and ground water	1,879,800	million m^3/year
Dependency rate	13%	n/a
Total renewable water resources per inhabitant (2013)	61,884	m^3/year
Dams (2010)	5,770	million m^3
Total	7,125,007.78	100%

(adapted from FAO, 2016)

As outlined in Table 6.2, the country is dependent on mostly surface water produced internally from precipitation as well as from glacial runoff from the Cordillera Blanca (Schauwecker et al., 2014).

Peruvian agricultural lands are comprised of 2 million agricultural producers (peasant families, agricultural enterprises, etc.) with 69 percent of this force located in the sierra (mountains), often under very difficult conditions of poverty and very small holdings (Huamanchumo et al., 2008). It is mostly its highly irrigated coastal land that drives and energizes the agricultural sector (Huamanchumo et al., 2008). Although Peru has long been reliant upon glacial runoff to buffer seasonal aridity, the impacts of climate change mean that glacial runoff is both less predictable and less consistent (Vuille et al., 2008; Carey, Huggel, Bury, Portocarrero, & Haeberli, 2010).

As an upper-middle income country, Peru has experienced remarkable economic growth within the last decade and has become one of South America's best-performing economies (World Bank, 2012). Effective trade policies coupled with solid economic reforms and increases in global demand for Peruvian agricultural exports render Peru an upper-middle income country. Peru's GDP growth has been significant for the country by averaging 6.3 percent between 2002 and 2010 (World Bank, 2012). This robust economic performance enabled recovery of Peru's income per capita, which increased by more than 50 percent after almost 30 years of stagnation. Agriculture alone represents almost 8 percent of the country's gross domestic product (World Bank, 2012). Because Peruvian livelihoods are so critically dependent upon agriculture, water has been key in determining landscape development. Water in Peru has facilitated economic and even political stability within the country by fostering economic growth (Rios, 2007).

With support from local producer associations, international development agencies as well as international financial institutions, a change in agricultural practices has taken place in Peru in recent years (OECD, 2007). This occurred notably with the shift towards more export-oriented crops such as asparagus. Such market opportunities come at a large cost to Peru's water and climate security. Unsurprisingly, this landscape transformation has had both positive and negative effects by creating and shifting a power balance between "winners" (economic security) and "losers" (water and climate security) within Peru. The country's continued path dependency and risk trade-offs have security implications for the stability of the country. The cumulative effect of both are discussed in detail in the following sections. The country is now facing a drastic decrease in water resources for irrigation and other needs. Through an examination of Peru's major export crop, asparagus,[3] the chapter highlights the need for a more holistic approach to resource governance, with increased climate and meteorological service information incorporated into hydrological services through joint actions of new technologies, increased water security, and better government regulation. In this manner, we argue that managing trade-offs must be considered as part of natural resource governance.

Governance and the water footprint of Peruvian asparagus

Within Latin America, Peru has reported the highest annual gains in agricultural productivity and it is likely that the growth of its agro-exports will only increase in the future Peru's success with exporters (Duff & Padilla, 2015). Although the National Institute of Natural Resources (INRENA) is responsible for the sustainable use of renewable natural resources including water, this body is in fact a public arm of the Ministry of Agriculture (Government of Peru, 2009). The major users of fresh water are agriculture (80 percent) with the balance split between domestic and industrial (18 percent) and mining (2 percent) (Government of Peru, 2009). Irrigated agriculture represents about 30 percent of cultivated land whereby more than two-thirds of the production go to exports. Despite its smaller size, Peru is the third leading country in food and agriculture exports, after Brazil and Argentina (Duff & Padilla, 2015).

World Bank reforms within Peru focused on exporting agricultural goods such as asparagus, table grapes and spices (World Bank, 2015). This monopoly of export crops has had devastating effects for small farmers. Human rights activists insist that agro-export companies have been aggressively buying up wells, pumping water heavily, and leaving small farmers with dry wells: "that's injustice and exclusion . . . [we] agree with having investment, but it can't be done in whatever way without any sort of control" (James, 2015). As evidenced in Figure 6.1, asparagus exports rates far surpass other agro-exports.

Peru is the world's largest exporter of green asparagus and second largest exporter of white asparagus. Asparagus is a high-value and labour-intensive perennial crop that is grown in countries that have a mild and relatively stable climate (Boriss & Brunke, 2006; Meade et al., 2010). With some of the world's highest yields for asparagus, Peru is the only country that can harvest asparagus year-round, ranging between two to three harvests per year per field (Azimi et.al., 2012; Meade et.al., 2010). The country's focus on asparagus was assisted by policies such as the Andean Trade Preference Act, which gave Latin American countries a

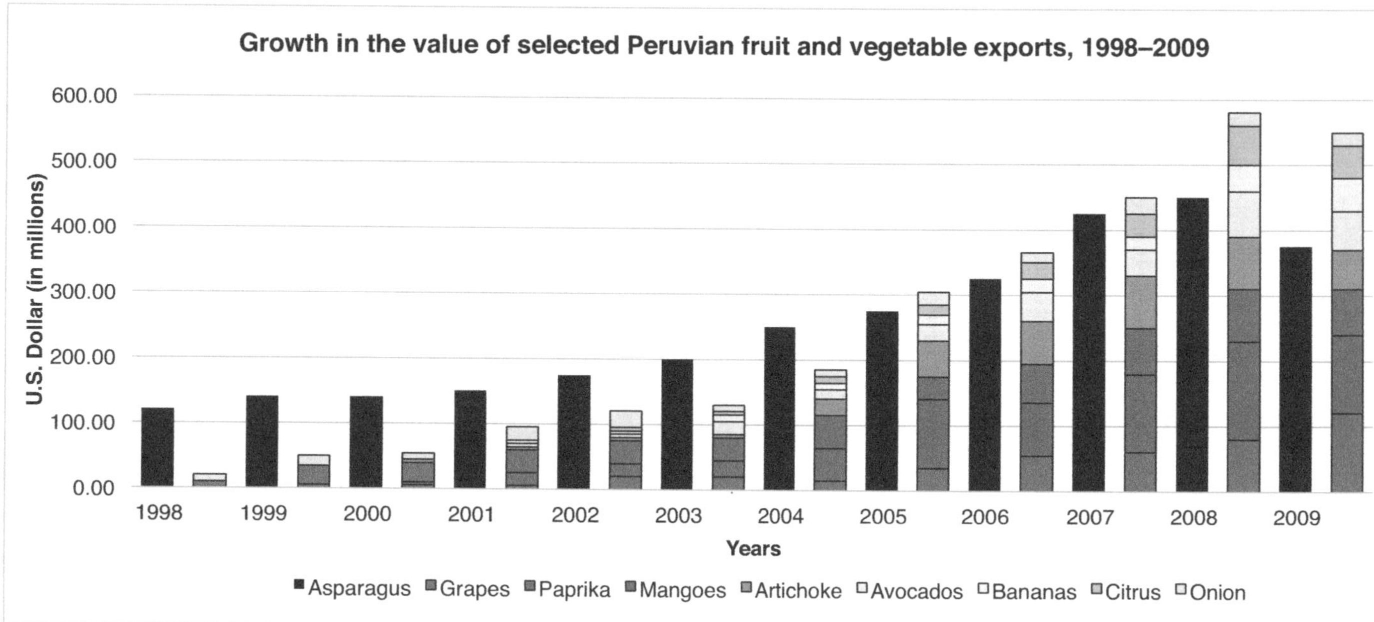

Figure 6.1 Growth in the value of selected Peruvian fruit and vegetable exports, 1998–2009
(IHS Markit Global Trade Atlas, 2010 and Meade, Baldwin, & Calvin, 2010)

reduction in tariffs, making their goods more cost competitive. Peruvian asparagus' tariff rate went down from 21.3 percent to approximately 5.6 percent (Shimizu, 2006). This crop in particular has been at the forefront of the investment boom in high value export crops within Peru. The sector is dominated by medium to large producing and exporting operators who see asparagus as a way to diversify an investment portfolio rather than as a livelihood strategy (Hepworth et al., 2010; OECD, 2007).

Between 1950 and 1970, production area for asparagus stood between 500–1,500 hectares. Since 1980, large increases in asparagus production allowed Peru to be one of very few countries that has continued to meet global demand (Benson, 2009). In 1990, the production area for asparagus increased by 1,233 percent, expanding to over 20,000 hectares with over 170,000 tons of production. In 2004, production increased to 193,000 tons with a decrease in area to 18,500 hectares (Benson, 2009; Shimizu, 2006). In 2006, there are reports that cultivated land for asparagus expanded to 22, 971 ha (Rios, 2007).

This rise in exports has been driven not only by low production costs within the country, but also because of the advantages offered by Peru's hydrological resources as well as its climate (Meade et al., 2010). Perhaps the most important reason for Peru's success with asparagus was a combination of two related factors. First, advances in water technology including drip irrigation in the 1980s made it possible to expand production to more arid locations (Shimizu, 2006). Second, the country's focus on large-scale irrigation schemes, largely subsidized by the government beginning in the 1990s, allowed irrigation further to expand production areas into more arid regions (Hepworth et al., 2010).

A major concern for asparagus is the variation in irrigation systems used in the main producing regions: in the north of the country (La Libertad), large-scale irrigation projects rely upon major rivers from the Andes mountains; in the south (Ica Valley), producers rely mainly upon groundwater (Nolte, 2010). Within Ica specifically, a highly arid eco-system, the annual asparagus water requirements are estimated as being as high as 147 million cubic meters (Hepworth et al., 2010). Most of the irrigation infrastructure here depends highly on groundwater fed systems. Within La Libertad, the Chavimochic irrigation project whose source is the Santa River, provides much need water (World Bank, 2016).

Although yields vary by region, the La Libertad area produces 13.4 tons/hectare, but yields in this area can reach up to 20 tons/hectares with the right conditions. We will focus on these two locations specifically because the regions of Ica and La Libertad represent 83 percent of national asparagus production (See Figure 6.2). Ica's production ranges between 39–44 percent and Ica's production ranges between 44–48 percent of the total annual production respectively (Azimi et al., 2012; Nolte, 2010).

Asparagus contributes heavily to the country's GDP with the country's top producing regions – Ica and La Libertad – outperforming the national average (Azimi et al., 2012). But despite these obvious financial incentives, the massive expansion of agricultural land used for asparagus farming has had significant ramifications on the country's water resources. First, the water footprint of the

Figure 6.2 Main asparagus-producing zones in Peru
(adapted from Geocurrents, 2016)

Table 6.3 Total annual water demand at sample agro-export farm (Hepworth et.al., 2010)

	Site 1 [Ha]	*Site 2 [Ha]*	*Crop need [m3/ha/yr]*	*Total [m3/yr]*
Asparagus	97	762	16,000	13,744,000 (79%)
Grapes	74	286	9,000	3,240,000 (19%)
Avocado	–	26	11,000	286,000 (2%)
Total				17,270,000

agro-export boom coupled with a rapid decline in water tables is worsening the situation through drying of domestic supply wells. Second, one giant asparagus farm can use as much water as an entire Peruvian city does on a daily basis (Dring, 2014). Third, as demonstrated in Table 6.3, the water requirements of asparagus far surpass other agro-exports. Fourth, because of agricultural practices, there is evidence of elevated levels of groundwater pollution, in particular, nitrates (Hepworth et al., 2010). Peruvian water resources are already under pressure due to agricultural productivity. Finally, but perhaps most importantly, with the added impact of climate change, water resources will only continue to be stressed.

Climate, surface water, and groundwater

Peru's topography means that its water resources are very unevenly distributed throughout the country, with recharge depending on glacial melt and orographic rainfall for both surface and groundwater (Chevallier et al., 2011). Peru and its water resources are highly sensitive to changes in its climate. Because of its proximity to the Amazon basin, coupled with the complicated relief of the Andes, and the complexity added by the Humboldt current, Pacific structures as well as El Niño, the whole region is highly sensitive and very difficult to model with high certainty (Condom, pers comm in Hepworth et. al., 2010). Global circulation model projections for maximum temperatures indicate that temperatures are increasing by at least 1° C to 2030 and up to 2° C in 2050. Unfortunately, the models are inconsistent regarding precipitation change (Senhami, 2009). Consequently, areas of Ica and La Libertad areas will experience extreme variability in temperature as well as some variability in precipitation up until 2030 (Jensen et al., 2015).

La Libertad and surface water

The sources of water for La Libertad is mainly fed by surface water from the Santa River that flows south to north through the Callejon de Huaylas valley under the shadow of the Cordillera Negra, and Cordillera Blanca. Glaciers in the Cordillera Blanca make up around 70 percent of annual flow in the Santa, and drain east to contribute to the headwaters of the Amazon (Aguirre, 2010). The La Libertad region is one of the areas most affected by climate change in Peru (USAID, 2012).

Glaciers are highly sensitive to changing climate regimes. Studies of climate data from stations in the tropical Andes indicate that temperature has increased in the region over the past four decades, on average up to 0.3° Celsius per decade (Bury, 2011). Peru, which hosts 70 percent of the tropical glaciers worldwide, has lost 43 percent of its glacial coverage by area since 1970 (Drenkhan, Carey, Huggel, Seidel, & Ore, 2015). As the area undergoes deglaciation, river flows increase to a peak flow, then decrease in volume over time while increasing in seasonal variability. It is estimated that peak flow has already been surpassed in some sub-catchments of the Santa River, which flows through several large Peruvian cities (Drenkhan et al., 2015).

Magrin et al. (2014) name loss of glacier mass and snowpack as the primary regional impact of climate change in Central and South America. This loss is attributable to climate change. In the past 30 years, the rate of glacial retreat has increased beyond previous predictions, threatening continued water access during the dry season and increasing flood risks during wetter seasons (Kohler & Maselli, 2009). Higher rates of glacial melt can initially increase stream flowrates, but after critical thresholds are passed, dry-season stream discharges trend downward. Studies of future Andean hydrological scenarios project annual watershed discharge losses of up to 30 percent over the next 50 years due to glacial retreat (Magrin et.al., 2014). Glaciers at lower elevations and tropical latitudes are particularly at risk (Fraser, 2012; Mark et al., 2010).

The impact of glacial recession modifies watersheds by forming wetlands and lakes – which may increase water availability in the short-term, but become dry and susceptible to changing climate regimes (Hepworth et.al, 2010). In effect, the impacts of changing climate regime upon the Santa River are severe. Deglaciation coupled with increased temperatures and erratic precipitation means that the watershed will lose its capacity to provide sustainable water access (USAID, 2012). Increased droughts followed by more intense and shorter precipitation periods are to be expected thereby reducing overall future availability of water (USAID, 2012). Irrigation demand in La Libertad is currently at 691.2 million m^3 and is expected to increase by almost 44.8 million m^3 with an increase in temperatures (Jensen et al., 2015). If La Libertad's water continues to be subsidized (Azimi et al., 2012), overexploitation coupled with changing climate regimes could create a highly tenuous situation within the region.

Ica and groundwater

In this section, we seek to understand the impacts of climate upon Peru's groundwater. The source of water for the Ica asparagus production region is the Ica Valley aquifer. This aquifer runs north to south approximately 75km in length (Government of Peru, 2016). Current aquifer recharge rates are far surpassed by consumption as outlined in Table 6.4. This aquifer is already running severe water deficits of up to 145 million m^3: with a capacity of 190 million m^3 of estimated usable reserves, annual use runs up to 335 million m^3 (Bullock, 2016).

Table 6.4 Ica Valley recharge rates

Year	*Ratio (consumption: recharge)*
1980	0.232
1990	0.161
1995	0.353
2000	0.300
2004	0.466
2007	0.437
2010	0.762

(adapted from Bullock, 2016)

More than 66 percent of Ica water usage comes from its groundwater (Bullock, 2016). Additionally, Ica's six largest growers consume over 78 percent of Ica's groundwater (Schmall, 2010).

Climate change will affect groundwater recharge rates; however knowledge of current recharge and levels is generally very poor, especially over long-term time horizons (Bates et.al., 2008; Green, Taniguchi, Kooi, Gurdak, & Allen, 2011). The direct effect of climate change on groundwater resources will depend upon the change in the volume and distribution of groundwater recharge. While uncertainty remains about projected impacts of climate change on diffuse groundwater recharge, increased groundwater abstraction will increase pressure upon already overexploited groundwater resources (Holman, Allen, Cuthbert, & Goderniaux, 2011; Taylor et.al., 2012).

Unfortunately, Peru's aquifer recharge rates are virtually unknown and more research is needed to better understand the relationship between groundwater and climate change (Vuille, 2013). The Ica region is a hyper-arid region characterized by high variability (de Grenade, 2014). Land management changes in irrigated agriculture on groundwater recharge rates, and thus on resource availability and quality, occur with abandonment of the traditional practice of spate irrigation in mountain-front areas (Global Water Partnership, 2012). In these areas, fields are deliberately flooded in the wet season to induce infiltration and increase aquifer dry-season storage (Global Water Partnership, 2012). This has occurred extensively in both the Ica Valley as well as La Libertad with the introduction of intensive asparagus cultivation (Garduno & Foster, 2010).

Climate change and natural resources governance in Peru

The impacts of climate change clearly affect future water availability in Peru. Peru recognizes its own vulnerability to climate change and how much its economic development has been contingent upon exploiting its water resources

(Government of Peru, 2016). In its Third National Communications Plan to the United Nationals Framework Convention on Climate Change, the Ministry of Environment clearly recognizes Peru's increasing vulnerability in the context of future change in both climate and its water resources (Government of Peru, 2016). The report describes how the agricultural sector is of critical importance to the nation since almost 5 percent (1.4 million out of its 31 million) of its citizens depend on agricultural livelihoods (ibid).

In Peru's National Strategy and Policy of Water Resources, there is recognition that while water is fundamental to its development, access to water resources information is simultaneously restricted across its various public institutions (Government of Peru, 2009). The report states that its objectives are to strengthen the role of institutions utilizing decentralized water management, while recognizing the boundaries of watersheds and not just political boundaries. Throughout all 85 pages of the strategy, climate change is mentioned only twice. In the first instance, the National Strategy stipulates the need to minimize the impacts of climate change as well as its productive sectors. In the second instance, the National Strategy briefly alludes to the need for its Basin Councils to identify vulnerable flood-prone areas and prevent drought impacts.

The Peruvian government recognizes that its overall vulnerability to climate change is shaped by many factors including poverty and low institutional capacity (Government of Peru, 2009; Government of Peru, 2016). It also recognizes that sectors critical to its economic growth, such as the agriculture, fisheries and water resources sectors are highly vulnerable to the impacts of climate change (Government of Peru, 2016). In its Regional Management Report on Climate Change (Government of Peru, 2013), the Ministry of Environment reviewed each region's climate preparedness through the development of the Regional Strategy to Climate Change (ERCC). The ERCC acknowledged again how the agricultural sector was its main priority and that in order to successfully employ climate-adaptation policies, the Ministry of Environment is critical in working across its 24 regions (Government of Peru, 2013). The Regional Report goes on to state that despite the recognition of agriculture, fisheries and the water resources sectors as highly vulnerable, a gap exists in reflecting these priorities across all levels of governance.

In the Ministry of Agriculture's assessment of climate-induced impacts upon its agricultural sector, the report describes how problems of water availability have led to conflicts over water access (Government of Peru, 2012). Seeking to remediate this while balancing the needs to maintain agricultural competitiveness in the context of climate change, MINAG attempts to incorporate climate change into all agricultural sectorial planning, for example through a national agricultural research plan for mitigation and adaptation that identifies the vulnerability of various agricultural projects. However, the report concludes by stating that Peru's exceptional conditions provide an enabling environment to sustain its extremely competitive agricultural sector. More consideration of climate change impacts upon the trade-offs between ensuring water for economic development and balancing competing needs is therefore essential.

Key takeaways and summary

We see the myriad impacts of the agro-export boom upon Peru's water resources. 85 percent of Peruvian water consumption is agricultural, and occurs mostly on the western slopes of the Andes down to the Pacific coast (Vergara et.al., 2007). Seasonal regulation of stream-flow and overall usage is critical to balance the need between providing irrigation to croplands with other competing needs. If, as projected, climate change impacts will cause increased seasonal stream-flow variability, the consequences for food production in the country will be severe. As glacial retreat, combined with changing precipitation and temperature, continue to accelerate, it will be essential to find more efficient ways of using water.

We also see that in Peru there are inextricable links between economic growth and water resources. With projected drier and more variable conditions, Peru could witness slower and more erratic economic growth. This creates a complicated situation between having sufficient water to allow for climatological impacts, as well as sufficient water for agriculture. It appears that while climate policy exists to realize the security implications of climate change, there is a tension between balancing this need with ensuring sufficient economic growth. The advent of decentralized governance, whilst ensuring a devolution of power from a centralized government, has also emphasized the need for economic growth above all else. In a context of water stress within Peru, ensuring sufficient water for agricultural productivity appears to be of the greatest concern.

Competing claims over water resources have increased over the last twenty years amidst a context of unclear and overlapping institutional and legal arrangements for water governance (Ore, Castillo, Orsel, & Vos, 2009). This unequal access to water resources and distribution systems is based on complex compositions of power and domination of business interest. Despite being the largest consumers of the country's water resources, the agricultural sector regards itself as threatened by increased competition from mining (Budds & Hinojosa, 2012). Without adequate governance, the resulting output could be sustained conflict between various competing business interests, government departments, and Peruvian citizens.

It is true that Peru has well-developed climate policy. Furthermore, its move towards decentralized governance to protect the rights of its citizens is also highly laudable. We are still left wondering whether decentralized governance provides benefits to enable long-term resilience to climate change impacts. In a USAID sponsored study (2016) examining this very question in Mali, the research team discovered that there was little technical capacity, information, or resources to sustain local level participation in climate adaptation. The same questions regarding the implications of decentralized governance upon Peruvian climate adaptation need to be examined in greater detail.

Nevertheless, we do know that there is a paucity of accurate climate and water resource information within the country and institutional silos remain prohibitive to information sharing (Government of Peru, 2009). It does not appear that the country's decentralized governance approach is addressing institutional silos; rather it is creating a data gap between water governance and climate adaptation.

Table 6.5 Using a value proposition approach to facilitate effective climate adaptation

1. **Policy:** create an enabling environment to support sustainable adoption of alternative technologies and business models to generate and share accurate climate and weather information;
2. **Finance:** ensure public finance incentivizes both public and private sector institutions to invest in improving climate information data generation/dissemination and technology;
3. **Partnerships:** create revenue-sharing agreements between public and private sector institutions to collaborate on generating, calibrating, packaging and distributing information and technology;
4. **Technology:** deploy cost-effective, accurate and easily maintained weather- and climate-monitoring systems;
5. **Incremental approach:** utilize previous steps to reach end-users, further engage with civil society, create effective cost-recovery mechanisms and monitor, evaluate and readjust approaches.

(adapted from Kurukulasuriya et.al., 2017).

This data gap means that without access to climate information, decision makers will be unable to anticipate, plan, respond or react to future climate change. In Table 6.5, Kurukulasuriya, Miller, and Rutaagi (2017) create a five-step value proposition to facilitate a more sustainable enabling environment to address the dichotomies between water governance and climate adaptation.

More effective communication and cooperation utilizing such a value proposition can address the disconnect between the agricultural, water resources, and climate sectors in Peru. Peru's decentralized governance must ensure that there are no long-term detrimental trade-offs between balancing the needs of its economic development priorities with those of its climate adaptation. By moving away from policies and practices that prioritize the agricultural sector, natural resource decision making must address the interconnected risks imposed by climate change.

Lebel et al. (2006) find that ecological feedbacks may constrain both governance and the capacity to manage change. Consequently, strengthening global, regional, and local climate and weather services is fundamental to manage and respond to the impacts of climate change (World Bank, 2016). Ensuring greater collaboration between Peru's National Service for Meteorology and Hydrology (SENAMHI) and the Ministries for Environment and Agriculture with local municipalities can decrease the severity of trade-offs occurring between the various sectors. World Bank sponsored research estimates that improving weather, climate and hydrological services could lead to a global reduction of US $2 billion annually in reduced asset losses and up to US$30 billion annually in global productivity of weather-dependent sectors (Hallegatte, 2012).

In addition to utilizing a climate services value proposition approach, Peru can modulate its risk by better understanding how trade-offs occur between its climate vulnerability and economic security. Peru must also understand its medium and long-term vulnerabilities caused by climate change by focusing beyond the current cross-sectoral ones. If this does not occur, negative trade-offs will ensue between stakeholder objectives and government priorities, livelihoods being the most negatively impacted. An unintended consequence of Peruvian water governance is the creation and reinforcement of vulnerabilities instead of the fostering resilience when considering the context of climate change. Thus, more holistic resource governance that takes into account climate change considerations is imperative.

Predicated upon this analysis of Peruvian asparagus and water resource governance, a brief summary follows:

- Changing climate regimes, coupled with Peru's desire to sustain its agricultural production of asparagus, are drivers of change that have numerous short- and long-term implications upon the country's water resources;
- In order to address these drivers of change, the country must:
  - Incorporate both climate and information about asparagus and other water intensive agro-exports into water governance decision making;
  - Recognize across all levels of governance, the vulnerability of the agriculture and the water resources sectors to changing climate regimes;
  - Inculcate a culture of climate services value-based action plans;
- If these gaps are not addressed, negative trade-offs could occur that increase Peru's exposure and sensitivity to climatic risks and compromise the country's economic security.

Notes

1 Since the 1960s, Peru has experienced political instability due to an uprising based on abolishing landlordism and has been in a protracted agrarian revolutionary war (Taylor, 2017). Scholars argue that the subsistence crisis of the 1980s (Taylor, 2017) occurred after sweeping agrarian reform both combined to create a context ripe for revolution. In effect, the hacienda system enabled revolutionary activity as the main insurgent group, the Shining Path, recruited from small land-holders who were not integrated into the capitalist market economy (McClintock, 1984). Additionally, critical mass support for the revolutionary effort was empowered by the dual burden of social alienation and economic deprivation within the highlands of Peru.

2 In the last couple of decades, Peru has been developing its industry and trying to develop its hydroelectric power as well as mining for natural resources. These industries are highly water intensive and often compete with demand with community needs, public health and the environment (Hepworth et.al., 2010).

3 Although Peru is increasingly exporting other export crops such as avocado and grapes, this rather recent trend means that not much data are available on these crops. Additionally, Peruvian asparagus has been often cited as an example of large water footprint export crops (see Azimi et. al., 2012; Meade et al., 2010; OECD, 2007). Finally, asparagus is one of Peru's most highly water-intensive agricultural products. For the combination of these reasons, this chapter will focus exclusively on asparagus.

References

Aguirre, M. (2010, April 8). Protecting the Rio Santa watershed in Peru. Retrieved from www.iucn.org/content/protecting-rio-santa-watershed-peru.

Alegria, J. F. (2006). The Challenges of Water Resources Management in Peru. Paper presented at the IAP Water Programme Regional Workshop for the Americas, Peru.

Allan, T. (1999). Water in International Systems: A Risk Society Analysis of Regional Problemsheds & Global Hydrologies. Paper presented at the Water resources and risk, Oxford University.

Azimi, K., Ferrero, C., Loucky, J., Olivares, H., & Rojas, A. (2012). Asparagus in Peru Microeconomics of Competitiveness. Retrieved from Boston: http://www.isc.hbs.edu/pdf/Student_Projects/2012%20MOC%20Papers/Asparagus%20Cluster-%20Final%20Report.pdf.

B.C. Bates, Kundzewicz, Z. W., Wu, S., & Palutikof, J. P. (Eds.). (2008). Climate Change and Water Geneva, Switzerland.

Beck, U. (2006). Living in the world risk society. *Economy and Society*, *35*, 329–345.

Benson, B. L. (2009). 2009 Update of the World's Asparagus Production Areas, Spear Utilization and Production Spears. Paper presented at the International Asparagus Symposium, Davis, CA. Berkes, F., Colding, J., & Folke, C. (Eds.). (2003). *Navigating Social-Ecological Systems: Building Resilience for Complexity and Change*. Cambridge, U.K.: Cambridge University Press.

Bodin, Ö., & Crona, B. I. (2009). The role of social networks in natural resource governance: What relational patterns make a difference? *Global Environmental Change*, *19*, 366–374.

Boelens, R. (2008). Water rights arenas in the Andes: Upscaling networks to strengthen local water control. *Water Alternatives*, *1*, 48–65.

Boelens, R., & Seemann, M. (2014). Forced engagements: Water security and local rights formalization in Yanque, Colca valley, Peru. *Human Organization*, *73*, 1–12.

Boriss, H., & Brunke, H. (2006). *Commodity Profile: Asparagus*. Davis, California: Agricultural Issues Center. University of California.

Budds, J., & Hinojosa, L. (2012). Restructuring and rescaling water governance in mining contexts: The co-production of waterscapes in Peru. *Water Alternatives*, *5*, 119–137.

Bullock, J. (2016). Development of Peru's Asparagus Industry. Retrieved from Leadership Academy for Development: https://scholar.harvard.edu/jbullock/publications/development-perus-asparagus-industry

Bury, J. (2005). Mining mountains: Neoliberalism, land tenure, livelihoods, and the new Peruvian mining industry in Cajamarca. *Environment and Planning*, *37*, 221–239.

Bury, J. T., B. G. Mark, J. M. Mckenzie, A. French, M. Baraer, K. I. Huh, M. A. Zapata, and R. J. Gomez. 2011. Glacier recession and human vulnerability in the Yanamarey watershed of the Cordillera Blanca, Peru. Climatic Change 105: 179–206.

Bury, J., Mark, B. G., Carey, M., Young, K. R., Mckenzie, J. M., Baraer, M., French, A., & Polk, M. H. (2013). New geographies of water and climate change in Peru: Coupled natural and social transformations in the Santa River Watershed. *Annals of the Association of American Geographers*, *103*, 363–374.

Bury, J. T., Mark, B. G., Mckenzie, J. M., French, A., Baraer, M., Huh, K. I., Lopez, G. (2011). Glacier recession and human vulnerability in the Yanamarey watershed of the Cordillera Blanca, Peru. *Climatic Change*, *105*, 179–206.

Carey, M., Huggel, C., Bury, J., Portocarrero, C., & Haeberli, W. (2010). An integrated socio-environmental framework for glacier hazard management and climate change adaptation: Lessons from Lake 513, Cordillera Blanca, Peru. *Climatic Change*, *112*, 733–767.

Chevallier, P., Pouyaud, B., Suarez, W., & Condom, T. (2011). Climate change threats to environment in the tropical Andes: glaciers and water resources. Regional Environmental Change, 11(1), 179-187. doi:10.1007/s10113-010-0177-6

del Castillo, L. (2006). *Personal Communication. Peruvian Centre for Social Studies (Centro Peruano de Estudios Sociales)*. Lima, Peru: Peruano de Estudios Sociales.

de Grenade, R. R. (2014). The Ica Valley: A place of little rain.

Drenkhan, F., Carey, M., Huggel, C., Seidel, J., & Ore, M. T. (2015). The changing water cycle: Climatic and socioeconomic drivers of water-related changes in the Andes of Peru. *WIREs Water, 2*, 715–733.

Dring, R. (2014, April 17). Asparagus: Draining Peru dry. Retrieved from http://sustainablefoodtrust.org/articles/asparagus-draining-dry/.

Duff, A., & Padilla, A. (2015). Latin America: Agricultural perspectives. In *Economic Research*. Raboresearch, Rabobank.

Duff, A., & Padilla, A. (2015). Latin America: agricultural perspectives. Retrieved from https://economics.rabobank.com/publications/2015/september/latin-america-after-the-commodity-boom/: https://economics.rabobank.com/publications/2015/september/latin%2Damerica%2Dagricultural%2Dperspectives/

Fajardo, S. (2013). *Losing the Hacienda: The Agrarian Reform's Effect on Landowners in the Peruvian Andes*. (Senior Thesis). Linfield College, McMinnville, OR.

FAO (Food and Agriculture Organization of the United Nations) (2016). Water resource sheet. AQUASTAT. Available from www.fao.org/nr/water/aquastat/sets/index.stm.

FAO (Food and Agriculture Organization of the United Nations) (2017). Peru. AQUASTAT. Retrieved from www.fao.org/nr/water/aquastat/sets/index.stm.

Fraser, B. (2012). Goodbye glaciers. *Nature, 491*, 180–182.

Foster, S., Shah, Tushaar (2012), Groundwater resources and irrigated agriculture: making a beneficial relation more sustainable, No H045034, IWMI Working Papers, International Water Management Institute, http://EconPapers.repec.org/RePEc:iwt:worppr:h045034

Garduno, H., & Foster, S. (2010). Sustainable groundwater irrigation: Approaches to reconciling demand with resources. In *GWP GW-MATE Strategic Overview Series SO-4*. World Bank. Washington DC, 1–20.

Government of Peru (1969). *Decree Law 17*. Informato Legal 258.

Government of Peru (2009). *Política y Estrategia Nacional de Recursos Hídricos*. (Autoridad Nacional del Agua). Lima, Peru.

Government of Peru (2012). Plan de Gestión de Riesgos y Adaptación al Cambio Climático en el sector Agrario. Retrieved from www.minagri.gob.pe/portal/download/pdf/especiales/plangracc/plangracc.pdf

Government of Peru (2013). Informe de Balance de la Gestion Regional Frente al Cambio Climatico en el Pais, Avances, Logros, Dificultades, Retos y Oportunidades. Retrieved from Lima, Peru: http://www.minam.gob.pe/cambioclimatico/wp-content/uploads/sites/11/2013/10/Informe-de-Balance-de-la-Gesti%C3%B3n-Regional-frente-al-Cambio-Climatico.compressed.pdf

Government of Peru (2016). *El Perú y el Cambio Climático, Tercera Comunicación Nacional del Perú a la Convención Marco de las Naciones Unidas sobre Cambio Climático*. (Comisión Nacional sobre Cambio Climático). Ministerio del Ambiente. Lima, Peru.

Green, T. R., Taniguchi, M., Kooi, H., Gurdak, J. J., & Allen, D. M. (2011). Beneath the surface of global change: Impacts of climate change on groundwater. *Journal of Hydrology, 405*, 532–560.

Grenade, R. R. D. (2014 November 18). The Ica valley: A place of little rain. Retrieved from www.watersecuritynetwork.org/the-ica-valley-a-place-of-little-rain/.

Hallegatte, S. A. (2012). Cost effective solution to reduce disaster losses in developing countries: Hydro-meteorological services, early warning, and evacuation. *Policy Research Working Paper* 6058. The World Bank, Washington, DC.

Hepworth, N. P., JC; Guemes Delgado B and P. Kjell. (2010). Drop by drop: Understanding the impacts of the UK's water footprint through a case study of Peruvian asparagus. Retrieved from Lima, Peru: https://static1.squarespace.com/static/5627894ce4b0a834a45c5cec/t/5787b70b4402430761241bcc/1468512017976/Drop+By+Drop+-+Understanding+the+impacts+of+the+UK%E2%80%99s+water+

footprint+through+a+case+study+of+Peruvian+asparagus.pdfHolman, I., Allen, D., Cuthbert, M. O., & Goderniaux, P. (2011). Towards best practice for assessing the impacts of climate change on groundwater. *Hydrogeology Journal*, *20*, 1–4.

Hordijk, M. (2005). Participatory governance in Peru: Exercising citizenship. *Environment & Urbanization*, *17*, 219–236.

Huamanchumo, J., Pena, Y., Silva, L., & Hendriks, J. (2008). Developing capacity in water users organizations: The case of Peru. *Irrigation and Drainage*, *57*, 300–310.

IHS Markit Global Trade Atlas (2010). Growth in the value of selected Peruvian fruit and vegetable exports, 1998–2009. Retrieved from www.ers.usda.gov/webdocs/publications/37036/8042_fts34501.pdf?v=41056.

International Trade Center (2004). Competitiveness Through Public-Private Partnership: Successes and Lessons Learned The National Strategic Export Plan. Retrieved from Montreaux, Switzerland: www.intracen.org/WorkArea/DownloadAsset.aspx?id=52343 The National Strategic Export Plan. In *Competitiveness Through Public–Private Partnership: Successes and Lessons Learned*. Montreaux.

IPCC (2014). *Climate Change 2014: Synthesis Report. Contribution of Working Groups I, II and III to the Fifth Assessment Report of the Intergovernmental Panel on Climate Change*. Geneva, Switzerland.

James, I. (2015). The costs of Peru's farming boom; thriving agribusiness, declining aquifers and conflicts over water. *The Desert Sun*.

Janssen, M. A., & Anderies, J. M. (2007). Robustness trade-offs in social-ecological systems. *International Journal of the Commons*, *1*, 43–65.

Jensen, R., Asudullah, A. M., Lasarte, A., Miralles-Wilhem, F., & Castillo, R. M. (2015). Climate Change Adaptation Case Study: Climate Change Impacts during Droughts on the City of Trujillo, Peru. Retrieved from Washington DC https://publications.iadb.org/handle/11319/7219Keith, R. (1976). Conquest and agrarian change: The emergence of the Hacienda system on the Peruvian coast. *The Hispanic American Historical Review*, *57*, 727–729.

Klaren, P. (2000). *Peru: Society and Nationhood in the Andes*. Oxford, USA: Oxford University Press.

Kohler., T., & Maselli, D. (Eds.). (2009). *Mountains and Climate Change: From Understanding to Action*. Bern, Switzerland: Geographica Bernensia.

Kurukulasuriya, P., Miller, A., & Rutaagi, R. (2017). Building a value proposition for climate and weather services. Retrieved from http://adaptation-undp.org/building-value-proposition-climate-and-weather-services.

Lebel, L., Anderies, J. M., Campbell, B., Folke, C., Hatfield-Dodds, S., Hughes, T., & Wilson, J. (2006). Governance and the capacity to manage resilience in regional social-ecological systems. *Ecology & Society*, *11*, 1–23.

Levin, S. A. (1998). Ecosystems and the biosphere as complex adaptive systems. *Ecosystems*, *1*, 431–436.

Liu, J., Dietz, T., Carpenter, S. R., Alberti, M., Folke, C., Moran, E., . . . Taylor, W. W. (2007). Complexity of coupled human and natural systems. *Science*, *317*, 1513–1516.

Lynch, B. D. (2012). Vulnerabilities, competition and rights in a context of climate change toward equitable water governance in Peru's Rio Santa Valley. *Global Environmental Change*, *22*, 364–373.

Magrin, G. O., Marengo, J. A., Boulanger, J.-P., Bucheridge, M. S., Castellanos, E., Poveda, G., . . . Vicun, S. (2014). *Climate Change 2014: Impacts, Adaptation, and Vulnerability. Part B: Regional Aspects. Contribution of Working Group II to the Fifth Assessment Report of the Intergovernmental Panel on Climate Change*. Cambridge, U.K.: Cambridge University Press.

Mark, B., Mckenzie, J., & Baraer, M. (2010). Climate change and tropical Andean glacier recession: Evaluating hydrologic changes and livelihood vulnerability in the Cordillera Blanca, Peru. *Annals of the Association of American Geographers, 100*, 794–805.

Matten, D. (2004). The impact of the risk society thesis on environmental politics and management in a globalizing economy principles, proficiency, perspectives. *Journal of Risk Research, 7*, 377–398.

Mayer, E. (2009). *Ugly Stories of the Peruvian Agrarian Reform*. Durham, NC: Duke University Press.

Mcclintock, C. (1984). Why peasants rebel: The case of Peru's Sendero Luminoso. *World Politics, 37*, 48–84.

Meade, B., Baldwin, K., & Calvin, L. (2010). *Peru: An Emerging Exporter of Fruits and Vegetables*. Washington DC: US Department of Agriculture.

Multisectorial, C. T. (2009). Política y Estrategia Nacional de Recursos Hídricos del Perú. Retrieved from Lima, Peru: www.ana.gob.pe/media/527865/política%20y%20estrategia%20 nacional_.pdf

Nelson, E., Mendoza, G., Regetz, J., Polasky, S., Tallis, H., Cameron, D. R., Shaw, M. R. (2009). Modeling multiple ecosystem services, biodiversity conservation, commodity production, and tradeoffs at landscape scales. *Frontier Ecological Environment, 7*, 4–11.

Nolte, G. E. (2010). *Annual Asparagus Report: Gain*. Washington, DC: US Department of Agriculture Foreign Agricultural Service.

Ore, M. T. (2005). *Agua, bien común y usos privados. Riego, estado y conflictos en La Achirana del Inca*. Lima, Peru: WALIR.

Ore, M. T., Castillo, L. D., Orsel, S. V., & Vos, J. (2009). *El Agua, Ante Nuevos Desafios*. Lima, Peru: Instituto de Estudios Peruanos.

Organisation for Economic Co-operation and Development (2007). Market Access and Private Standards: Case Study of the Peruvian Fruit and Vegetable Markets. Retrieved from Peru: http://www.oecd.org/officialdocuments/publicdisplaydocumentpdf/?doclanguage=en&cote=agr/ca/apm(2006)23/fina

Oxford Business Group (2016). Agricultural exports on the rise in Peru. Retrieved from www.oxfordbusinessgroup.com/overview/fresh-horizons-agricultural-exports-are-taking-place-traditional-foreign-currency-earners.

Pastor, F. R. D. C. (2013). *Water Politics: Governance, Conflict and Vulnerability in Andean Peru*. (Geography Thesis). Syracuse University.

Peters, B. G., Pierre, J., & King, D. (2005). The politics of path dependency: Political conflict in historical institutionalism. *The Journal of Politics, 67*, 1275–1300.

Pierson, P. (2000). Increasing returns, path dependence, and the study of politics. *The American Political Science Review, 94*, 251–267.

Reilly, T. K., Alvites, E., Tello, C., Velasco, K., & Villaverde, D. (2010). ABC de la Descentralización. In *Perú ProDescentralización*. Lima, Peru: USAID, 1–52.

Rios, L. D. (2007). Agro-industries characterization and appraisal: Asparagus in Peru. In *Agricultural Management, Marketing and Finance Working Document*. Rome, Italy: Food and Agricultural Organization of the United Nations.

Rogers, P. and Hall, A.W. (2003). Effective Water Governance. TEC Background Papers No. 7. Global Water Partnership

Salazar, B. (2013). El riego en el Peru: Falta mucho por hacer. *La Revista Agraria, 155*, 10–11.

Schauwecker, S., Rohrer, M., Acuna, D., Cochachin, A., Davila, L., Frey, H., . . . Vuille, M. (2014). Climate trends and glacier retreat in the Cordillera Blanca, Peru, revisited. *Global and Planetary Change, 119*, 85–97.

Schmall, E. (2010). *Peru Water Wars Threaten Agricultural Export Boom*. Reuters.

Servicio Nacional de Meteorología e Hidrología. (2009). Climate Scenarios for Peru to 2030. Retrieved from Lima, Peru: http://www.senamhi.gob.pe/usr/cmn/pdf/Resumen_Nacional_Ingles.pdf

Shimizu, T. (2006). Expansion of Asparagus production and exports in Peru. *IDE Discussion Paper*. Chiba, Japan.

Smith, J., Colan, V., Sabogal, C., & Snook, L. (2006). Why policy reforms fail to improve logging practices: The role of governance and norms in Peru. *Forest Policy and Economics*, *8*, 458–469.

Stearns, S. C. (1989). Trade-offs in life-history evolution. *British Ecological Society*, *3*, 259–268.

Taylor, R. G., Scanlon, B., Dols, P., Rodell, M., Beek, R. V., Wada, Y., . . . Treidel, H. (2012). Ground water and climate change. *Nature Climate Change*, *3*, 322–328.

Taylor, L. (2017) Sendero Luminoso in the New Millennium: Comrades, Cocaine and Couunter-Insurgency on the Peruvian Frontier. Journal of Agrarian Chang, 17:1, 106–121.

USAID (2012). *Follow the Water: Emerging Issues of Climate Change and Conflict in Peru*. Washington, DC.

USAID (2016). *Case-Study: Mali Governance and Climate Change Adaptation*. Washington, DC.

Vergara, W., Deeb, A. M., Valencia, A. M., Bradley, R. S., Zarzar, A., Grunwaldt, A., & Haeussling, S. M. (2007). Economic impacts of rapad glaciar retreat in the Andes. *Eos*, *88*, 261–268.

Vorosmarty, C. J., Green, P., Salisbury, J., & Lammers, R. B. (2000). Global water resources: Vulnerability from climate change and population growth. *Science*, *289*, 284–288.

Vuille, M. (2013). *Climate Change and Water Resources in the Tropical Andes*. Inter-American Development Bank.

Vuille, M., Francou, B., Wagnon, P., Juen, I., Kaser, G., Mark, B. G., & Bradely, R. S. (2008). Climate change and tropical Andean glaciers: Past, present and future. *Earth-Science Reviews*, *89*, 79–96.

World Bank (2012). Country Partnership Strategy for the Republic of Peru for the Period FY12-16. Retrieved from http://documents.worldbank.org/curated/en/177751468294360196/Peru-Country-partnership-strategy-for-the-period-FY12-FY16

World Bank (2015). The World Bank group's Peru portfolio and climate change. In *Derecho, Ambiente, y Recursos Naturales*. Washington, DC.

World Bank (2016). Chavimochic irrigation project, Peru. Retrieved from http://ppp.worldbank.org/public-private-partnership/library/chavimochic-irrigation-project-peru.

7 Water markets and the food–energy–water nexus in Australia

Michael Valli and Alexandra M. Girard

Introduction

Australia has abundant energy resources and productive agricultural land. Energy and food production, however, depend on water resources, which are scarce and variable on this dry continent. Water is the critical limiting factor to socio-economic development; and where water flows, power and prosperity follow. While water scarcity becomes increasingly problematic around the world, Australians have developed a unique governance framework for allocating water resources. Decades of policy reform established a market-based model, which created privately owned tradable water rights independent from land rights, while capping water extraction at a level consistent with environmental sustainability. This policy framework – the National Water Initiative – has been praised worldwide as a successful model for resource governance that should be emulated globally (NWC, 2011).

This chapter investigates how trading behaviour transfers water between users and how these flows reconfigure the relationship between food, water, and energy. Various stakeholders – water utilities, community and cultural groups, agriculture, mining, hydroelectricity, financial, and environmental interests – participated in the formation of water markets, and those same stakeholders now compete within the marketplace to buy or sell water. Policymakers assumed that trading drives water towards its highest productive use, as rational actors operate within an efficient water market structure (NWC, 2010). Yet, the diversity of stakeholders trading water raises some important questions about how the market is supposed to function. Should the environment be considered a 'consumptive water user'? Can the cultural value of water be accurately ascribed financial value through water prices (Tribe, Schelling, & Voss, 1976)? Does water trading create hidden socio-economic costs that offset aggregate productivity gains? Do government owned organisations follow 'rational' economic behaviour when management, mandate, and budgets are influenced by the shifting political landscape? Do water trading patterns conflict with environmental sustainability?

We argue that sustainable governance that allows water to flow to its highest value use is a generic narrative, which in a broad sense holds merit, but at a detailed level is riddled with caveats and contradictions (Pritchard & Valli, 2013).

The chapter therefore draws on empirical research from the southern Murray Darling Basin (MDB), the most active water market in Australia, and the first author's experience as a professional water trader. Unravelling complexities of water trading behaviour shows us that (1) farmers balance business decisions against lifestyle, family, environment, and community factors; (2) environmental water managers often trade water at unfavourable prices, transacting as political capital and departmental budgets allow it; (3) financial actors are circumscribed by their investment mandates and the overall calculus of risk management; and (4) superimposed on all stakeholders are the biophysical constraints of catchment hydrology, particularly the configuration of existing manmade water infrastructure. In short, water is a flow resource that cuts across jurisdictions, electorates, communities, economies, territories, cultures, and ecosystems. These contextual factors influence water trading behaviour, which creates a feedback loop as the movement of water resources reshapes both the natural and human landscape.

The following examines the historical context in which the market evolved. Environmental damage in the Murray Darling Basin provided the political impetus to reform water management along the principles of sustainable economic development, overturning a centralised state-led system in favour of a market-based one. The third section explores the classical economic analysis of water markets. Aggregate productive value of water resources is increasing as trade shifts water between different users, with important implications for the interrelationship of food, water, and energy. The fourth section presents a nuanced look at the market, repositioning stakeholders within the social and biophysical landscape of the Basin. Water remains an uncooperative commodity which resists market efficiency in numerous ways. The final section draws together key insights about the water market for researchers seeking vital but unexplored topics, and for policymakers hoping to replicate this governance framework outside Australia.

Development of water markets in Australia

In order to understand the perspective and power of the different actors involved in the water market, it is essential to comprehend the influence of the historical legacy and existing path dependency shaping the structural contexts of the market. This is particularly true because water management in Australia has always been tightly interwoven within socio-economic development discourses and water policies have therefore been central to Australian's socio-cultural, economic, and political preoccupations.

State-led development of water resources commenced with initial large-scale nation-building infrastructures in the1900s and continued uninterrupted to the 1970s (Cathcart, 2010). Global economic crises during that decade increased pressure on public finances within Australia and made governments unwilling or unable to finance new capital investment. Previously the unchallenged narrative of rural economic and social development resulted in the provision of irrigation services regardless of financial or environmental cost (Crase, O'Reilly, & Dollery, 2000; Musgrave, 1996). Yet unrestrained spending was no longer possible as

Australia entered a 'mature water economy' characterised by sharply rising incremental costs for water diversion and transmission; ageing or deteriorating infrastructure contributing to higher operation, maintenance and replacement costs; the exploitation of accessible dam sites; the intensification of competition between consumers; increasing relative scarcity; and increased inter-dependencies amongst water users (Randall, 1981). Moreover, the fragmented and uncoordinated State-based systems defined water rights with fixed annual allocation volumes, which guaranteed water over-extraction as governments pursued their economic development agenda.

As domestic water management problems became more acute, global policy discourse began shifting away from traditional social democratic views in favour of free market economics as the solution to restoring prosperity (Quiggin, 1996). As signatory to the Rio Convention, the Australian government agreed to implement its own framework for sustainable economic development, a process that commenced in 1990 and was later codified in the National Strategy of Ecologically Sustainable Development. The Strategy broadly aimed to fully integrate the wider socio-economic and environmental repercussions of all stakeholders' actions at the national, international, and biosphere levels (Commonwealth of Australia, 1992).

Ecological problems in the Murray Darling Basin (MDB) provided an immediate test for the new framework. Dryland salinity, invasive species, and water pollution were becoming persistent issues affecting irrigation water supply to significant areas of agricultural production in south-eastern Australia throughout the late 1980s (see Figure 7.1). Ecological problems evolved into political conflict as South Australia, which depends heavily on the Basin for its horticulture industry and urban water supply, blamed upstream states New South Wales and Victoria for causing downstream environmental harm through over extraction and poor water management. While environmental degradation concerned all State government, in reality, the holistic management of the basin, in terms of sustainable allocation, shared use, environmental health, and water security, was not made a priority (Gray, 2012). Political stalemate fractured in 1991–92 when giant algal blooms emerged throughout the Murray and Darling Rivers, highlighting the dire ecological health of the Basin, and elevated unsustainable water extraction to the national consciousness (Connell, 2007; Wahlquist, 2008).

The immediate policy response was to cap water extractions in the MDB as the initial step in achieving sustainable extraction levels by halting and then 'winding back' development pressures (Connell & Grafton, 2011). Moreover 'the Cap', as it became known, required that new irrigation development must only proceed by sourcing water from existing extractive users. Against that background and the broader National Competition Policy – a macroeconomic policy framework for deregulating sectors previously dominated by the state – water trading was discussed as the vehicle for reallocating resources between users within the constraints of the complex, existing water-sharing institutions and water rights systems (Crase, Pagan, & Dollery, 2004). The Council of Australian Governments (COAG) endorsed its Water Resource Policy in 1994, which outlined a

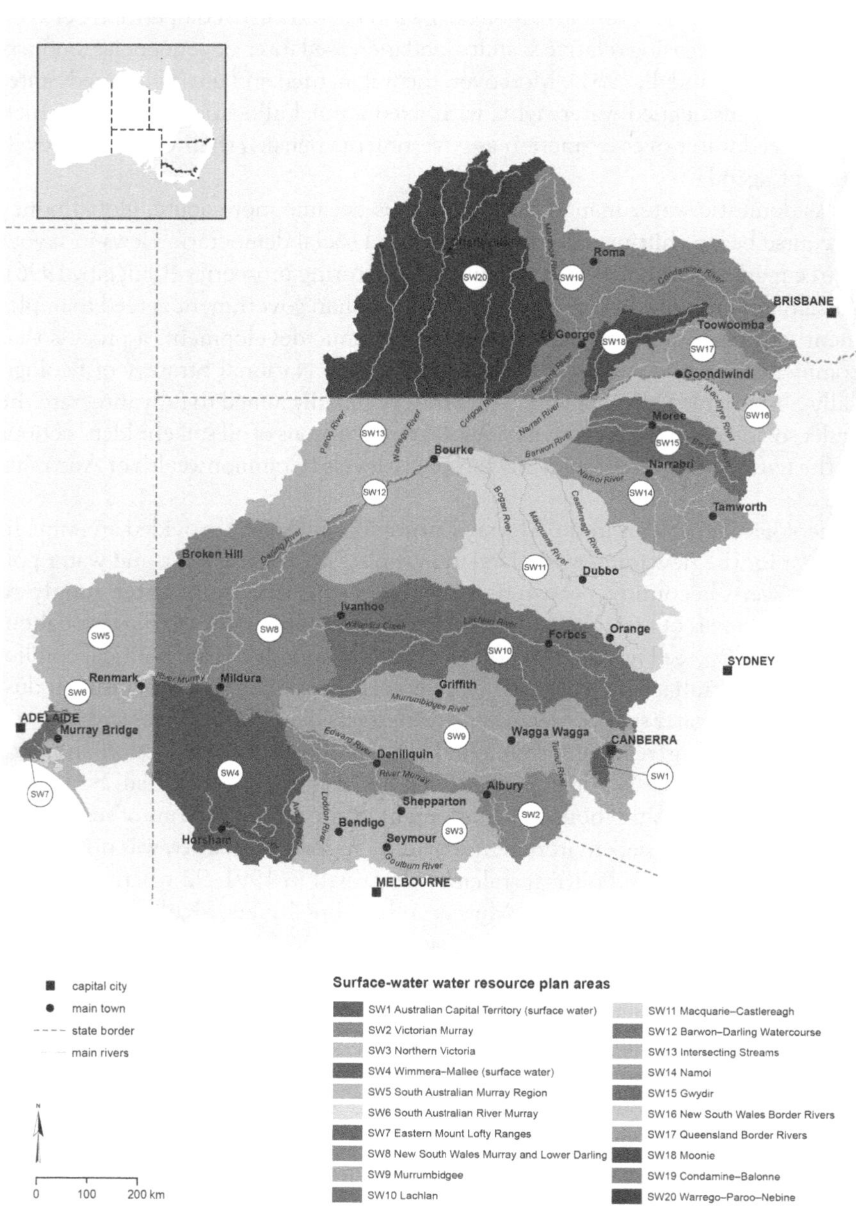

Figure 7.1 Surface Water Resource Plan regions in the Murray Darling Basin (Murray Darling Basin Authority, 2010)

comprehensive system of water property rights, separated from land title, with clear specification of ownership, volume, reliability, transferability and, if appropriate, quality (COAG, 1994). Ownership of pre-existing water rights was grandfathered into this new governance framework, which would be implemented in all catchments across Australia.

Theoretically, water trading promised a solution to intensifying competition between water users, unlocking economic efficiency, and improving environmental conditions. Markets would redistribute water from lower value to higher value users, raising the economic return of the resource (Frontier Economics, 2007). Markets would improve flexibility in managing irrigated farm businesses by freeing up capital or providing an alternative income stream for water sellers; and enabling water buyers to expand production efficiently (Bjornlund, 2004). Reallocating water in this manner would facilitate structural adjustment in the agricultural sector, enabling new farm businesses to replace failing ones more easily, and in some cases reducing rural poverty (Bjornlund, Wheeler, & Cheesman, 2011; Thobani, 1997). Markets would establish a price signal, which would discourage inefficient water practices and related environmental externalities like water logging and dryland salinity; shift production to the most suitable soil profiles and environmental conditions; and stimulate investment into water-saving technology or the retirement of degraded land (Quiggin, 2001).

Given the overwhelmingly positive expectations of the water markets, the reality of its operations proved different in several important ways. Crase et al. (2004) suggested water trading slowed the exit of less efficient producers, who could now sell water and deploy this income stream to support water-inefficient or unprofitable production. Policymakers equated *higher value use* with *higher water efficiency* in predicting the environmental benefits of water markets, a relationship which proved uncertain. Young and McColl (2003) argued strongly against the concept of *water efficiency* (input volume minus output volume) as a measurement of environmental benefits because water loss from one land parcel or production site is by definition returned to the watershed, and in some cases, contributes to ecosystem health. Encouraging water use efficiency and recycling could paradoxically lead to worse environmental outcomes.

More significantly, when water markets first commenced under the 1994 framework, governments decided to validate all existing water rights even where those rights had not previously been used for production. When the new water market established a price for water rights, those 'sleeper' licences (never activated) and 'dozer' licences (partially or intermittently activated) became valuable assets. Once activated by trade, these licences threatened to undermine the cap and the accompanying environmental benefits of attempting to halt extractive use at 1993–94 levels. Granting these dormant licences the same property rights status as active licences presented a financial windfall to less productive water rights owners, assuming licence idleness reflects productive value (Quiggin, 2006). This windfall further contributed to the persistence of otherwise unprofitable farm businesses, which further undermined the efficiency gains from the water market (Crase et al., 2004).

Frustration amongst all water stakeholders sparked another round of water policy reform. Unresolved problems with sustainable use limits, water property rights, risk sharing, and delayed policy implementation by State governments were addressed at COAG, 2004 with an updated water policy framework, the National Water Initiative (NWI). Then, political opportunism by the federal government and the intensifying Millennium Drought brought the NWI under greater scrutiny in 2007. Restoring over-allocated catchments in the Murray Darling Basin to sustainable diversion levels was a major sticking point in reviewing the NWI, and, as the federal election approached, water reform offered the conservative government an avenue to resolve frustrations amongst agricultural communities at water reform implementation and blunt the political success of the opposition party with urban green voters (Quiggin, 2011). The Howard government pronounced that water management problems were best addressed at the national level, invoking legal authority over the Basin through its external affairs powers in the Constitution. The Water Act (Commonwealth) 2007 and subsequent referral by State governments to the Commonwealth through the updated Murray Darling Basin Agreement, began a new whole-of-system approach to water management. Although the political strategy failed, the newly elected Rudd government enacted the water reforms and committed significant financial resources to its implementation.

Policymakers accepted that arbitrarily reducing water diversions to sustainable limits was politically impossible and, while legally sound,[1] undermining existing water rights in this manner would broadcast unacceptable sovereign risk to the domestic and international community (Crase et al., 2004). Commonwealth ministers therefore settled on two politically expedient, but financially costly, strategies to remedy over-allocation. Firstly, governments entered the water market to purchase water rights on behalf of the environment, so-called water buybacks (Owens, 2017). Secondly, irrigation infrastructure investment would reduce system transmission losses and transfer those savings to environmental use. Infrastructure investment applied to both public and private (on-farm) infrastructure, and is unsurprisingly the popular policy choice for recovering environmental water amongst the consumptive water users (Crase, O'Keefe, & Kinoshita, 2012). The Commonwealth Department of Environment would deploy an initial AUD8.9 billion budget (later increased to around AUD13 billion) while the newly formed Commonwealth Environmental Water Holder (CEWH) would manage the recovered water and other programs for environmental benefit. Consequently, the CEWH became the larger holder of tradable water rights in the Murray Darling Basin (see Figure 7.3).

Food, energy, and the water market: an orthodox perspective

Understanding the way markets are shaping the flow of the water in Australia is key to appreciate the complexity of the system with regards to the food, energy, water nexus. Applying a classical economics perspective shows a general pattern of water migrating to higher value production. For this, the Murray Darling Basin provides a great insight into the complexity of the regulatory system and the transactions happening within it at the catchment scale. Water is tradable within the boundaries of individual sub catchments of the Murray Darling Basin (see Figure 7.1). One exception is the southern Murray Darling Basin where the Murrumbidgee, Lower Darling, and rivers in northern Victoria are hydrologically connected to the Murray River. Connectivity allows water to be transferred between those river systems, so-called Inter-Valley Trade (IVT), and this consolidated water market includes more participants than any other region. Our discussion focuses on the southern Murray Darling Basin because its higher market activity and water resource consumption reveals deeper insights into water trading behaviour (National Water Commission, 2013).

Water Resource Plans determine how water is shared between the environment and consumptive stakeholders on a catchment-by-catchment basis (Owens, 2017). Water is assigned to the environment first in the form of 'passing flows', which intends to replicate the natural floods or dry spells for health of aquatic plants and animals. Rules specify timing and volume of storage releases or extractions from the water source, and while this water is not 'owned' by the environmental water holder, it is critically important to the ecosystem. Next available water is assigned to critical human needs and generally this water cannot be traded. Subsequent water is then assigned to consumptive users according to the quantity of water rights owned; this water is tradable. Shi (2006) clarifies the complex, multi-jurisdictional system of water rights and how available water resources are assigned to them. Ultimately, the 'water' being traded in the marketplace are claims on the available resources assigned to the consumptive pool, which broadly equates to water stocks stored in dams and flowing through the river channel. Total water availability for this consumptive pool varies considerably from year to year in the southern Murray Darling Basin (Figure 7.2).

Determining the economic value for water is a primarily a supply-demand proposition. Water is a seasonal natural resource, whose availability depends on catchment inflows via rainfall and catchment losses via evaporation or seepage. Unlike manufactured goods, water supply cannot simply increase to meet rising demand, but depends on the timing, volume, and location of rainfall. Climate also influences demand because rainfall is not confined to public dams and the river channel itself; it is spread across the entire catchment, so when it falls onto agricultural land or into private dams, it reduces the individuals' need to purchase water on the market and places downward pressure on prices. Weather is therefore the dominant short-term driver of water prices (Bjornlund & Rossini, 2007; Aither, 2016). In fact, the mere expectation of rainfall in weather forecasts can influences water prices (Mallawaarachchi, 2009).

Water ownership and use is dominated by agricultural production. Although Commonwealth Environmental Water Holder now owns a substantial water rights portfolio – by far the largest portfolio in the consumptive pool – this water remains 'outside' the market because current policy restricts trading environmental water in all but exceptional circumstances. Environmental recovery commenced in 2008–09 in earnest during the most acute phase of the decade-long Millennium drought and continues to this day (see Figure 7.3). Certainly, the

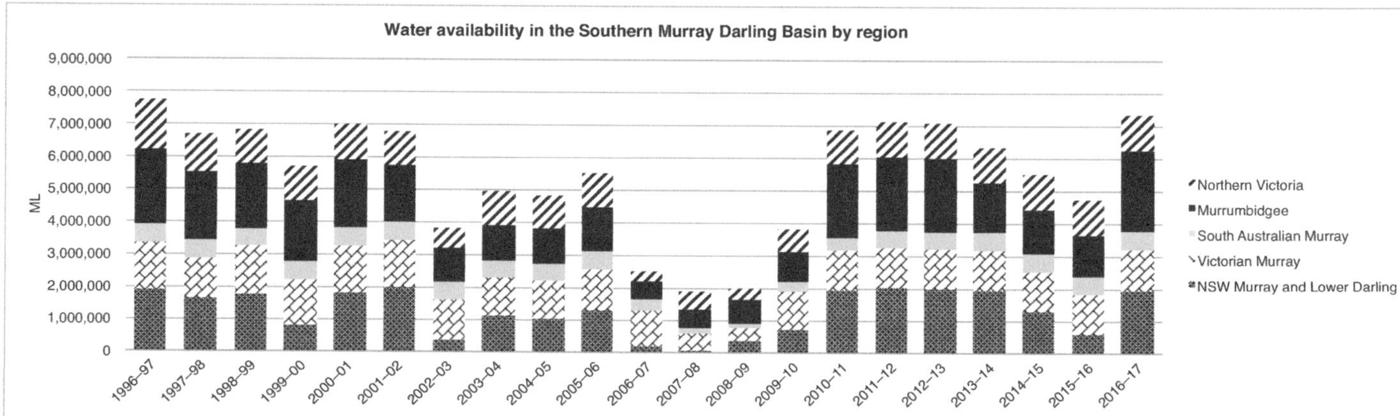

Figure 7.2 Water availability in the Southern Murray Darling Basin by Region

(Based on data from Australian Bureau of Statistics)

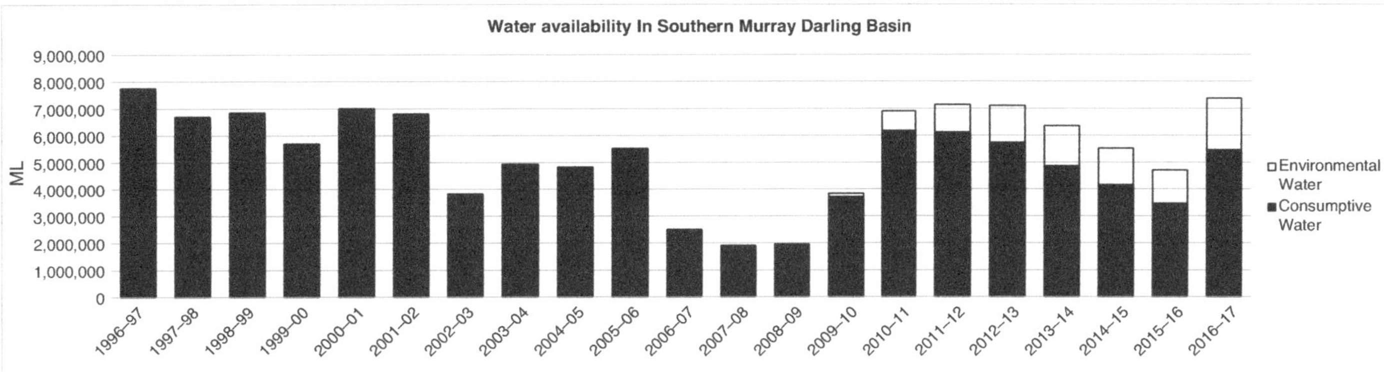

Figure 7.3 Consumptive and Environmental Water availability in the Southern Murray Darling Basin

(Based on data from Australian Bureau of Statistics)

presence of a large, price-insensitive buyer with known acquisition targets does impact water prices; however, these impacts are infrequent. Water markets are fundamentally a story about irrigated agriculture and water trading behaviour is strongly influenced by the economics of that industry.

For example, such decisions can be seen when evaluating the numerous trade-offs between rice or winegrape production. Maintaining winegrape production during water scarcity at the expense of rice production reduces the availability of a staple food item while sustaining a luxury discretionary one. Economic measurements of productivity disguise the materiality of the underlying process. Agriculture creates physical commodities and assigning water resources to one sector at the expense of another significantly impacts the agrifood system. The importance of this change becomes clear by investigating the wider political economy of agriculture.

One central tenet of water reform in Australia is that efficient markets drive water to its most productive use. Behind that statement is the economic principle that water users have different capacity to pay for water, which is usually expressed as the gross return per megalitre (AUD/ML) used in production (Qureshi, Shi, Qureshi, & Proctor, 2009). For example, in 2005 an average long grain rice producer in the Murrumbidgee region achieved a theoretical gross margin around AUD150/ML allowing for expenses and commodity prices (Singh, Mullen, & Jayasuriya, 2005). Cotton, dairy, horticulture, and even water utilities will have different gross margins per megalitre according to the costs structure of their particular industry. Rational economic behaviour assumes that rice growers will purchase water when prices are below AUD150/ML, and halt production in order to sell water whenever prices exceeds AUD150/ML. Other producers will purchase water from rice farmers at AUD150/ML up to their own gross margin per megalitre. As water becomes more scarce, and therefore expensive, lower value producers become sellers and higher value producer become buyers, redistributing water from one group to the other (Appels, Douglas, & Dwyer, 2004).

The Millennium drought in the mid-2000s highlighted this dynamic on a grand scale (see Figure 7.4). Severe water scarcity and record prices coincided with steep declines in water use for lower value producers (rice, cotton, dairy) while water consumption of higher value commodities (fruit/nuts, grapes, vegetables) remained relatively stable. Consequently, the percentage of available water flowing to those higher value commodities increased substantially. At the macroeconomic scale, gross value of irrigated agricultural production (GVIAP) remained steady throughout this period despite reduced aggregate output (Mallawaarachchi & Foster, 2009). Rice, cotton and dairy production recovered after the drought, however a longer term adjustment away from those sectors continues, which policymakers consider as validation for establishing water markets (NWC, 2012).

Another aspect to consider is the energy consumed throughout the agriculture value chain. Clearly, the energy required to transfer bulky wine bottles across vast distances to international customers is different to rice transported short distances for low-intensity processing for domestic customers. Yet inside the farm gate agricultural production in Australia is itself highly mechanised and energy intensive.

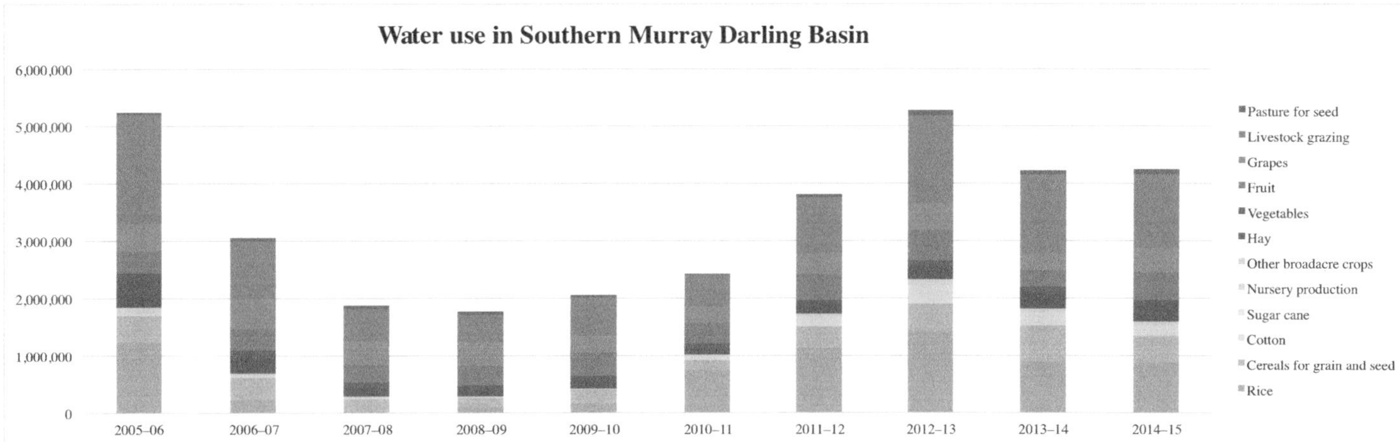

Figure 7.4 Water use by commodity in the southern Murray Darling Basin

(Source: RIRDC, 2016)

Water is heavy and so irrigation farms consume significant energy resources; not least because powerful pumps are necessary to extract water from rivers and aquifers and distribute it around the farm. Irrigation systems differ between commodities – drip, flood, pivot, channel – with different implications for on-farm energy consumption and water efficiency such that qualitative differences loop back to environmental sustainability issues. For example, rice is the most water intensive crop in Australia, which typically requires 12ML/ha, compared to winegrapes and other horticulture, which consumes around 4ML/ha. Ultimately, selecting irrigation infrastructure involves trade-offs between the market value of water savings, relative electricity expenses, infrastructure capital costs, crop yield benefits, and the biological constraints of the particular crop. The key point is that water markets enable rapid shifting of water between different production systems; intensify the competitive forces on all water users thereby increasing the pace of structural change in the agricultural sector; and dynamically making and remaking the complex interactions between food, water, and energy.

Water market activity in the southern MDB: a heterodox perspective

Markets provide an incentive structure for water resource consumption. Yet the assumptions that the 'market' represents a mechanical system whereby water inevitably flows towards higher value users needs to be understood as a generic narrative, which in a broad sense holds merit, but breaks down upon detailed examination. An accurate picture emerges by situating water markets within wider social and ecological structures, in which individual water traders fit within evolved arrangements characterised by complex interactivity with different actors with different agency, essentially a heterodox economics view of the market. The 'structured markets' perspective emphasises the operation of water markets within agent-structure feedback loops at various temporal and spatial scales. Water trading behaviour is more complex than the narrow pursuit of maximising return per megalitre; price and aggregate measurements of productivity value obscure individual idiosyncrasies and non-rational behaviour. The following section presents several factors that challenge the narrative of water market efficiency and its interplay with food and energy.

Physical geography

One key limiting factor to market efficiency is the mismatch between abstractions of the Basin and physical water flows 'on the ground'. Environmental management debates are dominated by images and discourses about the Murray Darling Basin as a geographic whole, yet the creation of one bounded space potentially obscured other ones. For example, natural watercourses, represented as blue lines on the map, are dry riverbeds during certain conditions and major rivers that intersect on maps may not do so on the ground (see Figure 7.1). Indeed, mapping 'surface water resource plan areas' in the Murray Darling Basin inevitably hides the complex hydrological connectivity between surface and groundwater flows. The Basin is often stylised as a pristine natural ecosystem, but is actually a patchwork of channels, weirs, locks, pumps, and dams modifying the pre-existing watercourse. Hydrology is controlled to such an extent that maximum and minimum river flows now occur exactly opposite to the dry and wet climate seasons; and resource managers control the precise river flow and height anywhere in the Basin via centralised, computer systems. The system resembles an integrated water distribution network – part natural watercourses and part manmade infrastructure – which operates to deliver water across the Basin.

Idealised models of water trading tend to assume frictionless ability of market actors to take positions. While water trading is an abstract process where debit and credit lines are entered into individual accounts on the water trading register, delivering water to an individual extraction site is constrained by the physical dimensions and configuration of the network. Proximity to the watercourse, channel capacity, pump size/rate, topography, and access to on-farm water storage influence individuals' actions to use and trade water as much as prices. For example, transforming water into a tradable commodity has accelerated the socio-economic expansion and contraction of different regions. Water delivery systems and the configuration of agricultural land are immobile, so when markets rapidly relocate water resources from one location to another, it devalues the fixed capital embodied in the neglected land and water infrastructure. Devaluation is most acute in regions where irrigators are exposed to the competitive water marketplace, but are unable to adapt their business.

Frontier Economics (2007) observed this process unfolding in the Victorian Murray Valley where smaller land parcels on shared irrigation channels cannot achieve profitability through scale; where the delivery network prevents new irrigation techniques; or where soil is less suitable or does not permit different crop production. Consequently, water shifted away from those regions to greenfield sites, which offer direct river pumping, new irrigation layouts and economies of scale. Whatever the driver of change, all branches of shared irrigation networks require a minimum maintenance cost and minimum water flow to prevent infrastructure damage. Steep declines in irrigation demand at one branch will then raise the incremental cost for the remaining water users. So-called 'stranded assets' problem adds financial pressure to declining regions and may force water resource managers to terminate delivery to that channel. One project, the Northern Victoria Irrigation Renewal Project (NVIRP), fought this trend by spending AUD2 billion to modernise, rationalise, and decommission channels within the Goulburn Murray Irrigation System (NVIRP, 2012). Essentially NVIRP uses public funding to repair an externality caused by normal water market activity. The enormous budget for this single program offsets much of the productivity gains attributed to the introduction of water trading and sits awkwardly with the assumption of water market efficiency.

Physical geography influences behaviour at the microeconomic level because agricultural production internalises the rhythm and flow of nature within an individual firm. Over time, agricultural production locks into path dependency whereby 'choices made in the past become embodied in specific bodies of equipment, firm assets and competences and labour skills as well as the organisational habits that shape economic action' (MacKinnon, 2008). Farm managers must therefore take decisions with numerous and varied sources of uncertainty, mindful of past and present constraints, all of which could result in the catastrophic loss of initial capital investment.

Water risk emerges because irrigators must lock in their water budget before water becomes available. In other words, the spatial and temporal mismatch between water use (fixed according to plan biology) and water availability (variable according to weather) exposes irrigators to water availability (and price) fluctuations. Water trading decisions always look towards the future and vary with (sometimes irrational) individual interpretations of it. Indeed, water trading activity may reflect the continuous repositioning of water risk as much as sourcing water for production (Loch, Bjornlund, Wheeler, & Connor, 2012). Moreover, once crops are planted the financial risk/reward calculus changes from generating profit to minimising loss. Annual crop producers (rice, cotton, wheat) are somewhat elastic to water prices because adjusting planted area within and between seasons is possible without significant financial loss. Permanent plantings (grapes, fruit, nuts) are highly inelastic to water prices because trees or vines take years to mature and the large capital investment to establish trees will be lost with any interruption to irrigation supply (Wheeler, Bjornlund, Shanahan, & Zuo, 2008). During water scarcity, irrigators pay almost any price to secure the marginal megalitre of water to reach harvest.

Environmental water use

Dramatic government intervention in the water market – in the shape of Environmental Water Holders – obstructs rational behaviour as these visible and price insensitive buyers pursue their objective. The Murray Darling Basin Plan targets 2750 GL water rights acquired for the environment from the consumptive pool. Although the Commonwealth recovery program is given the most scrutiny, State governments also have environmental water holders, which commenced operations before the Commonwealth. State environmental recovery programs acquired water rights on a smaller scale and in the same manner as private water traders – by private treaty and negotiation – and therefore escaped criticism of distorting prices and crowding out private traders from the marketplace. In contract, the size and methodology of Commonwealth program became a focal point in the political campaign against excessive environmental recovery targets in the Basin Plan.

The Commonwealth Environmental Department is responsible for acquiring water rights and the Commonwealth Environmental Water Holder manages this water for environmental benefit. Acquisition of water rights from private owners in the consumptive pool occurs via two methods: (1) by entering the market directly; and (2) by offering funding to irrigators to upgrade their irrigation infrastructure to more efficient technology, who then transfer water rights to the

CEWH equivalent to the projected water savings. Infrastructure funding is typically distributed through delivery partners such as State governments, industry groups, water brokers, and water utilities.

Direct market activity commenced with a series of 'expression of interest' rounds in 2008. Potential water sellers nominated their preferred price with the Commonwealth assessing those offers in the context of 'value for money' and the overall environmental watering strategy. Normal water market turnover and liquidity dropped significantly during each tender process as potential sellers withdrew their offers from the wider market in order to test price limits of this large, price-insensitive buyer. As the Commonwealth earned a reputation for paying inflated prices – whether deserved or not – it elevated the water prices across the entire market and crowded out all other private buyers, who desperately needed water to ease the impacts of the ongoing drought. Government purchases during the drought dominated market activity during the drought, accounting for half of all transactions in northern Victoria from 2009 to 2012 (DELWP, 2015), suggesting environmental purchases were done at inflated prices. Government purchases were indiscriminate in their location and exacerbated the problems of 'stranded assets' on the irrigation network, which a more coordinated process with infrastructure operators may have avoided. For those reasons, the Productivity Commission (2010) found 'tenders are sound purchasing mechanisms where active markets for water entitlements do not exist. But where active markets do exist, acquiring water directly from those markets is likely to be more efficient'. As taxpayers, irrigators were also alarmed at the CEWH purchasing water at peak drought prices because political capital and departmental budget allowed it. When rains returned and water prices fell, impairment to those government-owned water rights was substantial.

Infrastructure modernisation was adopted in response to criticism of direct purchases. On-farm infrastructure investment amounts to a public subsidy to private water users, but is intended to minimise the socio-economic impacts and externalities of purchasing water shares in the open market. Yet the task of measuring water savings is complex and the costs are high. Dairy Australia estimated water recovery via infrastructure grants cost the CEWH around AUD3,700/ML compared to prevailing water prices between AUD2000/ML in the open market (RMCG, 2013). Spotting an opportunity, irrigators apply for infrastructure grants without owning the required water to transfer back to the government; if successful, those irrigators purchase water in the open market at lower prices and immediately exchange them for the higher implicit water value in new irrigation technology. Since their windfall is guaranteed, these irrigators bid strongly for water shares and temporarily drive up the price. Again, the Productivity Commission (2010) found that 'subsidising infrastructure is rarely cost effective in obtaining water for the environment, nor is it likely to be the best way of sustaining irrigation communities'.

Outlooks for the program itself are uncertain since the current expenditure rate will exhaust its budget before the 2750 GL recovery target is met. Regardless of whether and when this aspiration is met, it remains the case that the entry of the

Commonwealth Government into the MDB water market is a major destabilising force. As Wheeler, Bjornlund, Zuo and Shanahan (2010) have suggested, the level of government involvement in environmental water purchases: 'may well disrupt their intended function of redistributing water between competing private irrigators, and such disruptions might result to slow down the ability of existing irrigators to restructure their farm business to be viable in today's economic and political climate'.

Political intervention

Water markets are not organic or spontaneous phenomenon, but arise from persistent and determined action by the state. The Murray Darling Basin is fragmented by political action and it is unsurprising that political contests, which created tradable water rights, continue to influence efficient market operations today. Indeed, the National Water Initiative permits trade restrictions by the minister in order to manage the influence of environmental, geographical, hydrological, and socio-cultural factors (Gray, 2012; Owens, 2017).

One recent example is the Victorian government cap on the water traded out of irrigation districts in northern Victoria at 4 percent per annum to alleviate the socio-economic impacts of water resources leaving those regions *en masse*. Water users with extraction sites outside those districts or non water users (NWU) with no associated extraction site were prevented from buying once the 4 percent limit was reached. Trade restrictions inevitably reduce the pool of market participants and therefore artificially increase or decrease prices depending on market conditions. Enforced between 2007–14, the 4 percent limit applied during a period of downward pressure on water prices where most irrigators sold water to exit farming, alleviate debt or free working capital following the Millennium Drought. NWUs (typically net buyers) were prevented from trading precisely when irrigators needed maximum buyer activity, resulting in artificially lower prices once the 4 percent limit was triggered each year. Arguably this political intervention harmed rather than helped irrigators adjusting to change.

Water trading is permitted between different catchments in the southern MDB; however the specific features of water rights in each catchment differ according to their State legislation. Differences in water rights encourage traders to transfer water into Victoria at the end of each financial year to exploit favourable storage arrangements. Victorians argued this behaviour diluted their water property rights, leading their State government to make another intervention in the water market in 2012. The Victorian water minister imposed limits on the inter valley water transfers from New South Wales and South Australian trading zones into Victoria; and inter valley water transfers from the Goulburn, Loddon, Campaspe rivers to the Murray river within Victoria.

Unintended consequences of political intervention again became evident over time: while the restriction secures storage rights of Victorian water rights owners, it also causes higher water prices paid. Differences in water rights sometimes result in higher water availability (and lower prices) for New South Wales water rights and, when the restriction is active, prevent Victorian water users from accessing cheaper water across the border. Trade restrictions of this type are continuously criticised for reducing water market efficiency and impeding the overall objectives of water reform by the Australian Competition and Consumer Commission (ACCC, 2013).

Farmers and rational behaviour

Undermining the efficient functioning of the water market is the relationship between farmers' behaviour and multilayered socio-cultural, political, and geographical contexts. The efficient markets hypothesis in Murray-Darling water policy tends to treat farmer decision-making somewhat as a black box. The implied assumption is that farm businesses are profit-maximising entities in possession of clear strategic objectives on how to manage their assemblages of assets and resources. On this basis, farmers are assumed to make decisions about their water needs in coherent and consistent pursuit of profit-making optimality.

Contemporary research on farmer behaviour interrogates the extent of these assumptions. It needs to be recognised that farmers are not a homogenous social category, but differ across a range of socio-cultural, demographic and economic fronts. This diversity casts considerable variety into the question of how farmers manage their businesses. Firstly, there is substantial evidence that farmers tend to subsume profit goals within a broader framework of utility maximisation (Amador, Sumpsi, & Romero, 1998). For example, farmers operate their business with multiple objectives in mind, including income, lifestyle, family, environment, succession arrangements and reputation within the community (Kuehne, Bjornlund, & Loch, 2010). For family-based farming establishments, the convoluted navigation of farm and family aspirations within the ebb and flow of life cycle considerations clearly plays a vital role in structuring farmer behaviour (Frontier Economics, 2007). Secondly, even if farmers construct their aspirations around profit-maximising goals, there is extreme uncertainty with regards to how to arbitrate the temporal scales of profitability. In farming, there are difficult-to-compute trade-offs concerning short-, medium-, and long-term strategy setting. Within farming communities, these are frequently managed through what sociologists would label 'conventions' (activities which are considered acceptable within a community) (Biggart & Beamish, 2003). Accepted wisdoms of what entails 'being a good farmer' (Burton, 2004) shape the ways in which farmers structure their desires to meet goals within different time frames. Again, these can be mediated strongly by life cycle factors – decisions to forgo higher profit to 'keep the land in better shape' may be circumscribed by the intricacies of intergenerational and succession issues within individual farm families. The factors previously described underline the shortcomings of simple conflations between farmer behaviour, efficiency, and market signals. Farmers with water rights in the Murray Darling subsequently trade water in highly nuanced, multidimensional and individually specific ways.

Financialisation of water

Towards the end of the Millennium Drought, newspaper articles described a new category of water trader. Headlines like 'Investors dip their toes in water rights' and 'Liquid gold' announced the growing activity of investment funds participating in the water market (Snow & Jopson, 2010; Boreham, 2011). Investors operating in the agricultural sector is nothing new and corporate farm businesses traded water alongside family operations since the market opened. What distinguished these new firms was their investment in water as a stand-alone asset and as their primary business activity.

Financial actors are at first glance perfect agents of profit maximisation; yet investors and speculators may obstruct objectives of the National Water Initiative to manage water resources for 'socially and economically beneficial outcomes in a manner that is environmentally sustainable' (COAG, 2004). Investors enjoy a significant advantage in the water market. Generally speaking, these firms do not consider social or environmental issues in their water trading behaviour; their access to capital and technical expertise allows them to exploit opportunities unavailable to most market participants; and as non-landowners (non water users) they are free of geographic limitations associated with water delivery, which constrain other stakeholders. Indeed, inefficiencies in the water market simply creates price anomalies and profit opportunities for professional water traders. For example, water allocation prices often vary between regions or between brokers in the connected southern MDB water market, which a professional water trader will capture by purchasing low in one place and selling high in another. One effect of this arbitrage is to smooth out differences in water prices. Since market efficiency is synonymous with perfect information (and therefore no opportunity for arbitrage), professional traders are viewed as a necessary and desirable aspect of water market maturity (Bjornlund & Rossini, 2008). Yet Das (2011) argues forcefully that modern finance does not create value, but merely shuffles resources into new configurations in order to squeeze efficiencies from existing assets. Arbitrage in the water market is therefore a zero-sum game where speculative traders take profits that would otherwise stay with productive water users, who engage in genuine value creation through agriculture and other production.

Institutional investors arriving into the water market increases the quantity of money chasing the same finite pool of water rights; all things being equal the price must rise. Moreover, as non water users, investors typically own water rights in order to lease them to water users at attractive rates. By contributing to rising prices and seeking maximum lease returns, investors intensify the competitive pressures on water users to raise productivity; to generate more output with less water. Investors are therefore an important agent for driving water towards its highest value user, but with an important disclaimer: the water productivity gains do not flow to water users, but are captured by the investors themselves and siphoned off to financial centres around the world.

Water reform in Australia promises socio-economic benefits, yet rising aggregate water productivity and who receives its economic benefits are two different questions. If investors or speculators in urban financial centres capture the economic benefits of water reform, it casts doubt on the original justification that water markets would benefit rural communities and the agricultural sector. On the other hand, financial actors increase market liquidity and may drive the development of innovative water products such as options or futures, which greatly benefit waters users in their water risk management. Financialisation of water, like wider the food– energy–water nexus in Australia, remain an unexplored research topic available to the new generation of geographers.

Conclusion

Water markets in Australia are an important case study for research, policy, and corporate stakeholders grappling with natural resource governance challenges around the world. This chapter highlights the difference between a top-down, theoretical perspective of market activity versus the real, messy, on-the-ground practices of individuals and business. Commodification and marketisation of water resources in Australia relies heavily on the assumption that an efficient marketplace will shift water resources towards highest value users. Reality is far more complex since market agents operate within a dynamic social, historical, and biophysical structure, which influences their water market activity as much as the narrow pursuit of profit maximisation.

Water markets are fundamentally a governance system for reallocating resources between competing stakeholders. Movement of water via the marketplace influences complex and interconnected feedback loops between food, water, and energy; mapping water trading behaviour is therefore critical to understanding the food–energy–water nexus in Australia. This nuanced analysis of water trading activity should moderate the potential social, economic, and ecological outcomes of market-based water management.

Note

1 Water resources ultimately vest with the Crown and their status as private property therefore remains unclear (Gray, 2012).

References

Aither (2016). *Supply-side drivers of water allocation prices. Identifying and modelling supply-side drivers of water allocation prices in the southern Murray-Darling Basin*. Department of Agriculture and Water Resources.

Amador, F., Sumpsi, J. M., & Romero, C. (1998). A non-interactive methodology to assess farmers' utility functions: An application to large farms in Andalusia, Spain. *European Journal of Agricultural Economics*, *25*, 92–109.

Appels, D., Douglas, R., & Dwyer, G. (2004). Responsiveness of water demand: A focus on the southern Murray-Darling Basin. *A Productivity Commission Staff Working Paper. Melbourne*.

Australian Competition and Consumer Commission (ACCC) (2013). ACCC Water Monitoring Report 2011–12. Retrieved from www.accc.gov.au/system/fles/Water%20Monitoring%20Report%202011–12.pdf.

Biggart, N. W., & Beamish, T. D. (2003). The economic sociology of conventions: Habit, custom, practice, and routine in market order. *Annual Review of Sociology*, *29*(1), 443–464.

Bjornlund, H. (2004). Formal and informal water markets: Drivers of sustainable rural communities. *Water Resources Research*, *40*(9), 1-12. doi:10.1029/2003WR002852.

Bjornlund, H., & Rossini, P. (2007). Fundamentals determining prices in the market for water entitlements: An Australian case study. *International Journal of Water Resources Development*, *23*(3), 537–553.

Bjornlund, H., & Rossini, P. (2008). Are the fundamentals emerging for more sophisticated water market instruments? *Proceedings from the 14th Annual Conference of the Pacific Rim Real Estate Society*. Kuala Lumpur, Malaysia.

Bjornlund, H., Wheeler, S., & Cheesman, J. (2011). Irrigators, water trading, the environment and debt: Buying water entitlements for the environment. In R. Q. Grafton & D. Connell (Eds.), *Basin Futures: Water Reform in the Murray-Darling Basin* (pp. 291–302). Canberra: Griffin Press.

Boreham, T., (2011). Investors dip their toes in water rights. *The Australian Newspaper*. Retrieved from www.theaustralian.com.au/business/investors-dip-their-toes-in-water-rights-theres-certainly-enough-of-it-around-but-should-you-buy-it/story-e6frg8zx-1226039898967.

Burton, R. J. (2004). Seeing through the 'good farmer's' eyes: Towards developing an understanding of the social symbolic value of 'productivist' behaviour. *Sociologia Ruralis*, *44*(2), 195–215.

Cathcart, M. (2010). *The Water Dreamers: The Remarkable History of Our Dry Continent*. Melbourne, Australia: Text Publishing.

Commonwealth of Australia (1992). *National Strategy for Ecologically Sustainable Development: December 1992*. ACT, Australia: Australia for Australian Government Publishing Service. Retrieved from www.environment.gov.au/about-us/esd/publications/national-esd-strategy on 28 April 2016.

Connell, D. (2007). *Water Politics in the Murray-Darling Basin*. Leichhardt, NSW: The Federation Press.

Connell, D., & Grafton, R. Q. (2011). Water reform in the murray-darling basin. *Water Resources Research*, *47*(12), 1-9. doi:10.1029/2010WR009820.

Council of Australian Governments (COAG) (1994). Communiqué, meeting of COAG in hobart 25 February 1994. *Attachment A: Water Resource Policy*. Retrieved from www.environment.gov.au/system/files/resources/6caa5879-8ebc-46ab-8f97-4219b8ffdd98/files/policyframework.pdf.

Council of Australian Governments (COAG) (2004). Intergovernmental agreement on a national water initiative. *Canberra, ACT*. Retrieved from www.nwc.gov.au/__data/assets/pdf_file/0008/24749/Intergovernmental-Agreement-on-a-national-water-initiative.pdf.

Crase, L., O'Keefe, S., & Kinoshita, Y. (2012). Enhancing agrienvironmental outcomes: Market-based approaches to water in Australia's murray-darling basin. *Water Resources Research*, *48*(9), 1-7, doi 10.1029/2012WR012140.

Crase, L., O'Reilly, L., & Dollery, B. (2000). Water markets as a vehicle for water reform: The case of New South Wales. *Australian Journal of Agricultural and Resource Economics*, *44*(2), 299–321.

Crase, L., Pagan, P., & Dollery, B. (2004). Water markets as a vehicle for reforming water resource allocation in the murray darling basin of Australia. *Water Resources Research*, *40*(8), 1-10, doi 10.1029/2003WR002786.

Das, S. (2011). *Extreme Money: Masters of the Universe and the Cult of Risk*. London: FT Press.

Department of Environment, Land, Water and Planning (DELWP) (2015). Victorian Water Trading: 2014–15. *Annual Report*. Parliament of Australia.

Frontier Economics (2007). *The Economic and Social Impacts of Water Trading: Case Studies in the Victorian Murray Valley*. Barton, Australia: Rural Industries Research and Development Corporation. Australian National Water Commission.

Gray, J. (2012). The legal framework for water trading in the Murray-Darling Basin: An overwhelming success? *Environmental and Planning Law Journal*, *29*(4), 328–349.

Kuehne, G., Bjornlund, H., & Loch, A. (2010). *Why Do Farmers Make Non-Profit Decisions? Investigating Decisions Made During Drought*. Barton, Australia: Rural Industries Research and Development Corporation.

Loch, A., Bjornlund, H., Wheeler, S., & Connor, J. (2012). Allocation trade in Australia: A qualitative understanding of irrigator motives and behaviour. *Australian Journal of Agricultural and Resource Economics*, *56*(1), 42–60.

MacKinnon, D. (2008). Evolution, path-dependence and economic geography. *Geography Compass*, *2*, 1449–1463.

Mallawaarachchi, T., & Foster, A. (2009). *Dealing with Irrigation Drought: The Role of Water Trading in Adapting to Water Shortages in 2007–08 in the Southern Murray-Darling Basin*. (Research report 09.6 to the Department of the Environment, Water, Heritage and the Arts). Canberra: Australian Bureau of Agriculture and Resource Economics.

Murray-Darling Basin Authority (2010). *Guide to the Proposed Basin Plan: Technical Background*. Canberra: Murray-Darling Basin Authority.

Musgrave, W. (1996). The irrigation industry in the Murray-Darling Basin and aspects of its reform. In J. Pigram (Ed.), *Security and Sustainability in a Mature Water Economy: A Global Perspective* (pp. 43–63). Armidale: Centre for Water Policy Research.

National Water Commission (NWC) (2010). *The impacts of water trading in the southern Murray: Darling Basin: An economic, social and environmental assessment*. Canberra: National Water Commission.

National Water Commission (NWC) (2011). Strengthening Australia's water markets. Retrieved from www.nwc.gov.au.

National Water Commission (NWC) (2012). Impacts of water trading in the southern Murray – Darling Basin between 2006–07 and 2010–11. Retrieved from www.nwc.gov.au/__data/assets/pdf_file/0019/10783/681-NWC_ImpactsofTrade_web.pdf.

National Water Commission (NWC) (2013). *Australian Water Markets Report 2012–13*. Canberra: National Water Commission.

NVIRP (2012). Northern victoria irrigation renewal project. *Annual Report 2011–12*. Tatura, VIC: Goulburn-Murray Rural Water Corporation.

Owens, K. (2017). *Environmental Water Markets and Regulation*. London, New York: Routledge.

Pritchard, W., & Valli, M. (2013). What kind of a market for what kind of water? Geographical perspectives on Murray-Darling water reform. In H. Bjornlund, C. A. Brebbia, & S. Wheeler (Eds.), *Sustainable Irrigation and Drainage IV: Management, Technologies and Policies* (pp. 51–62). Southampton: Wit Press.

Productivity Commission (2010). Market mechanisms for recovering water in the Murray-Darling Basin. *Final Report*. Australian Government Productivity Division.

Quiggin, J. (1996). *Great Expectations: Microeconomic Reform and Australia*. Sydney: Allen & Unwin.

Quiggin, J. (2001). Environmental economics and the Murray-Darling river system. *Australian Journal of Agricultural and Resource Economics*, *45*(1), 67–94.

Quiggin, J. (2006). *Water Reform in Australia: Strengths and Limitations of Market Based Mechanisms*. Submission to the Productivity Commission research study on Rural Water Use and the Environment: The Role of Market Mechanisms.

Quiggin, J. (2011). Managing risk in the Murray-Darling Basin. In R. Q. Grafton & D. Connell (Eds.), *Basin Futures: Water Reform in the Murray-Darling Basin* (pp. 313–326). Canberra: Griffin Press.

Qureshi, M. E., Shi, T., Qureshi, S. E., & Proctor, W. (2009). Removing barriers to facilitate efficient water markets in the Murray-Darling Basin of Australia. *Agricultural Water Management*, 96(11), 1641–1651.

Randall, A. (1981). Property entitlements and pricing policies for a maturing water economy. *Australian Journal of Agricultural Economics*, *25*(3), 195–220.

RMCG (2013). *Cost Benefit Analysis of Farm Irrigation Modernisation: Final report*. Bendigo, VIC: RMCG.

Rural Industries Research and Development Corporation (RIRDC) (2016). *Contemporary Trends and Drivers of Irrigation in the Southern Murray-Darling Basin*. Barton, Australia: RIRDC.

Shi, T. (2006). Simplifying complexity: Rationalising water entitlements in the southern connected river Murray system, Australia. *Agricultural Water Management*, 86(3), 229–239.

Singh, R. P., Mullen, J. D., & Jayasuriya, R. T. (2005). *Farming Systems in the Murrumbidgee Irrigation Area in NSW* (Economic Research Report No. 10). Yanco, NSW: Department of Primary Industries.

Snow, D., & Jopson, D. (2010, September 3). Liquid Gold. *Sydney Morning Herald*. Retrieved from www.smh.com.au/environment/water-issues/liquid-gold-20100903-14ueu.html.

Thobani, M. (1997). Formal water markets: Why, when and how to introduce tradable rights in developing countries. *The World Bank Research Observer*, *12*(2), 161–179.

Tribe, L. H., Schelling, C. S., & Voss, J. (Eds.). (1976). *When Values Conflict: Essays on Environmental Analysis, Discourse, and Decision*. Cambridge, MA: Ballinger Publishing Company.

Wahlquist, Å. (2008). *Thirsty Country: Options for Australia*. Sydney: Allen & Unwin.

Wheeler, S., Bjornlund, H., Shanahan, M., & Zuo, A. (2008). Price elasticity of water allocations demand in the Goulburn: Murray Irrigation District. *Australian Journal of Agricultural and Resource Economics*, 52(1), 37–55.

Wheeler, S., Bjornlund, H., Zuo, A., & Shanahan, M. (2010). The changing profile of water traders in the Goulburn-Murray Irrigation District, Australia. *Agricultural Water Management*, 97(9), 1333–1343.

Young, M. D., & McColl, J. C. (2003). Robust reform: The case for a new water entitlement system for Australia. *Australian Economic Review*, 36(2), 225–234.

8 Institutional bricolage to address sustainability challenges in the South African sugarcane industry

A case study of the SUSFARMS® initiative in the Midlands area of KwaZulu-Natal, South Africa

Jessica Cockburn, Vaughan Koopman, Laura M. Pereira and Jaci van Niekerk

Introduction

In light of complex rural development challenges in the KwaZulu-Natal province of South Africa, as well as global market demands for more sustainable sugar farming practices, new governance arrangements for sustainable production are emerging in the South African sugar industry. Using the lens of institutional bricolage, this chapter critically explores how these new governance arrangements were constructed by drawing on social learning theory and expansive learning methodology.

The chapter begins with an introduction to sustainability concerns in global sugar supply chains. It then provides an overview of the South African sugar industry, the agricultural development and sustainability challenges faced in South Africa, and the local sugar industry response to these sustainability challenges. The second part of the chapter focuses on the case study. First, we introduce the SUSFARMS® tool (Sustainable Sugarcane Farm Management System (SASA, 2012), which was developed to guide decision-making and better management for sustainability at the farm level. We then describe a mediated expansive learning process which was initiated to address some of the implementation challenges of the SUSFARMS® tool. This is followed by a description of the SUSFARMS® 2018 Collaboration (Binedell, Cockburn, & Koopman, 2014), a novel multi-stakeholder governance platform in the Midlands area of KwaZulu-Natal. We then examine the SUSFARMS® initiative (both the development of the 'SUSFARMS® tool' and the 'SUSFARMS® 2018 Collaboration') through the lens of institutional bricolage. We demonstrate how the mediated expansive learning process enabled local actors to be institutional bricoleurs who could re-model institutional structures and processes to address challenges they faced in championing the implementation of the SUSFARMS® tool. We elicit lessons learnt that can be applied to other natural resource governance initiatives seeking to address sustainability challenges in complex agricultural value chains. We conclude with a critical assessment of the SUSFARMS® initiative and consider its suitability as an institutional mechanism to deal with the sustainability and agricultural development challenges facing the South African sugar industry.

Sustainability challenges and responses in sugar supply chains

The global sugar production system faces complex sustainability challenges ranging from consumer food and health concerns to social and environmental impacts related to sugarcane cultivation (Ahmad, 2014). The environmental footprint of sugar cultivation (beet and sugar cane) includes biodiversity loss as a result of land clearing for plantations, intensive water usage, and applications of synthetic fertilisers, herbicides and pesticides (Cheesman, 2004; WWF, 2005). Critics have pointed out the socio-economic impacts of sugarcane, highlighting land grabs for cane cultivation, low wages and limited benefits for workers, insecure job tenure, and concerns about workers' health and living conditions (Stedman-Edwards, 2004; Weiss, 2007; Oxfam, 2013; Martiniello, 2015; Visser & Ferrer, 2015). The biodiversity-related challenges facing the South African sugar industry are similar to those previously described.

In response to these challenges, and in line with increasingly rigorous responsible business practice norms, corporate sustainability and social responsibility have become ever more prominent in sugar business discourse (Jenkins, Baptista, & Porth, 2015). As this discourse gains traction, organisations, which produce and procure sugar, are seeking ways to measure and manage their interactions with suppliers, investors, and civil society to bring about more engaged sustainability initiatives. Leading global brands, for example, have recently announced their sustainability principals and targets for procurement of agricultural produce.

An important development in this move towards more sustainable global sugarcane supply chains is the establishment of Bonsucro in 2005 as a global multi-stakeholder, non-profit organisation dedicated to reducing the environmental and social impacts of sugarcane production (Moura & Chaddad, 2012). Initially coordinated and hosted by the World Wide Fund for Nature (WWF), Bonsucro is a collaboration between sugar retailers, investors, traders, producers, and NGOs which seeks to improve the sustainability of sugar production globally. It has developed the Bonsucro Standard to mitigate environmental and social impacts of sugarcane production and primary processing, while balancing the need for economic viability (Bonsucro, 2015).

Likewise, sugarcane farmers and milling companies in South Africa are responding locally to these sustainability pressures and discourses in the market. The SUSFARMS® initiative in South Africa is such an example and is used as a case study in this chapter to illustrate how a diverse group of actors have reorganised outmoded institutional practice in novel ways to meet new accountability and sustainability challenges in their agricultural supply chain. In so doing, they are positively influencing not only their own businesses, but also having a broader influence on both national and international sugar industry sustainability policy and practice.

The South African sugar industry: an overview

Globally, South Africa is ranked 15th out of approximately 120 sugar producing countries, and is considered one of the leading cost-competitive producers of sugar (SASA, 2015). The country has six milling companies that operate 14 sugar mills that are supplied with sugarcane grown in the two eastern provinces of KwaZulu-Natal and Mpumalanga (Figure 8.1). Across these two provinces, sugarcane comprises approximately 50 percent of field crop gross farming income, thus making a significant contribution to rural economic development in these two provinces (SASA, 2015). The South African sugar industry has dedicated units and staff resources that provide expert advice and support on matters of environmental sustainability (SASA, 2015). The development and endorsement in 2009 of the Sustainable Sugarcane Farm Management System (SUSFARMS®) as the primary tool for ensuring responsible practice in sugarcane cultivation at farm level has been cited as a significant commitment by the industry to address sustainability concerns (Binedell et al., 2014).

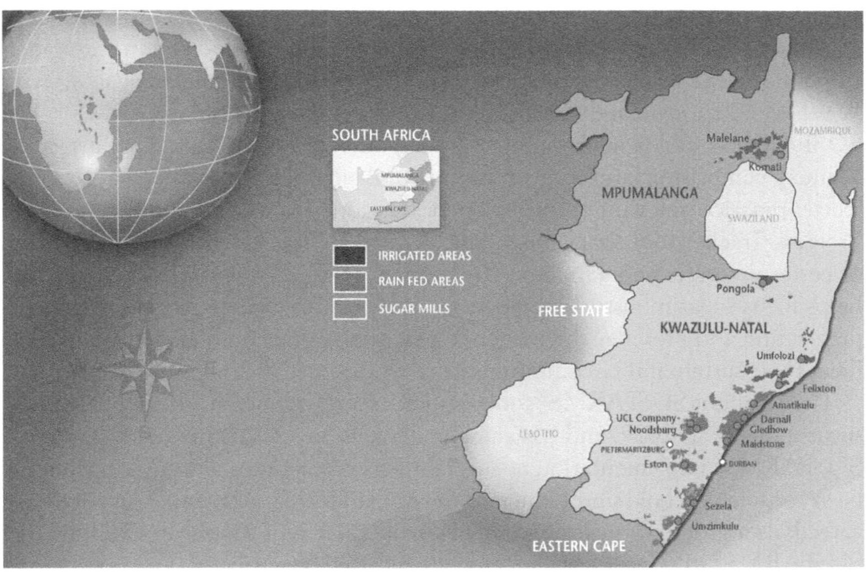

Figure 8.1 Geographical location of the South African sugar industry. The SUSFARMS® 2018 Collaboration covers the production areas for three sugar mills in the Midlands region of KwaZulu-Natal: UCL Company, Noodsberg and Eston

(SASA, 2015).

Case study: development of the 'SUSFARMS® tool' and the 'SUSFARMS® 2018 Collaboration'

Information sources for this case study were based on the masters research projects of two of the authors (Cockburn, 2013; Koopman, 2014) along with project documentation from the SUSFARMS® 2018 Collaboration (Binedell et al., 2014). The information sources included data collected from surveys, interviews, workshops, and focus group discussions along with archival documents of the Collaboration in the form of meeting minutes and project reports.

The next section introduces the SUSFARMS® decision-making and management tool, the first step in the SUSFARMS® initiative. A description of the recently formed SUSFARMS® 2018 Collaboration, developed to improve implementation of the SUSFARMS® tool and to further adapt it to local and global sustainability challenges and market pressures, follows.

SUSFARMS®: a decision-making and farm management tool

It has been argued that better management practices for sustainability are commonly developed by scientists in isolation from farmers, resulting in low adoption rates of new practices. The SUSFARMS® journey however was a bottom-up initiative, driven proactively by the farmers in the KwaZulu-Natal Midlands area in a novel collaboration with WWF. In 1998, a group of leading commercial farmers in the Midlands cane growing area acknowledged that a system for environmental management was needed (Maher & Schulz, 2003). This grew out of a local interest in conservation and development issues, driven by strong leadership among the farmers. These early efforts lead to the farmers engaging with WWF in 2002, ultimately resulting in the development of SUSFARMS®. The SUSFARMS® tool is made up of two components (Figure 8.2). The first is the Manual, comprising farm level environmental, agronomic, social, and economic 'better practice' relevant to sugarcane production. The second component is the Progress Tracker tool that farmers use to quantify the extent of their adoption of better practices and adherence to legislation. The Progress Tracker tool also helps focus sugar industry extension services on areas where farmers are underperforming and provides the evidence base that value chain customers need to meet their sustainability commitments.

The resultant SUSFARMS® tool was officially recognised by the South African sugar industry in 2009, and has since undergone several refinements. Currently SUSFARMS® implementation by sugar farmers is voluntary, although in what has been considered a bold step, farmers in the Midlands area, with miller support, have agreed, as a group, to make it compulsory in their mill supply areas. Uptake of SUSFARMS® in other sugarcane growing regions has not been as successful as in the Midlands area, a concern for the industry and for those involved in championing this tool.

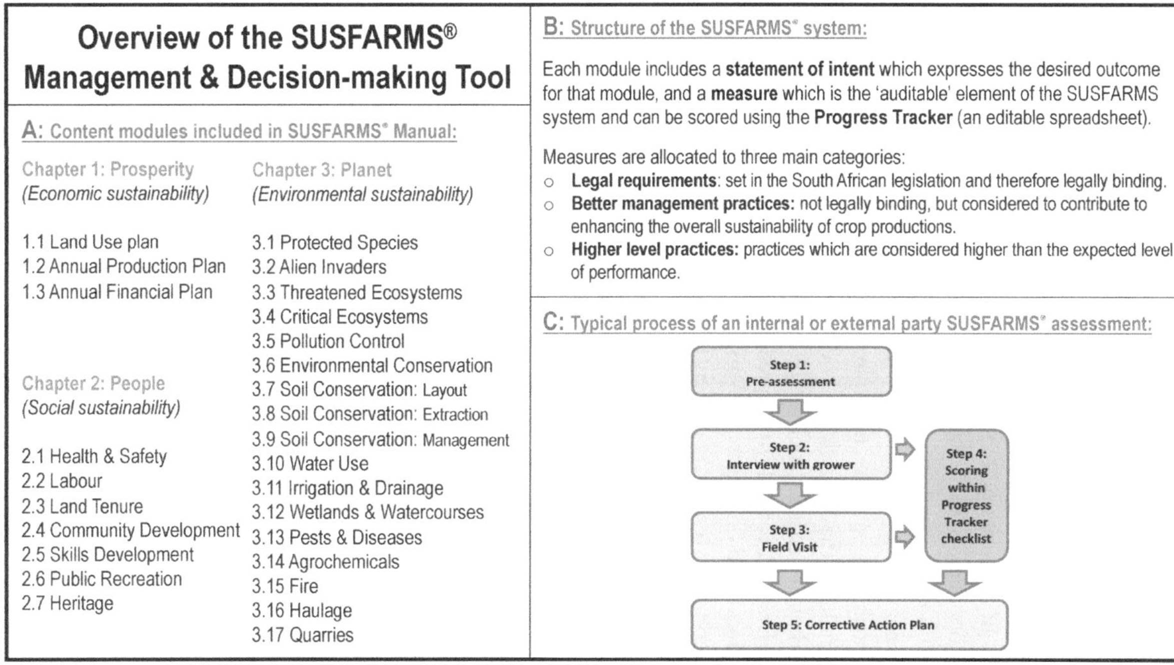

Figure 8.2 Overview of the SUSFARMS® tool: A: The content of the Manual is divided into three chapters, each with a number of modules. B: The tool is structured according to 'statements of intent', which are accompanied by measures for assessment and verification. C: A SUSFARMS® assessment follows a five-step process, and is verified through a Progress Tracker in which scores for each measure are captured.

SUSFARMS® implementation challenges: insights from a facilitated expansive learning process

Through the SUSFARMS® journey (Figure 8.3), it became evident that the sugar farmers in the Midlands are not only innovative and proactive; willing to learn and change their practices, but that there is strong leadership among the farmers and a history of working in partnerships (e.g. with WWF). However, effective implementation of the SUSFARMS® tool at farm level was considered low, even in the Midlands. A process of identifying hurdles and solutions to the implementation of SUSFARMS®, emphasising the learning process, was initiated with farmers and extension officers in 2011 by one of the authors in fulfilment of a Master's Degree in Agriculture (Koopman, 2014). Engeström's (1987, 2001, 2007) notion of expansive learning was chosen to guide this process of enquiry on the basis of two important notions: (1) that agency is gained through the process of actors meaningfully engaging with distributed expertise, multiple viewpoints, and disputed knowledge, and (2) the recognition that knowledge and the process of learning in complex multi-agency workplace settings (such as the sugar value chain) is not necessarily stable, readily understood, or made up of predefined answers.

In workshops modelled on Engeström's interventionist research methodology, participant farmers and extension staff identified six key hurdles to SUSFARMS® learning and practice which broadly occurred at two scales: (1) issues relevant to broader sugar industry operations and institutional arrangements that were beyond the immediate influence of participants (now addressed through the Collaboration), and (2) issues that are within the sphere of influence of farmers and extension officers. The hurdles to implementation of SUSFARMS® are briefly discussed below:

Lack of attention and planning to extension staff's knowledge and learning needs

This was exacerbated by the scarcity of extension skills in the current market place, augmented by many of the experienced extension staff being near retirement age.

Farmers typically ask for more practical learning opportunities from extension staff, but tend to attend in low numbers when such opportunities are offered

Participants identified several reasons for this. Firstly, research information is often not presented in an easy-to-use format and learning approaches often do not match the way farmers prefer to learn. Time shortage was also mentioned as a hurdle. Both farmers and extensionists viewed environmental sustainability learning as being 'additional' or over and above, rather than being integral or fundamental components of agronomic learning or knowledge.

Farmers' aversion to administration and paperwork

Farmers are generally resistant to participating in SUSFARMS® audits and to providing documented evidence of better practice (through the Progress Tracker Tool).

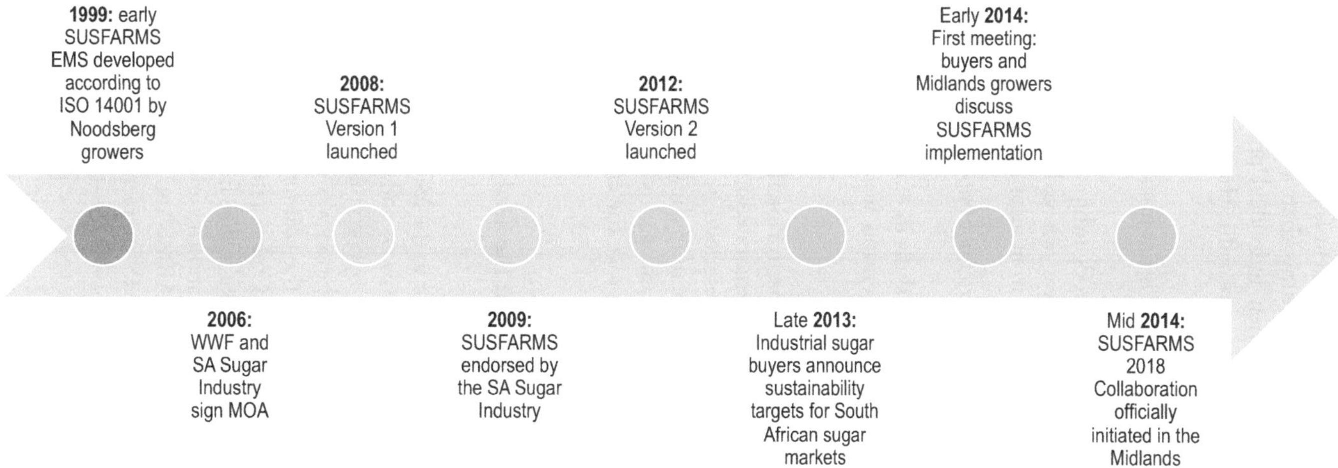

Figure 8.3 Timeline of the SUSFARMS® journey: from early activities driven by the Midlands farmers, through to the sugar industry endorsement in 2009 and most recently, the launch of the SUSFARMS® 2018 Collaboration

(EMS=Environmental Management System, ISO=International Standards Organisation, MOA=Memorandum of Agreement).

Farmers' need for practical and economic evidence before accepting change
There is a general reluctance to finance better management practices when no evidence of direct financial benefit is available. Farmers tend to worry more about their immediate returns than about long-term sustainability of their practices, or how their individual practice contributes towards landscape scale sustainability, especially when yields and incomes tend to be under pressure.

Strategic leadership on environmental issues from top industry structures was not evident to farmers on the ground
The message of environmental custodianship and support for SUSFARMS® and the pressure to implement was received from 'outsiders' (e.g. WWF), rather than by extension services from sugar industry structures.

Limitations inherent in the current extension model and farmers' need for more frequent extension contact
In addition to the lack of relevant skills in extension staff, participants also felt the efficacy of industry extension was hampered by structural and budgetary constraints.

This study underlined some of the institutional hurdles to the adoption of SUSFARMS®. These hurdles were incorporated into the participatory planning and strategising of the Collaboration described in the next section, and formed part of the motivation for the development of the Collaboration, particularly for WWF, who were involved in the study and wanted to see progress on the implementation of practices which would improve the environmental sustainability of sugarcane farming.

The SUSFARMS® 2018 Collaboration: the need for a novel governance platform

When the largest industrial buyer of sugar in the Midlands expressed an interest in engaging with WWF and the Midlands growers and millers on sourcing responsibly produced sugar from the Midlands, the opportunity for broad supply chain collaboration was recognised. The Collaboration was initiated at a workshop in the Midlands, in June 2014, and set 2018 as the year for reaching sustainability objectives and targets, hence the name. The Collaboration includes miller and farmer stakeholders from the Midlands area, stakeholders from the sugarcane supply chain and industry, as well as a social NGO and the environmental NGO, WWF. In recognition of sluggish uptake of SUSFARMS® at farm level, the purpose of the Collaboration is to facilitate SUSFARMS® implementation and verification on Midlands farms to achieve two broad aims: (1) to satisfy local grower commitments to using SUSFARMS® as a management tool to enhance responsible production, efficiency, and legal compliance, and (2) to help Midlands farmers and millers reach responsible practice targets agreed to with their buyers of sugar. The Collaboration is intended as a pilot for the implementation of SUSFARMS® across the

South African sugar industry. The Collaboration arose out of the main industrial buyer of Midlands sugar's announcement of specific volumetric targets for sustainably produced sugar in late 2013.

Participants in the Collaboration committed to an open, transparent, and participatory process based on good governance principles. In the early stages of the collaborative process, time was taken to develop a shared understanding of all participants' perceptions of sustainability and responsible practice, and their visions for the Collaboration. This shared vision was used to guide development of the objectives of the Collaboration (Box 8.1). It was recognised that several on-going South African sugar industry initiatives aligned with the goals of the Collaboration. These include the on-going updating of the SUSFARMS® Manual and Progress Tracker tool by the South African sugar industry, the development of an adapted isiZulu (local language) version of SUSFARMS® for small-scale farmers and the development of sustainability functions of senior managers within the milling companies.

Box 8.1 Jointly-developed objectives of the SUSFARMS® 2018 collaboration

1. Ensure farm-level verification of SUSFARMS® Progress Trackers.
2. Benchmark and align SUSFARMS® to other relevant systems, e.g. Bonsucro Production Standard.
3. Improve the implementation of SUSFARMS® Progress Trackers: fast-track implementation, ensure compliance and develop a system for data capture and analysis.
4. Support local skills development for land use planning to support SUSFARMS® implementation.
5. Customise SUSFARMS® for small-scale farmers.
6. Strengthen support systems and learning opportunities to enable the implementation of best management practices.
7. Ensure that governance systems and resources are in place for long-term implementation of SUSFARMS 2018 objectives.

Challenges in building the collaboration

The implementation of SUSFARMS® by farmers demands an unprecedented level of integrated action on the part of traditionally competing actors in the sugar supply chain. Key among these in the Midlands are the main industrial buyers of refined sugar, cane farmers, the South African sugar industry, and millers. The process of gathering all the relevant stakeholders together into a single, collaborative group was not without challenges. Keeping the group focused on understanding and discussing the risks and benefits of shared action, setting work plans, and deciding on joint activities remains a challenge.

One of the key obstacles was the issue of infringing on the functions or mandates of other formally recognised institutions within the South African sugar industry. For example, the industry has a natural resource manager, a knowledge manager, and an extension manager, all three of whom are mandated to work together on the development and implementation of SUSFARMS® – which overlaps directly with the objectives of the Collaboration. To reduce the perception that the Collaboration was taking over these mandates, all three managers were included in the Collaboration from the start (once initial discussions between farmers and buyers had affirmed the need to include senior sugar industry representatives). It was clarified that the Collaboration sought to supplement and support rather than to challenge or override existing mandates. A second obstacle was the tension between producers and buyers around the notion of rewarding responsible practice with price premiums. Farmers felt that implementing responsible practices often comes at a cost, and they expected to be compensated for this with a price premium. This sensitive issue was discussed openly in meetings of the Collaboration, and the buyers were clear from the start that premiums would not be a given upfront. Investment from the buyers was initially focused on providing the technical support needed for implementation rather than rewarding farmers with a premium. Thirdly, farmers repeatedly called for additional capacity and resources to deal with responsible practice issues and the implementation of SUSFARMS® at the local level and felt this was not forthcoming from the South African sugar industry or sugar buyers.

An enabling condition that reduced conflicts around the development of this platform, was the long-standing relationship, built on trust, between WWF and the various industry bodies and individuals responsible for sustainability issues and SUSFARMS® implementation. This ensured on-going and open communication about the Collaboration. WWF were selected early in the process of setting up the Collaboration as the 'most neutral' partner to facilitate linkages between other partners. In this way, WWF acted as a trusted intermediary or 'bridging organisation' in setting up the Collaboration, a role recognised as important in innovative governance approaches to complex natural resource management challenges (Olsson et al., 2006). A final factor that may have contributed to the relatively smooth process by which the Collaboration was set up was the firm commitment from the main sugar industrial client to support openly the Collaboration and act as a broker with higher-level sugar industry bodies on linking sustainability and procurement issues.

The SUSFARMS® 2018 Collaboration as a case of institutional bricolage

Institutional bricolage is a concept that has emerged out of the field of critical institutionalism (Cleaver & de Koning, 2015). It can be defined as 'the ability of people to creatively combine old and new institutions which appropriately serve their purposes' (Van Oosten, 2013). Institutional bricolage is different from institutional change, in that it focuses on the human agency of the 'bricoleurs', i.e.

the ability of people to actively change institutions themselves to suit a particular purpose or in response to changing needs (Cleaver, 2001; Van Oosten, 2013; Cleaver & de Koning, 2015).

The concept of institutional bricolage has been applied as a means of understanding a variety of complex natural resource governance challenges and the role of local actors in addressing these. For example: communal use of grazing land and water resources in the Usangu Basin in Tanzania is seen as a socially-embedded practice and institutional bricolage (Cleaver, 2002; Cleaver, 2012); agricultural water governance reforms at the local level in Kyrgyzstan and Tajikistan as institutional bricolage influenced by path dependencies (Sehring, 2009); local adaptation of imposed political and economic institutions by the Serer of the Siin region of Senegal as bricolage through the interweaving and transformation of informal and formal institutions (Galvan, 2004); and community forest management in the Amazon in Bolivia and Ecuador, which illustrates that institutions in practice work differently than imagined through local people's agency as bricoleurs (De Koning, 2014). These cases illustrate how institutional bricolage happens by the creation of new institutions, which are a dynamic mixture and a messy patching of the 'modern' and 'traditional', 'formal' and 'informal' (Cleaver, 2002) and can be likened more to 'institutional DIY' than engineering or design (Cleaver, 2012).

In a critical institutionalist approach to understanding institutional bricolage, actors are seen as able to 'strategize, innovate and negotiate in their engagement with institutions and management of natural resources', with agency being shaped by power dynamics (Cleaver & de Koning, 2015, p. 8). Such power dynamics include the societal allocation of resources, the power inherent in social norms, rules and functioning, and the power to challenge and subvert current institutional structures (Cleaver & de Koning, 2015). In the case of SUSFARMS®, both the development of the tool as a bottom-up initiative, and the subsequent building of the SUSFARMS® 2018 Collaboration, will be examined through the lens of institutional bricolage by paying particular attention to human agency of the local farmers in the Midlands as 'bricoleurs' and how they have interacted with the power dynamics inherent in the formal institutional structures of the South African sugar industry.

We propose that the mixing of formal standards institutions at the global level (e.g. Bonsucro certification) with local level governance systems, i.e. local farmers' associations and the SUSFARMS® 2018 Collaboration, is a form of institutional bricolage. The development of the SUSFARMS® tool and the Collaboration can be considered to perform institutional bricolage as outlined in the next section.

Development of the SUSFARMS® tool at the local farmer group level

The development of the SUSFARMS® tool in the Midlands area, which is now endorsed across the South African sugar industry as the standard for sustainable farming, facilitated and funded by WWF, initially went strongly against industry norms of practice, which were focused on profitability as the key outcome of farming. WWF drew on the theory of expansive social learning, which is congruent with the notions of institutional bricolage, to guide the process of developing SUSFARMS® as a new institution governing natural resource management in sugarcane farming. Through this facilitated learning process, particular attention was paid to building human agency at the local level by enabling access to additional knowledge resources and social networks of local farmers who were thus empowered to engage with global bodies such as Bonsucro and push the adoption of SUSFARMS® at higher industry levels. In this way, the local actors were able to engage with powerful players in the industry and bring about institutional change e.g. through getting SUSFARMS® endorsed industry-wide.

Development of the SUSFARMS® 2018 Collaboration: building a collaborative platform to further develop the SUSFARMS® tool and improve implementation

Although the South African sugar industry already has an organisational structure and shows significant, formal institutional development (as previously described), the Collaboration was seen as a necessary novel platform for collaboration between global sugar buyers and local sugar farmers. This may be understood as a move by farmers in the Midlands area to circumvent the control that the South African sugar industry holds over sugar markets through the formal institutions set up by the Sugar Act (Government of South Africa, 1978). Thus, it is an institutional innovation within the sugar industry that shows a multi-scale governance structure, as participants from local, to regional, and global scales are equally involved. We argue that this self-organisation and re-arrangement of existing institutions in order to respond to sustainability challenges emerging in the sugar industry, from the local to the global level, is an example of institutional bricolage. Indeed, such multi-scale engagement with institutional change processes is a key characteristic of institutional bricolage (Cleaver & de Koning, 2015).

Shifting power to re-focus attention and resources on local-level sustainability challenges

Through the SUSFRAMS 2018 Collaboration, local Midlands farmers have been able to reach across institutionalised, hierarchical power structures in the sugar industry to bring attention to the needs of farmers at the local level for increased support towards implementing the SUSFARMS® tool. This was done by bringing senior industry role players (such as the South African sugar industry natural resources manager, the knowledge manager etc.) around the table to voice their concerns about the lack of financial and human resources dedicated to local farmer groups for the implementation of SUSFARMS®. This is an example of bricolage as it seeks to re-direct societal allocation of resources to the needs of local resource managers (farmers) through strategising and negotiations.

Shifting SUSFARMS® from a voluntary standard to a locally-agreed upon compulsory process

At industry level, SUSFARMS® is still a voluntary standard, and farmers are free to choose whether they want to implement SUSFARMS® and complete a Progress Tracker. However, the farmers from the Eston, Dalton, and Midlands supply areas have taken the unprecedented step of agreeing that the Progress Tracker submission be compulsory. This is a signal from the leadership in the area that they are committed to the implementation of sustainable practices and are willing to collaborate with the requirements of sugar buyers to produce a sustainable product following agreed-upon standards. Local millers are enforcing this arrangement (they will not accept a load of sugarcane into the mill if the farmer has not submitted a Progress Tracker). To date, all three mill-supply areas have had 100 percent submission of Progress Trackers. Of course, completion and submission of a Progress Tracker is no guarantee of compliance to SUSFARMS®, but it does indicate a formative commitment from farmers and millers to begin implementing SUSFARMS®. This has required an adaptation of established institutions (i.e. compliance with all requirements of the miller before cane delivered to the mill will be accepted) to take this new approach on board.

Lessons for novel natural resource governance initiatives

The SUSFARMS® case study has demonstrated how a group of stakeholders have worked together over long periods of time to modify and adapt the SUSFARMS® tool and develop an enabling institutional platform to improve the sustainability of sugarcane farms. This was not a simple, linear process: it required a mixture of a special set of enabling conditions (described in the lessons that follow) with a mediated expansive social learning process to build collective agency for people to make a difference in local governance arrangements to accommodate sustainability issues.

The SUSFARMS® initiative (setting up the tool itself and building the Collaboration) provides three lessons for the success of resource governance initiatives more broadly.

The important role of a trusted intermediary

Bringing the right people around the table may require a trusted intermediary or 'broker' to navigate the complex interactions between people and organisations that would not traditionally have worked together or trusted each other (Long, Cunningham, & Braithwaite, 2013; Olsson et al., 2006; Westley et al., 2013). In the case of the SUSFARMS® Collaboration, this role was played by WWF and the main sugar industrial client, who helped build links, establish long-term relationships and build trust amongst the diverse and otherwise disconnected stakeholders.

Taking advantage of 'windows of opportunity'

Making use of 'windows of opportunity' is critical in effecting sustainable governance interventions. In this case, changing market pressures from consumers seeking assurance on sustainability of products was one of the drivers for the sugar industrial clients developing their 2020 sustainability targets, driving them to seek sources of 'sustainable sugar' at the local producer level. Windows of opportunity, when they present themselves, can be effectively leveraged if work has already been done 'on the ground' beforehand. The participation by local farmers in the WWF-facilitated development process of the SUSFARMS® tool and more recently in the Collaboration, heightened these farmers' awareness and understanding of local and global sustainability concerns and some of the multi-stakeholder processes used to address them. This meant they were better prepared for engaging global players such as Bonsucro, the industrial sugar clients, and WWF in a multi-stakeholder process that gave rise to the Collaboration when the time was appropriate. They could also share their learning from such global collaborations with other local farmers and role-players to impress upon them the importance of being involved in crafting their own sustainability responses. Preparedness to be able to take advantage of windows of opportunity, through the enactment of human agency, is as important as knowing when that correct window has arrived in order to grasp an opportunity (Olsson, Folke, & Hahn, 2004; Westley et al., 2011).

Commitment, leadership, and social cohesion at the local level

Through their involvement in developing the SUSFARMS® tool and more recently, the Collaboration, the farmers and millers in the Midlands emerged as leaders in sustainability practices in the South African sugar industry and can be considered 'institutional entrepreneurs' (See Westley et al., 2011) or champions. Strong leadership and a sense of community among these groups seems to have contributed significantly to the success of the initiatives, as the local stakeholders were already unified around issues of sustainability and development and were thus 'ready' when the international players approached them for collaboration.

Human agency in action: using expansive learning approaches to support institutional bricolage

The SUSFARMS® initiative has shown that by engaging multiple internal and external stakeholders, using a mediated expansive learning approach to guide the process, local actors' knowledge and human agency was strengthened. This in turn enabled and sustained the bringing together of a diversity of stakeholders around the table to re-model institutional relationships, co-construct objectives and activities, to leverage additional resources and to challenge institutionalised power dynamics towards achieving their sustainability commitments. However, it leaves open the question of some remaining sustainability challenges in the sugar industry that may necessitate more than novel governance structures to address adequately (e.g. land restitution issues, concerns about the conditions of farm workers' conditions, meaningful agrarian reform, and empowerment of small-scale growers).

Critically assessing the contribution of the SUSFARMS® 2018 Collaboration to sustainable natural resource governance

As mentioned in the introduction to the previous case study, two of the authors have been involved closely with this case over a number of years as facilitating practitioners and researchers. This raises questions about biased perspectives about the case; thus a critical assessment of the case is necessary. The development of the SUSFARMS® tool and the Collaboration together hold significant promise for addressing the sustainability challenges of the South African sugar industry at the farm level, including addressing concerns about the environmental impacts of sugarcane and questions around social development in rural contexts. This is evidenced by the commitment shown by local, regional and global stakeholders to participate in the Collaboration. Furthermore, lessons of the SUSFARMS® case may well hold broader relevance to other agricultural sectors in South Africa that face similar institutional reform and sustainability challenges. Even though the SUSFARMS® tool and the Collaboration are not explicitly designed to deal with the broader rural development challenge in the sugar industry, we consider here some of the potential drawbacks for sustainability that may emerge from further development of the sugar sector, and some of the shortcomings of the SUSFARMS® initiative to date.

What are the shortcomings of the SUSFARMS® initiative?

- It can not guarantee that all farmers will implement all guidelines provided equally.
- It has not guaranteed environmental and social sustainability of sugarcane farming in SA more broadly because of its limited geoagrphic implementation in the KZN Midlands.
- It has not addressed issues pertaining to land redistribution, although this is being addressed elsewhere in the South African sugar industry.
- It has yet to be adapted to suit the needs of small-scale farmers, and benefits to small-scale farmers in improving their viability and their role as significant decision-makers in the industry, are not yet apparent.
- It does not interrogate the meaning of sustainability through the lens of public health and the food system at large. In this sense, SUSFARMS® still promotes 'Business as Usual' in the sugar industry, despite growing concerns about the health-related impacts of sugary diets.

What has the SUSFARMS® initiative done?

- It has clearly laid out the environmental, social, and economic sustainability challenges facing sugarcane farming (both from a best practice and a legal perspective), and has provided guidelines for farmers, primarily large-scale commercial farmers, on how to address these. By endorsing this standard across the industry, and communicating this to all farmers, the South African sugar industry has ensured that no farmer can claim ignorance.

- It has provided a tool for farmers to monitor their compliance not only with best practice guidelines for environmental, social, and economic sustainability practices, but also with the relevant South African law (through the Progress Tracker).
- It has provided a mechanism and an institutional tool for the implementation, compliance monitoring, and assurance of sustainable sugarcane farming, which will improve access to possible 'niche markets' for sustainably produced sugar from compliant farmers and millers.
- It has shifted power dynamics between local farmers and senior industry actors through the development of the Collaboration, which brings stakeholders who would otherwise be separated by hierarchical, institutional barriers, together.

The impact of SUSFARMS® is yet to be identified in terms of how it will address the sustainable development challenges facing the industry, particularly concerns relating to land restitution and meaningful empowerment of small-scale growers (Dubb, Scoones, & Woodhouse, 2016; James & Woodhouse, 2016) and working conditions for farm labourers (Castel-Branco, 2012). Furthermore, the relevance of SUSFARMS® for smallholder sugar farmers, who only supply 9 percent of sugar in the country, but who are nevertheless important actors in agrarian reform in South Africa, needs to be interrogated. The diversity of farmers' knowledge and skills must be taken into account, including traditional knowledge systems. If smallholder farmers are to be fully incorporated into the SUSFARMS® model, the adaptation of the SUSFARMS® tool to their contexts should receive high priority in order to be relevant and accessible. A further extension of the bricolage process would be to incorporate these forms of knowledge into the tool, thereby providing a more inclusive and flexible platform.

Conclusion

The SUSFARMS® tool and the SUSFARMS® 2018 Collaboration have been developed through research and multi-stakeholder engagement. They provide a comprehensive set of guidelines, along with a collaborative governance platform, for sugarcane farmers to improve the sustainability of their farming operations. However, unless farmers throughout the industry actually implement guidelines and are held to account, SUSFARMS® will remain a manual on the bookshelf for the majority and will not bring about the necessary change in the sugar industry.

Thus, continuing to achieve the goals of the Collaboration to improve implementation and further adapt SUSFARMS® to local sustainability challenges and global market pressures is crucial. The Midlands farmers have already showed strong, visionary collective leadership by engaging with multiple stakeholders both within, and outside, the industry to shift the SUSFARMS® tool firstly into being widely recognised by local industry and international role players. Secondly, it has changed from being a voluntary measure to a requirement for farmers to send their cane to the mill and thirdly, by satisfying their main sugar buyer's need for responsibly produced sugar. The tool and Collaboration are examples of institutional bricolage at the local level.

To conclude, in terms of the lessons that can be learnt from the Collaboration for other natural resource governance initiatives, it is critical to appreciate that the process of achieving the Collaboration was not simple nor linear: it required a combination of a special set of enabling conditions with a mediated expansive social learning process to build agency of people around a table towards collectively making a difference. The value of a trusted intermediary, the importance of being ready to make the most of 'windows of opportunity' and the significance of human agency along with social cohesion, commitment, and strong leadership at the local level to bring about enabling conditions for change have been illustrated in this case. Most importantly, overcoming the messiness of wicked problems requires a diverse set of inputs that no one actor can do alone.

References

Ahmad, N. B. (2014). The international sugar trade and sustainable development: Curtailing the sugar rush, 39 N.C. *Journal of International Law & Commerce*. Reg. 675, Retrieved from http://digitalcommons.pace.edu/lawfaculty/968/.

Binedell, M., Cockburn, J., & Koopman, V. (2014). Midlands sustainable sugar supply Chain collaboration. *South African Sugar Journal*, 98(4), 86–87.

Bonsucro (2015). Bonsucro Production Standard V 4.1.1. Better Sugarcane Initiative Ltd., London. Retrieved from http://bonsucro.com/site/wp-content/uploads/2013/02/Bonsucro-Production-Standard-4.1.1.pdf.

Castel-Branco, R. (2012). The dilemma of growing sugarcane in KwaZulu-Natal. Retrieved from www.theafricareport.com/Soapbox/the-dilemma-of-growing-sugar-cane-in-kwazulu-natal.html.

Cheesman, O. (2004). *Environmental Impacts of Sugar Production*. Surrey, UK: CABI Publishing.

Cleaver, F. (2001). Institutional bricolage, conflict and cooperation in usangu, Tanzania. *IDS Bulletin*, 32(4), 26–35.

Cleaver, F. (2002). Reinventing institutions: Bricolage and the social embeddedness of natural resource management. *The European Journal of Development Research*, 14, 11–30.

Cleaver, F. (2012). *Development Through Bricolage: Rethinking Institutions for Natural Resource Management*, Oxford: Earthscan.

Cleaver, F., & de Koning, J. (2015). Furthering critical institutionalism. *International Journal of the Commons*, 9(1), 1–18.

Cockburn, J. (2013). *Implementation of the push-pull strategy for Eldana saccharina control on sugarcane in KwaZulu-Natal, South Africa*. Unpublished Masters Thesis. Potchefstroom: North-West University.

de Koning, J. (2014). Unpredictable outcomes in forestry: Governance institutions in practice. *Society & Natural Resources*, *27*, 358–371.

Dubb, A., Scoones, I., & Woodhouse, P. (2017). The political economy of sugar in Southern Africa: Introduction. *Journal of Southern African Studies*, 43(3):447–70.

Engeström, Y. (1987). *Learning by Expanding: An Activity Theoretical Approach to Developmental Research*. Helsinki, Finland: Orieanta-Konsultit.

Engeström, Y. (2001). Expansive learning at work: Toward an activity theoretical reconceptualization. *Journal of Education and Work*, *14*(1), 133–156.

Engeström, Y. (2007). Enriching the theory of expansive learning: Lessons from journeys toward coconfiguration. *Mind, Culture, and Activity*, *14*(1–2), 23–39.

Galvan, D. (2004). *The State Must Be Our Master of Fire. How Peasants Craft Culturally Sustainable Development in Senegal*. Berkeley: University of California Press.

Government of South Africa (1978). Sugar Act 9 of 1978. Pretoria, South Africa: Government of South Africa. Retrieved from www.gov.za/documents/sugar-act-7-apr-2015-1159.

James, P., & Woodhouse, P. (2017). Crisis and differentiation among small-scale sugar cane growers in Nkomazi, South Africa. *Journal of Southern African Studies*, 43(3):535–549

Jenkins, B., Baptista, P., & Porth, M. (2015). *Collaborating for Change in Sugar Production: Building Blocks for Sustainability at Scale*. Cambridge, MA: The CSR Initiative at the Harvard Kennedy School and Business Fights Poverty.

Koopman, V. (2014). *Exploring expansive learning in sustainable agriculture: A case study of commercial sugarcane farmers and extensionists in KwaZulu-Natal Midlands and South Coast*. Unpublished Masters Thesis. Pietermaritzburg: University of KwaZulu-Natal.

Long, J., Cunningham, F., & Braithwaite, J. (2013). Bridges, brokers and boundary spanners in collaborative networks: A systematic review. *BMC Health Services Research*, *13*, 1–13.

Maher, G. W., & Schulz, L. (2003). An environmental management system for sugarcane in the Noodsberg area of South Africa. *Proceedings of the South African Sugar Technologists Association*, *7*, 189–195.

Martiniello, G. (2015). *'Don't stop the mill': South African sugar, agrarian change and outgrowers adverse incorporation in the Kilombero Valley, Tanzania*. Retrieved from www.plaas.org.za/bicas: BRICS Initiative for Critical Agrarian Studies (BICAS).

Moura, P. T., & Chaddad, F. R. (2012). Collective action and the governance of multistakeholder initiatives: A case study of Bonsucro. *Journal on Chain and Network Science*, *12*(1), 13–24.

Olsson, P., Folke, C., & Hahn, T. (2004). Social-ecological transformation for ecosystem management: The development of adaptive co-management of a Wetland Landscape in Southern Sweden. *Ecology and Society*, 9(4), 2.

Olsson, P., Gunderson, L. H., Carpenter, S. R., Ryan, P., Lebel, L., Folke, C., & Holling, C. S. (2006). Shooting the rapids: Navigating transitions to adaptive governance of social-ecological systems. *Ecology and Society*, *11*(1), 18.

Oxfam (2013). Sugar Rush: Oxfam briefing note. Retrieved from www.oxfam.org/en/grow/policy/sugar-rush.

SASA (2012). *SUSFARMS®: Sustainable sugarcane farm Management System* (2nd ed.). Mount Edgecombe: South African Sugar Association.

SASA (2015). *Sugar Industry Directory 2014–2015*. Mount Edgecombe: South African Sugar Association.

Sehring, J., (2009). Path dependencies and institutional bricolage in post-Soviet water governance. *Water Alternatives*, *2*(1), 61–81.

Stedman-Edwards, P. (2004). Review of commodities: Rice, cane sugar, and cotton. Retrieved from http://wwf.panda.org/about_our_earth/all_publications/index.cfm?14553/From-Negotiations-to-AdjustmentReview-of-Commodities-Rice-cane-sugar-and-cotton.

Van Oosten, C. (2013). Forest landscape restoration: Who decides? A governance approach to forest landscape restoration. *Natureza & Conservação, 1*(2), 119–126.

Visser, M., & Ferrer, S. (2015). Farm workers' living and working conditions in South Africa: Key trends, emergent issues, and underlying and structural problems, Report commissioned by the pretoria office of the international labour organization.

Weiss, T. (2007). *The Global Food Economy: The Battle for the Future of Farming*. London: Zed Books.

Westley, F., Olsson, P., Folke, C., Homer-Dixon, T., Vredenburg, H., Loorbach, D., Thompson, J., Nilsson, M., Lambin, E., Senddzimir, J., Bannerjee, B., Galaza, V., & van der Leeuw, S. (2011). Tipping toward sustainability: Emerging pathways of transformation. *Ambio, 40*(7), 762–780. doi:10.1007/s13280-011-0186-9.

Westley, F. R., Tjornbo, O., Schultz, L., Olsson, P., Folke, C., Crona, B., & Bodin, Ö. (2013). A theory of transformative agency in linked social-ecological systems. *Ecology and Society, 18*, 27.

WWF (2005). Sugar and the environment. Zeist, Netherlands. Retrieved from http://awsassets.panda.org/downloads/sugarandtheenvironment_fidq.pdf: World Wildlife Fund: WWF Global Freshwater Programme.

9 Corporate water risk and return

Alex Money

Introduction

Shortage of water infrastructure is a critical juncture for the management of nexus resources including food and energy. According to McKinsey, the world's estimated need for water infrastructure investment between 2013 and 2030 is US$ 11.7 trillion (Dobbs & Pohl, 2013), rising linearly with GDP and population growth. Meanwhile, the OECD projects that the average annual world infrastructure expenditure on water between 2020 and 2030 will need to be US$ 1,037 billion, or 1.03 percent of world GDP, and more than the combined expenditure on road, rail, telecoms, and electricity, over the same period (OECD, 2006). Financing this investment in the post financial crisis environment presents particular difficulties, given the deterioration in many public sector balance sheets (O'Brien & Keith, 2009), increased risk aversion by lenders and financial intermediaries, and the relatively small proportion of institutional asset allocation to infrastructure investment from the private sector. On the other hand, this proportion of institutional capital allocation is increasing as investors diversify their portfolios (Clark, 2000; Hagerman, Clark, & Hebb, 2007; Hebb, 2006, 2007), and interesting new models for financing urban infrastructure based on profit-driven motivations rather than the creation of public goods are emerging (e.g. Hebb & Sharma, 2014). This chapter builds on the literature in this area to propose how and why profit-motivated firms might finance substantial investments in water-related infrastructure via financial products that are a hybrid between market and hierarchy (i.e. firm), even where the firm may not be the primary beneficiary of these investments.

A recent report claimed that unpriced natural capital cost the global economy over US$7 trillion per annum, 13 percent of nominal GDP (Trucost Plc, 2013), with US$2 trillion accruing to water consumption. Natural capital accounting is one of several approaches being used to further the notion that current rates of corporate water use are unsustainable. The argument is that water use presents a large, potentially uncapped liability to firms and their shareholders as a consequence of physical, reputational, regulatory, and litigation risk (Barton, 2010) affecting a company's licence to operate (Sarni, 2011). The corporate water risk concept, emphasises, inter alia, the lack of adequate infrastructure, uncertainties related to populist politics, inadequate information flows and weak regulatory frameworks, and the variability and unpredictability of water demand due to demographics, urbanisation, and climate variability (Hope & Rouse, 2013). Approaches used to give salience to the concept of corporate water risk include surveys such as those by the World Economic Forum, which has consistently ranked water supply crises as one of the business world's top five risks (Howell, 2013). In addition, heuristic devices have been developed by various third parties with the objective of helping companies model, manage, and disclose their water risk (e.g. Global Reporting Initiative, 2011; WBCSD, 2013; World Resources Institute, 2010).

Despite all this, the mechanisms that exist for firms to identify, evaluate, mitigate, and govern the risks they face in ensuring sustainable access to water supplies and services remain at best embryonic. Corporate water use is characterised by resource dependence (a function of reliance on a scare resource that cannot be substituted) and organisational interdependence (a function of complex supply chains (Money, 2014a)), and the disclosure by firms of the risks they face is generally narrow in scope and generic in content. This notwithstanding, there is an assumption amongst key stakeholders such as a firm's shareholders, that sufficiently robust internal processes exist within firms to ensure these risks are managed effectively. These stakeholders often assume that the lack of disclosure by firms is more a function of managers' unwillingness to share what they perceive to be commercially sensitive information (Money, 2014b). These are assumptions that derive from a sense of 'tacit knowledge' (Gertler, 2003), which influences the relationship between shareholders and firms. Further, there are few pure market structures for the intermediation of water supply and services. In part, this is likely a function of the unique social, cultural, and economic attributes associated with water use, including human rights to water that have been developed within the literature (e.g. McCaffrey, 1992) and ratified under international law (United Nations General Assembly, 2010). In short, it is often the case that senior executives at many firms do not have substantive plans to manage future water-related risks, and equally that many shareholders in these firms will unconsciously accept this situation as the status quo (Money, 2014a, 2014b).

That said, it is clear that the issue of corporate water risk *is* salient to some shareholders. For example, Norges Bank Investment Management (NBIM), one of the world's largest sovereign wealth funds, sets out in writing what it expects from portfolio companies in terms of water management. Equally, corporate water risk is undoubtedly highly salient to senior management at many firms. However, without an effective framework to evaluate water risk issues, responses are symbolic rather than substantive (Money, 2014c), and risk being interpreted as 'blue-washing', or a mere public relations exercise.

In this chapter I apply transaction cost theory, resource dependency theory, and real options theory to propose a framework based on the use of hybrid governance structures that intermediates this gap between symbolism and substance. Hybrid governance structures in this context include equity investment (e.g. a joint venture between a firm and a utility), long-term binding contracts, routines for mutual monitoring, and prearranged procedures for third-party mediation or arbitration in the event of a dispute, without immediate recourse to a court of law. The hybrid/alternative structure may involve many more actors than the buyer and seller of the service; it might include, for example, other water consumers, regulators, government agencies, taxpayers, civic society groups, or NGOs.

The purpose of this chapter is to provide a conception of economic return that goes beyond the usual financial parameters that are applied to decide whether a project (such as a water infrastructure plant) is built or not. I therefore propose the idea of 'corporate water return' as a corollary to corporate water risk. I argue that firms may take actions that are outside the frame of business as usual if the action is predicated on a competitive advantage that the firm enjoys and if it provides shared net benefits both to the company and its external stakeholders. The paper offers its five-factor framework (5FF) to describe this process, using as examples the water infrastructure deficit challenge outlined previously, and the concept of corporate water return. This chapter also offers a robust justification for *managing*, not only measuring and reporting, water risk. If such concepts can gain traction it offers hope that private sector capital can play a more meaningful role in achieving long-term sustainable growth, not only with relation to water but more broadly to other types of infrastructure, including those related to energy and food production.

The chapter is conceptual rather than empirical in its framing – the implications of the outcomes proposed apply at global scale. It applies transaction cost theory to frame what actions are typically conducted by the firm, and to help identify the circumstances in which transactions take place within or outside a firm's governance structure. The construct of corporate water return is also developed in this chapter, along with a series of propositions to understand corporate water risk and corporate water return through the theoretical lenses presented in the second section. The third section sets out the five-factor framework, using the exemplar of water infrastructure to illustrate the framework. The final section includes a discussion of the framework's ability to catalyse change in the financing of sustainable infrastructure and its limitations, areas for future research, and conclusion.

Conceptual foundations: transaction cost theory, resource dependency theory, and real options theory

Transaction cost theory emerged in the twentieth century in response to the question of why firms even existed if market transactions were the most efficient basis to organise economic activity; and given that firms did exist, on what basis did managers decide which activities to organise within firms, and which to leave to the market (Crook, Combs, Ketchen, & Aguinis, 2012). Oliver Williamson, who formalised what has become transaction cost theory (TCT) (Williamson, 1975, 1979, 1985), defined transaction costs as expenses that are distinct from the cost of production, and arise from identifying qualified exchange partners, negotiating contracts, creating dispute resolution mechanisms, monitoring performance,

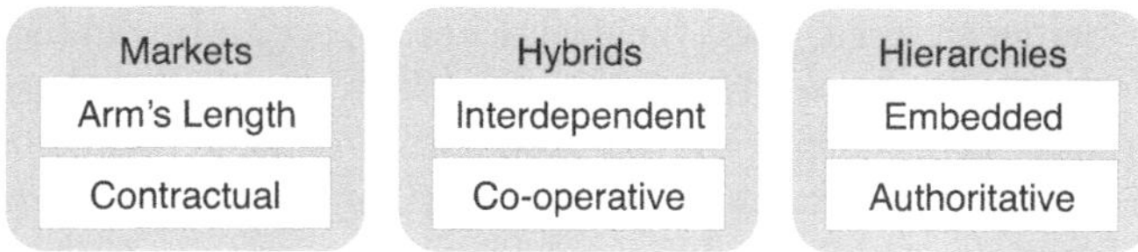

Figure 9.1 TCT: structures for transactions

and adapting to changing conditions (Williamson, 1991). TCT proposes that transactions across workgroups can be organised into one of three structural alternatives: markets, hybrids, or hierarchies; from which managers should select in order to minimise transaction costs (Williamson, 1991; See Figure 9.1). The costs themselves arise due to imperfect information that restricts the identification of qualified exchange partners, price discovery, and the ability to anticipate future contingencies (Simon, 1945). This creates the potential for maladaption (Williamson, 1999), where exchange partners may be unwilling or unable to adapt to changing circumstances (Klein, Crawford, & Alchian, 1978).

In terms of the structural alternatives, market transactions are simple arms' length transactions between buyers of goods and services, and sellers. The price and quantity are market determined, and the transaction may be supported by a formal contract. Hybrid transactions involve two or more exchange partners, but typically involve long-term, greater-than-market commitments such as cooperation agreements, franchises, development partnerships, and joint ventures (Kale & Sing, 2009; Siegel & Zervos, 2002). In hierarchies, transactions that could take place across many exchange partners instead take place within one firm. TCT presents markets and hierarchies as polar structural alternatives, as defined in terms of ownership autonomy, incentive intensity, administrative controls, and adaptation (Williamson, 1985, 1991). In summary, TCT posits that a relationship exists between the structural alternatives of markets, hybrids and hierarchies that managers choose from, in response to the transactional attributes of asset specificity, uncertainty and frequency (Williamson, 1985). It provides a framework that can explain changes in the level of integration in response to perceived exchange hazard and adaptation risk.

In the past half century, the predictive boundaries of TCT have been variously challenged. For example, TCT and resource-based theory view assets differently. Resource-based theory locates the value and competitive advantage of a firm in its resources and capabilities, including physical assets and the human capability to best use these assets (Barney, 1991). Under TCT, high asset *specificity*, the unique value of an investment only for a particular transaction, rather than an investment, results in market failure. As a result, managers move toward hierarchical structures to reduce exchange hazards. Under resource-based theory, managers bundle *strategic* assets, those assets essential for the firm's future value creation within a hierarchy in order to create and unlock value. However, specific assets and strategic assets are not equivalents and there has been little empirical examination of whether strategic assets have a greater impact on hierarchical integration than assets that are merely specific[1] (Crook et al., 2012). It has thus been suggested that asset specificity, the 'big locomotive' of TCT (Williamson, 1985, p. 56) is less useful in predicting integration than the resource-based view, which considers assets to be strategic if they are valuable, rare, and not easily substituted. This apparent relationship is of particular interest in the context of a firm's use of water resources and services, and will be revisited shortly.

A more direct challenge to TCT comes from real options theory, which assumes a different relationship between the transactional attribute of uncertainty, and structural alternatives to manage this uncertainty. Specifically, real options theory proposes that, instead of pushing managers to greater integration and hierarchy, they respond to uncertainty by desiring greater flexibility (McDonald & Siegel, 1986). This can be to the extent that managers rationally choose a structure that is suboptimal, for example by deferring irreversible decisions on investments until uncertainty is reduced, even if these investments would deliver a positive net present value (Folta & O'Brien, 2004). For managers, sunk costs imply commitment-intensive choices (Ghemawat, 1991) and may present significant barriers to entry (Baumol & Willig, 1981; Caves & Porter, 1977; Clark & Wrigley, 1995). In short, greater uncertainty may lead managers to prefer market structures over hierarchy, which is at odds with the outcomes predicted by TCT. Testing for the relationship between volume uncertainty and preference by managers for hybrid structures over hierarchies, a positive and significant correlation has been identified (Crook et al., 2012, p. 70). However, in the presence of behavioural uncertainty, where the tasks being performed are complex, opaque, and difficult to evaluate, managers prefer increased integration (Crook et al., 2012, pp. 70–71). This points to a nuanced relationship between uncertainty and the management structure chosen to address it, which is also of particular interest in this chapter.

Looking to a different perspective, real options theory suggests that managers opt to defer making attractive but irreversible investment decisions in the face of uncertainty (Folta & O'Brien, 2004). They may also, in certain circumstances, take a 'growth option', and make early investments under uncertain conditions (Amram & Kulatilaka, 1998), if they judge that there are early mover advantages in doing so (Kulatilaka & Perotti, 1998). Indeed, greater uncertainty may actually add value to the growth option, encouraging investment that delivers new firm capabilities (Kogut & Kulatilaka, 2001), technological advantages, brand recognition, or other comparative advantages over later movers (Lieberman & Montgomery, 1988). This is consistent with the integrative structural preference predicted by TCT.

However, when seeking first mover advantages, managers are more likely to exercise their growth options if they are able to minimise the size of their irreversible investments, or sunk costs (Clark & Wrigley, 1997a; Folta & O'Brien, 2004). One way that managers might do this is by using more flexible arrangements such as market contracts and hybrids, even where these structures imply a higher transaction cost than a hierarchy (Kogut, 1991; Schilling & Steensma, 2002), provided the value of being a first mover exceeds the transaction cost. In this configuration, real options theory challenges the predictions of TCT. Instead of hierarchies, it is hybrid structures that allow managers to make limited commitments and to reduce the risks associated with volume and technological uncertainty.

This section sought to provide an overview of some of the complexities that emerge when comparing different theoretical understandings about how a firm makes decisions and conducts transactions in uncertain conditions. The next section explores some of the implications of these theoretical perspectives within the applied concept of corporate water risk. In the process of doing this, the chapter will introduce corporate water return as a corollary to corporate water risk. Five propositions are presented in the next section to draw out key features of managing corporate water risk and water return as they could be managed by firms behaving in accordance with the theory presented previously.

Corporate water return

Corporate water risk is an exemplar of the nuanced transaction hazard that managers face when deciding between the structural alternatives of market, hybrid, and hierarchy. Transactions that involve water resources (such as abstraction and storage) or services (such as treatment and disposal) incorporate assessments of asset specificity, uncertainty, and frequency. However, a more complete understanding of the transaction hazard is rendered if assets are also considered from a resource-based view, i.e. being value-enhancing, limited in supply, and difficult to substitute (Barney, 1991). Moreover, unpacking the TCT attribute of uncertainty to its components of volume, technology, and behaviour (Williamson, 1985) also affords interesting insights from a real options theory perspective, as it points to a conflict where managers seek greater integration in response to behavioural uncertainty, but more freedom in response to volume uncertainty. These sub-attributes may be positively correlated: for example, more uncertainty about the volume of water demanded might make suppliers more opaque about their pricing or provisioning strategies. This implies that the uncertainty effect is not monatomic (e.g. Folta & O'Brien, 2004), and that intra-attribute 'wars of primacy' may take place as managers wrestle to determine the appropriate structural alternative for the transaction. That is to say, it is hard to tell which attribute of uncertainty; for example, the volume of water available, the technology to treat water effectively, or how much people will consume, will be prioritised by managers and thus determine which approach they take to manage it.

As discussed previously, real options theory suggests that in an environment of increased volume uncertainty, managers may prefer more market-based structures in order to preserve flexibility. However, if the increased uncertainty is behavioural, there is empirical evidence that managers prefer more integrated structures (Crook et al., 2012). Consider the provision of water supply and services for a multinational firm that seeks to expand in an infrastructure-scarce but rapidly growing part of the world. This could for example be in Sub-Saharan Africa, where real GDP more than doubled between 1995 and 2010 (International

Monetary Fund, 2013). The cost of addressing the continent's infrastructure deficit is US$75 billion per annum, or 12 percent of Africa's GDP, of which there is currently a funding gap of US$ 35 billion per year (World Bank, 2014). With GDP growth of 6 percent in Sub-Saharan Africa forecast for 2014, compared to world output of just 3.6 percent (International Monetary Fund, 2013), the IMF anticipates that the region will account for more of the global economy than at any time in the last 30 years. This provides the basis for at least an initial assumption that there are early mover advantages to be gained by firms exercising a 'growth option' and investing in water infrastructure, despite behavioural uncertainty about how this water will be used and regulated and *because of* the uncertainties associated with rapidly increasing volume demand related to e.g. socioeconomic development. This leads us to:

> *Proposition 1: Corporate Water Return can be understood in terms of the value of a 'growth option' that affords early mover advantages to firms that invest in water infrastructure under conditions of uncertainty*

In essence this is the counterpoint to the corporate water risk concept, which emphasises, inter alia, the lack of adequate infrastructure, uncertainties related to populist politics, inadequate information flows and weak regulatory frameworks, and the variability and unpredictability of demand due to demographics, urbanisation, and climate variability (Hope & Rouse, 2013). As a result, firms opt to defer investing in water infrastructure, *despite* early mover advantages, and because of behavioural uncertainties related to e.g. political and regulatory risk:

> *Proposition 2: Corporate Water Risk can be understood in terms of the value of a 'deferred option' that affords flexibility and protects firms from investing in water infrastructure under conditions of behavioural uncertainty*

According to a report prepared by McKinsey, global spending on basic infrastructure currently amounts to US$2.7 trillion per annum, compared to an optimal $3.7 trillion per annum that is required to close the output gap of lost economic activity because of the shortage of infrastructure (World Economic Forum, 2013). In as much as a portion of this investment gap is a 'deferred option' where managers of firms decide not to make infrastructure investments despite their delivering a positive net value, it can be deduced that this situation exists because the risk of behavioural uncertainty outweighs the early mover advantage of volume uncertainty.

> *Proposition 3: Where the net difference between Corporate Water Return and Corporate Water Risk is a negative value, managers will not integrate the transaction within the firm, but instead seek structural alternatives*

As has been discussed, the structural alternatives according to TCT are markets and hybrids. In terms of markets, the literature in relation to the market provision of water supply and services has emphasised the importance of water pricing that delivers sustainable cost recovery to the supplier, within a transparent governance structure that provides for the poor (Rouse, 2013). To the limited extent that managers of firms comment externally about the appropriateness of market structures for water supply and services, they emphasise the importance of 'stewardship' (e.g. Cofino, 2013), which broadly equates to advocacy of incentive-based models of activity such as market structures, with minimal distortions (e.g. from pricing subsidies) that are consistent with TCT:

> *Proposition 4: Where managers seek market structures to manage Corporate Water Return and Corporate Water Risk, they will advocate arms-length 'stewardship' models based on incentives, consistent with transaction cost theory*

Indeed, in a 'pure' market structure, it is likely that most of the costs accruing to water use correspond directly to the level of consumption, rather than as discrete transaction costs. However, there are few pure market structures for the intermediation of water supply and services. As highlighted above, this is likely a function of the unique and varied social, cultural, and economic attributes associated with water use (e.g. McCaffrey, 1992). Moreover, water pricing rarely includes any amortisation of the original capital expenditure costs of infrastructure, often because this was covered by the public purse. Water rates typically include allocations for repair and maintenance but exclude a provision for replacement.

> *Proposition 5: For transactions where Corporate Water Return and Corporate Water Risk are salient but cannot be fully integrated, managers will always prefer to participate in hybrid structures over market structures*

Proposition 5 treats water as a strategic asset, as understood by the resource-based view. It introduces the idea of issue salience within the strategic cognition process, where managers evaluate return and risk in terms of their relative materiality, specificity, and urgency (Money, 2014c). No assumption is made here as to the level of integration between the polar structures of market and hierarchy. However, it posits that where full integration is not appropriate, the preference by managers for a hybrid structure over a market structure is unconditional.

The following section offers an empirical framework to test these propositions.

A Five-Factor Framework (5FF)

What follows is a theory-building framework that has been deliberately structured to facilitate both before and after empirical analyses of actions taken by managers at firms to manage water risk and return. It is therefore important to present the framework with appropriate context, and an illustrative scenario is used for this purpose.

ACME Plc is a multinational firm within the food and beverage sector. Its portfolio includes packaged foods and powdered and liquid beverages. The firm is listed on the London Stock Exchange. Its annual revenues are roughly US$20 billion, as is its market capitalisation, and it is a member of the FTSE 100 index. Its share ownership is geographically diverse, but it is concentrated amongst institutional holders rather than retail investors. The firm also uses debt to finance its activities, and is rated 'A2' (upper-medium grade, low credit risk), similar to its peers. ACME currently has distribution networks and retail presence in over 100 countries, and a manufacturing presence in 30 countries. At present, 40 percent of sales are derived from Europe, 30 percent from the Americas, 20 percent from Australasia, and 10 percent from the Middle East and Africa (MEA).

ACME's managers anticipate annual sales growth of 4 percent over the next decade, which would increase group revenues to US$30 billion. However, they also believe that by the end of that period, MEA will account for 20 percent of sales, while Europe's share will fall to 30 percent. These dynamics reflect the expected demand growth rates within the respective regions, and imply compound sales growth of over 10 percent per annum in MEA, compared with less than 2 percent in Europe. The firm's managers recognise that very few of their manufacturing plants are based in MEA, and the facilities in place are insufficient to accommodate an increase in sales from US$2 billion per annum to US$6 billion per annum over the next decade.

Given the importance of MEA growth to the firm's overall performance (it would account for 40 percent of the group's total increase in sales value), ACME's managers make it a strategic priority to increase the firm's manufacturing capacity in the region. As part of this programme, a new finishing factory for several of the firm's key lines is commissioned at a site on the outskirts of Lusaka, Zambia. The decision is shaped by the site's proximity to road, rail, and air transport networks, and to space available to expand. It is also shaped by Zambia's relatively stable political regime, strong civic institutions, and a legal system based on English common law. ACME's managers decide to build sufficient capacity into the plant to meet customer demand not just in Zambia, but in several adjacent markets.

The factory is modern and efficient but relatively water intensive, given the nature of ACME's product portfolio producing food and beverages. A rolling 12-month agreement has been concluded with the municipal water utility in order to ensure that the firm has sufficient supply at an agreed price. However, it is clear that the municipal utility will struggle to cope with the increasing water demand from its rapidly growing, urbanising customer base, without significant infrastructure investment that is currently not available. Zambia's infrastructure funding gap is estimated at US$500 million per annum, 40 percent of which is in the water and wastewater sector (Foster & Dominguez, 2011).

The implications are clear. Rising demand and relatively inelastic supply is likely to result in maladaption (Williamson, 1999) as the utility finds itself unable to meet ACME's requirements. This creates transaction costs for ACME, as behavioural uncertainty rises. Apart from the utility itself, various actors can contribute to this uncertainty, including regulators, politicians, formal civic groups, and activist local residents. This narrative around 'licence to operate', the social tacit approval for the economic activity to continue, comes with significant economic and reputational risk – and is widely rehearsed in the existing corporate water risk literature (e.g. Barton, 2010; Sarni, 2011). The point of interest is *corporate water return*. Irrespective of the increased transaction costs associated with behavioural uncertainty, ACME carries latent benefits in the growth options that it can exercise to exploit its early mover advantage.

The question is whether a relational structure, an institutional mechanism for transactions as discussed in the previous section, exists where these latent benefits are greater than the transaction costs so that ACME would exercise its growth option. In this example, if the firm could operate within a structure where it was sure that it would not face constraints on water supply and services as it expanded, this structure would have a value to the firm. The question would then be what the value was, and if it exceeded the cost of the structure.

To better understand how such a structure is identified and evaluated the Five-Factor Framework (5FF) presented below can be used. It helps to provide a way through the complex decisions described thus far between the ways risk is understood by managers within firms and how based on this understanding managers then see and work within different options to manage this risk. In the situation below you could picture ACME in the context presented working through this Framework. Figure 9.2 sets out a 5FF under which such a structure can be identified and evaluated.

1 **Action**: The first condition of the 5FF is that the firm is considering an action (such as an investment in water supply and services) that has lower transaction costs when it occurs *outside* the hierarchy. If the desired outcome is possible by action taken within the integrated firm structure at a lower transaction

Figure 9.2 A Five-Factor Framework (5FF)

cost than any alternative structure, then logically the action would be taken within the firm. In the parlance often used by managers of firms, this action would be 'business as usual'. For the action to take place outside the hierarchy, the transaction cost associated with an alternative structure has to be lower than the cost associated with an integrated transaction. In the parlance of managers, this is an action that is 'beyond the fence line'. In ACME's case, it is assumed that the transaction costs to the firm of securing a dedicated, irrevocable right to sufficient water at the Lusaka site are too high for that action to take place. These costs may be purely economic, but may also relate to the firm's licence to operate.

2 **Advantage:** The second condition of the 5FF is that, in taking an action, the firm lowers the transaction costs associated with the activity, due to a comparative advantage that the firm brings to the transaction. That is, the firm's involvement results in lower transaction costs for the activity than are otherwise prevailing. For example, assume that the utility that supplies ACME wishes to increase its capacity to store and distribute water within its customer catchment area. The investment required for such a capital-intensive activity is significant, and the utility considers borrowing the money by issuing bonds. The current sovereign credit rating for Zambia is 'B1' (Moodys, 2014) and the Zambian utility is likely to be rated below this – assume one grade lower, i.e. 'B2'. Recall that ACME is rated 'A2', which is nine notches higher than the utility. Based on current market rates, for the utility to borrow e.g. US$250 million over ten years, lenders might require an effective interest rate of 8 percent per annum. For ACME to borrow the same amount over the same period, lenders might require annual interest of just 3 percent. This differential offers scope for a meaningful reduction in transaction costs, due to ACME's comparative advantage over the utility.

3 **Alternative:** The third condition of the 5FF is that there is, in practice, an alternative structure to the status quo, under which the firm (e.g. ACME) engages in the transaction (e.g. underwriting a US$250 million investment in water infrastructure) with another actor (e.g. the utility) on a Pareto-improved basis (e.g. lower financing costs benefiting the utility as well the utility's customers, including ACME). This structure may have some or all of the characteristics expected of hybrids in TCT, such as equity investment (e.g. a joint venture between ACME and the utility), long-term binding contracts (e.g. regarding supply and payment terms), routines for mutual monitoring (e.g. ACME monitors the utility's progress in cost recovery; the utility monitors the volume of ACME's water use; the regulator monitors the utility's abstraction and ACME's discharge), and prearranged procedures for third-party mediation or arbitration in the event of a dispute, without immediate recourse to a court of law. The alternative structure may involve many more actors than the buyer and seller of the service; it might include, for example, other water consumers, regulators, government agencies, taxpayers, civic society groups, or NGOs.

4 **Attribution:** The fourth condition of the 5FF is that, beyond a Pareto improvement, the benefits accruing to the lower transaction cost are attributable to the respective beneficiaries. It is moreover a necessary condition that the beneficiaries include at least one other actor than the firm (or concert party), although the firm itself does not have to be a beneficiary. The attributed benefits need not be financial, nor must they be quantified. However, the benefits must be defined, discrete, and direct. Such benefits for example could be related to reputation, forms of social respectability, value, and other issues related to social licence to operate. The necessity of attribution is a function of hybrid structures. In a hierarchy structure, controls exist to manage and resolve disputes, and managers exercise broad discretion to direct behaviour. In a market structure, powerful incentives exist to ensure that buyers and sellers are satisfied. However, a hybrid structure lacks the authority of control by fiat, while long-term contractual arrangements erode the power of incentives. As a result, the efficacy of monitoring and managing relationships between transacting parties and other actors becomes critical. Attribution is key to this: again, in the parlance, 'you cannot manage what you cannot measure'.

5 **Awareness:** The final condition of the 5FF is that the beneficiaries of the action, beyond the transacting parties themselves, are aware that they are better off as a consequence of the transaction taking place. In the case of ACME, beneficiaries would include the institutional investors who hold shares in the firm and share in its profits. The transaction helps to increase the flow of earnings and dividends that result from ACME's additional sales to the MEA region. It is important the investors are aware of the benefits, as they are the principals who ultimately authorise managers, acting as their agents, to take such actions. In the case of the utility, the beneficiaries may be proximate domestic water consumers who have improved access to water supply and services at a price that is affordable thanks to the lower cost of infrastructure funding. It is important that beneficiaries are aware because their support may be necessary in granting a 'licence to associate', or create innovative hybrid structures that deliver Pareto improvements that may not have been realised under market or hierarchical regimes. The 'licence to associate' in terms of delivering corporate water return is the counterpoint to the 'licence to operate' on which the corporate water risk literature is focused.

Discussion

The 5FF posits a framework to understand and evaluate intermediating actions taken by managers to lower transaction costs and deliver desired outcomes. Although it is conceptual in structure, the framework facilitates ex ante and ex post empirical evaluation of these actions from different perspectives such as transacting parties, shareholders and other stakeholders. While the 5FF sets out minimum criteria that must be met for each condition, it is also possible to conceive a relatively high (or low) contextual applicability of the framework, where conditions are more (or less) comfortably met.

Let us take each condition in turn by way of a brief example. First, where the differential in transaction costs for a firm is substantial if the action is internalised (high applicability), the firm is more likely to look at alternative structures than where the differential is modest (low applicability). Second, where the comparative advantages that the firm brings to the action are greater and/ or more numerous, the rationale for the transaction increases (high applicability). Third, where alternative structures to hierarchy and market already exist and are well established, e.g. prevalence of cooperative agreements, joint ventures, monitoring agencies etc., the applicability of the framework is higher. Fourth, where the benefits accruing from the transaction are bigger, more widely spread, and more clearly apparent to the respective parties, the framework gains stability and reinforcement. Finally, where the ultimate beneficiaries can make the clearest link between their benefit and the original action, the lower the resistance may be to deployment, and the higher the corresponding applicability.

The 5FF is a cyclical, self-reinforcing framework. Other things being equal, the benefit of an action that reduces a firm's transaction costs and raises its profitability are that its value should increase and its creditworthiness should improve. This in turn contributes to a higher credit rating. For example, if ACME is upgraded to 'A1', the interest rate it is required to pay on future borrowing should fall, compared to its current 'A2' rating. In effect, this increases the firm's comparative advantage, potentially making more hybrid transactions attractive as the marginal cost falls. These additional actions further improve creditworthiness and lowered borrowing costs, creating a virtuous cycle of Pareto-improved value creation. Meanwhile the investment in infrastructure – which would otherwise not have taken place – delivers multiplier benefits through job creation, creating tax receipts and income that can be used to reinforce the mechanisms necessary for monitoring and managing hybrid governance structures. In theory, this helps to ensure they are fit for purpose, disputes are resolved more swiftly, and the benefits accruing from the process are delivered more completely. Separate to this, the improvement created in water supply and services infrastructure has benefits in terms of productivity (e.g. fewer days lost to illness) and welfare (e.g. improved health outcomes). Investments should also create opportunities in terms of skills training and education – productivity benefits that are not the exclusive domain of water.

However, if the benefits of such an approach are so manifest, why is it not in widespread use? The realities are that the governance and enabling structures necessary are embryonic and evolving. Challenges are most evident in terms of public investment planning, policy uncertainty, and development capacity. However, the fact that challenges exist is not, in itself, a sufficient reason to discard alternative models to catalyse change.

According to the research provider Dealogic, the market for green bonds increased by 500 percent year on year in 2013 (Bolger, 2014), mainly issued by supranational organisations such as the World Bank and EIB. However, corporate activity is rapidly increasing and in the first three months of 2014, issuance was two-thirds of the total issued for 2013 (Edwards, 2014). The market has also been developed through the publication of voluntary guidelines by a consortium of leading banks, which categorise potential eligible projects, including water infrastructure. It is reasonable to assume that such publications contribute to lowering transaction costs by improving the flow of information, and encourage the development of structural alternatives. In absolute terms, the capital committed to green bonds in 2013 was US$11bn, a small number compared to the estimated infrastructure investment gap (Bolger, 2014). However, if the recent trend rate is extrapolated, green bond issuance could quickly have a larger impact. The 5FF predicts that such investments create a positive feedback loop, and it is conceivable that capital allocations to infrastructure via hybrid structures achieve critical mass within the decade. This would be an immensely exciting outcome, given the economic and societal benefits that would accrue from, for example, major improvements in access to improved water and sanitation services in less developed nations.

In March 2014, the consumer goods firm Unilever announced the issuance of the first ever green sustainability bond (Unilever Plc., 2014) in the sterling market. The amount raised was £250 million, at a fixed interest rate of 2 percent. The firm also described a 'Green Sustainability Bond framework' that would provide 'clarity and transparency around the use of proceeds with a yearly reporting structure to reconcile use of funds'. The criteria for investments include operational reductions of water used by 50 percent for new factories and 30 percent for retrofitted factories. The firm also set out a pipeline of projects in which the bond's proceeds would be invested, including factories in South Africa, China, Turkey, and the USA. There are various aspects of the firm's decision that are consistent with the propositions in this paper, and are predictable from the 5FF. Unilever is prosecuting its 'Sustainable Living Plan' which targets a doubling of the size of the business, while reducing the overall environmental footprint. It can be speculated that this target creates transaction costs that the firm would seek to reduce via hybrid structures. Given the company's 'A1' credit rating, it enjoys a comparative advantage over many sovereign borrowers. The 'Green Sustainability Bond framework' provides a basis for attributing the benefits, while the yearly reporting structure should keep stakeholders aware of new developments.

This paper has used corporate water risk and return as a limited conceptual example for the 5FF. Given water's unique economic, social, and cultural attributes, this raises the legitimate question as to whether the framework has broader validity or relevance beyond water. The idea that an action by a firm can precipitate a benefit (or value) on a mutual (or shared) basis that would otherwise not exist is not, of course, a new or original concept. Instrumental stakeholder theory (Donaldson & Preston, 1995) has explored value creation for shareholders through stakeholders (Freeman, Wicks, & Parmar, 2004). The literature on how firms can create 'shared value' for stakeholders as well as shareholders highlights 'three distinct ways to do this: by reconceiving products and markets, redefining productivity in the value chain, and building supportive industry clusters at the company's locations' (Porter & Kramer, 2011, p. 65). As much as the 5FF incorporates these dimensions, it is complementary to, rather than a substitute for, the outputs of Porter and Kramer's (2011) shared value solutions. However, while the creation of shared value is a necessary component of the 5FF's process, it is not the motivation for the action. If transaction costs for a firm are lower if the transaction takes place within a firm's structure, then the 5FF predicts that the transaction will be internalised, even if an alternative structure creates shared value. As such, it is at odds with the proposition that 'the purpose of a corporation must be redefined as creating shared value, not just profit per se' (Porter & Kramer, 2011, p. 63). The 5FF makes no normative statement about what the purpose of a corporation must be in relation to society. Instead it seeks to build theory on the motivations and conditions under which managers may engage with hybrid structures which, inter alia, have societal implications.

In conceptualising corporate water return and risk through the lens of TCT and real options theory, the 5FF attempts to proxy some of the applied dynamics under which managers make decisions (Donaldson & Lorsch, 1983; Hayes & Garvin, 1982). Much as the literature on corporate water use has hitherto focused on the risk side of the equation, so too has most of the empirical research on real options theory focused on the option to defer, rather than to grow (Folta & O'Brien, 2004). Given that both options increase in value with increased uncertainty, the disproportionate interest in options to defer (and corporate water risk) over options to grow (and corporate water return) suggest that more attention to the latter is justified. While there is nothing substantive within the academic literature on corporate water return, it is curious that even in practitioner reports – where commercial incentives exist for a wider framing of the issue – the conceptualisation of opportunity as the counterpoint to risk is very narrow, and restricted principally to the economic benefits for firms of greater efficiency. It is hoped that the 5FF will catalyse new and innovative approaches to conceptualising and realising the value of options to grow, which are likely to increase in an environment of rising uncertainty. These approaches should have far wider benefits than the economic dimensions highlighted here; and have broader implications for food and energy infrastructure.

Conclusion

The 5FF predicts that a positive feedback loop exists when a firm exercises its growth option by taking action outside the hierarchical structure. However, adherence to the following conditions is necessary:

- First, that the firm must benefit from lower transaction costs as a result of taking the action outside the firm structure rather than within the hierarchy.
- Second, that in taking an action the firm brings a comparative advantage that lowers transaction costs.
- Third, that a structural alternative to the polar positions of hierarchy and market exists.

- Fourth, that the benefits accruing from lower transaction costs are attributable to the respective beneficiaries' participation.
- Fifth, that the beneficiaries of the action, beyond the transacting parties themselves, are aware that they are better off as a consequence of the transaction taking place.

This chapter does not attempt to qualify the relative importance of each of the five conditions set to the overall applicability of the 5FF. This would be a useful area of future research, particularly with reference to empirical analysis and statistical significance testing. And by embedding a Pareto improvement within the conditions, the chapter also does not allow for anyone to become worse off as a consequence of the action.[2]

Certainly, the stakes are high enough. It has been estimated that the 'output gap' between rising infrastructure needs and insufficient provision is at least US$ 1 trillion per year (World Economic Forum, 2013). Public–private partnerships (PPPs), which can tap the private sector's financial resources and skills in delivering infrastructure, are frequently described as the obvious and optimal solution to closing the gap. However, insufficient preparation, low quality feasibility studies, a poor balance between risk allocation and regulation, and an inadequately accommodating environment are cited as reasons why PPPs have not delivered against their potential. The 5FF does not obviate these hurdles, but provides a different perspective by which they can be conceptualised and addressed. Preparation requires an experienced team, good governance and adequate funding. Feasibility studies require sophisticated modelling, supplemented with accurate data. Balancing risk allocation and regulation requires clear signalling and adaptive responses. A conducive environment requires a solid legal framework and societal communication. The 5FF is based on a hybrid governance structure where the requisite elements to catalyse these investments exist.

If the key stakeholders in the process, notably the transacting parties, governance and monitoring agents, and the ultimate beneficiaries (e.g. shareholders and citizens), can be aligned in the pursuit of an objective, there is reason to be optimistic that the output gap can be closed. The onus lies first with firms to reconceptualise risk in terms of options to defer and options to grow. Second, it lies with utilities and others to take a longer-term view of their infrastructure requirements. And third, it lies with governments to develop the environment needed where complex hybrid structures can thrive. But this is not simply a wish list. The responsibility of firms – as profit-seeking entities, accountable to their owners – ranks first among equals. The current status quo may reflect problems of agency, but the momentum for change builds as the reality of insufficient infrastructure in a growing, developing, urbanising, and resource-scare world come home to roost.

All of these issues have deep and wide-ranging implications for the governance of resources other than water, notably food and energy. Even allowing for forecast error, the annual US$ 1 trillion 'value gap' between the demand for new infrastructure and its supply represents a dramatic market failure. The consequences are lower economic activity and sub-optimal welfare. Closing the gap would lift more people out of poverty more quickly, and PPP hybrid structures that engage firms with their counterparts to deliver this infrastructure would be valuable and desired. There are clear and present challenges to be addressed, and it is self-evident that many critical enabling conditions will need to be more developed as a precondition for this framework to achieve practicable salience. This underpins the exhortations made in this chapter for further research in developing an empirical base; testing the sensitivities of the factor conditions; and widening the frame of reference beyond water, infrastructure, and investment.

Finally, it is important to emphasise that while this is a theory-building paper, various real-world developments in recent months offer strong support for the predictive implications of the 5FF, and the propositions that underpin it. The market for 'green bonds' – closely analogous to the action proposed in the example herein – has increased five-fold in the past year, and demand is accelerating sharply as the conditions required to embed hybrid structures are strengthened. This already reflects new investment in the order of tens of billions of dollars per annum. These may be the dawning days of a new era in post-crisis sustainable environmental financing, during which theories such as the 5FF can play a role in securing a better future for the generation to come.

Notes

1 It has also been proposed that transaction costs are insufficiently demarcated for empirical analysis when compared to sunk costs (Clark & Wrigley, 1995).

2 This may conceivably be an unrealistic criterion in an applied context, and again would be an interesting area for empirical analysis.

References

Amram, M., & Kulatilaka, N. (1998). Real Options: Managing Strategic Investment in an Uncertain World. Harvard Business School Press

Barney, J. (1991). Firm resources and sustained competitive advantage. *Journal of Management, 17*(1), 99–120.

Barton, B. (2010). *Murky Waters? Corporate Reporting on Water Risk*. Boston, MA. Publisher Ceres.

Baumol, W. J., & Willig, R. D. (1981). Fixed costs, sunk costs, entry barriers, and sustainability of monopoly. *The Quarterly Journal of Economics*, 96(3), 405–431.

Bolger, A. (2014, March 11). Climate focus drives demand for "green" bonds. *Financial Times*. Retrieved from www.ft.com/cms/s/0/94d1e2e6-a84c-11e3-a946-00144feab7de.html#axzz2xjODOspL.

Clark, G., & Wrigley, N. (1995). Sunk costs: A framework for economic geography. *Transactions of the Institute of British Geographers*, *20*(2), 204–223.

Clark, G., & Wrigley, N. (1997a). Exit, the firm and sunk costs: Reconceptualizing the corporate geography of disinvestment and plant closure. *Progress in Human Geography*, *21*(3), 338–358. doi:10.1191/030913297674898950.

Cofino, J. (2013). Nestlé's Peter Brabeck: Our attitude towards water needs to change. *Guardian Online*, 1. Retrieved from www.theguardian.com/sustainable-business/nestle-peter-brabeck-attitude-water-change-stewardship.

Crook, T., Combs, J., Ketchen, D., & Aguinis, H. (2012). Organizing around transaction costs: What have we learned and where do we go from here? *Academy of Management Perspectives*, *27*(1), 63–79. doi:10.5465/amp.2012.0008.

Dobbs, R. and Pohl, H. (2013) Infrastructure Productivity: How to save US$1 trillion a year. McKinsey Global Institute.

Donaldson, G., & Lorsch, J. (1983). *Decision Making at the Top: The Shaping of Strategic Direction*. New York: Basic Books.

Donaldson, T., & Preston, L. E. (1995). The stakeholder theory of the corporation: Concepts, evidence, and implications. *Academy of Management Review*, *20*(1), 65–91.

Edwards, B. (2014, April 2). Investors warm to green bonds. *Wall Street Journal*. Retrieved from http://online.wsj.com/news/articles/SB1000142405270230384780457947705187222857.

Folta, T. B., & O'Brien, J. P. (2004). Entry in the presence of dueling options. *Strategic Management Journal*, *25*(2), 121–138. doi:10.1002/smj.368.

Foster, V., & Dominguez, C. (2011). Zambia's infrastructure: A continental perspective. *Policy Research Working Papers*. The World Bank. doi:10.1596/1813–9450–5599.

Freeman, R. E., Wicks, A. C., & Parmar, B. (2004). Stakeholder theory and "the corporate objective revisited." *Organization Science*, *15*(3), 364–369.

Gertler, M. S. (2003). Tacit knowledge and the economic geography of context, or the undefinable tacitness of being (there). *Journal of Economic Geography*, *3*(1), 75–99.

Ghemawat, P. (1991). *Commitment: The Dynamics of Strategy*. New York: Free Press.

Global Reporting Initiative (2011). G3.1 Sustainability Reporting Guidlines. GRI.

Hagerman, L., Clark, G., & Hebb, T. (2007). Investment intermediaries in economic development: Linking pension funds to urban revitalization. *Community Development Investment Review*, *3*(1), 45–65.

Hayes, R., & Garvin, D. (1982). Manging as if tomorrow mattered. *Harvard Business Review*, *60*(3), 70–79.

Hebb, T. (2006). *Pension Funds and Urban Revitalization California Case Study B: Real Estate CalPERS' California Urban Real Estate Initiative*. Oxford: Oxford University Press. Retrieved from http://urban.ouce.ox.ac.uk/research.php.

Hebb, T. (2007). *CalSTRS Case Study: Diversity as an Investment Framework: CalSTRS Targeted Investment Strategy*. Oxford: Oxford University Press. Retrieved from http://urban.ouce.ox.ac.uk/research.php.

Hebb, T., & Sharma, R. (2014). New finance for America's cities. *Regional Studies*, *48*(3), 485–500. doi:10.1080/00343404.2013.843163.

Heide Caves, R. E., & Porter, M. (1977). From entry barriers to mobility barriers: Conjectural decisions and contrived deterrence to new competition. *The Quarterly Journal of Economics*, *91*, 241–261.

Hope, R., & Rouse, M. (2013). Risks and responses to universal drinking water security. *Philosophical Transactions: Series A, Mathematical, Physical, and Engineering Sciences*, *371*(2002), 20120417. doi:10.1098/rsta.2012.0417.

Howell, L. (2013). *Global Risks 2013* (Vol. 8). Switzerland: World Economic Forum.

International Monetary Fund (2013). Sub-saharan Africa: Keeping the pace. Retrieved from www.imf.org/external/pubs/ft/reo/2013/afr/eng/sreo1013.pdf.

Kale, P., & Sing, H. (2009). Managing strategic alliances: What do we know now, and where do we go from here? *Academy of Management Perspectives*, *23*(3), 297–326. doi:10.5465/AMP.2009.43479263.

Klein, B., Crawford, R., & Alchian, A. (1978). Vertical integration, appropriable rents, and the competitive contracting process. *Journal of Law and Economics*, *21*, 297–326.

Retrieved from http://heinonlinebackup.com/hol-cgi-bin/get_pdf.cgi?handle=hein.journals/jlecono21&section=19.

Kogut, B. (1991). Joint ventures and the option to expand and acquire. *Management Science*, *37*(1), 19–33.

Kogut, B., & Kulatilaka, N. (2001). Capabilities as real options. *Organization Science*, *12*(6), 744–758.

Kulatilaka, N., & Perotti, E. C. (1998). Strategic growth options. *Management Science*, *44*(8), 1021–1031.

Lieberman, M. B., & Montgomery, D. B. (1988). First-mover advantages. *Strategic Management Journal*, 9(S1), 41–58.

Strategic Management Journal, *23*(6), 535–550. doi:10.1002/smj.247.

McCaffrey, S. C. (1992). Human Right to Water: Domestic and International Implications, A. Georgetown International Environment Law Review., 5, 1.

McDonald, R., & Siegel, D. (1986). The value of waiting to invest. *The Quarterly Journal of Economics*, *101*(4), 707–727.

Money, A. (2014a). Corporate Water risk: A critique of prevailing best practice. *Journal of Management and Sustainability*, *4*(1), 42–59. doi:10.5539/jms.v4n1p42.

Money, A. (2014b). Corporate Water risk: Investor tolerance of the status quo. *Journal of Management and Sustainability*, *4*(1), 60–75. doi:10.5539/jms.v4n1p60.

Money, A. (2014c). *Evaluating Issue Salience through Corporate Water Risk*. Smith School of Enterprise and the Environment, University of Oxford. Working paper

Moodys (2014). Rating: Zambia, Government of. Retrieved from www.moodys.com/credit-ratings/Zambia-Government-of-credit-rating-806356896.

O'Brien, R., & Keith, A. (2009). The geography of finance: After the storm. *Cambridge Journal of Regions, Economy and Society*, *2*(2), 245–265. doi:10.1093/cjres/rsp015.

OECD (2006). *Infrastructure to 2030: Telecom, Land, Transport, Water and Electricity*. OECD Publishing

Porter, M., & Kramer, M. (2011). Creating shared value. *Harvard Business Review*, 89(1–2), 62–77.

Rouse, M. (2013). *Institutional Governance and Regulation of Water Services: Second Edition* (2nd ed.). International Water Association Publishing, London UK

Sarni, W. (2011). *Corporate Water Strategies*. London: Earthscan.

Schilling, M. A., & Steensma, H. K. (2002). Disentangling the theories of firm boundaries: A path model and empirical test. *Organization Science*, *13*(4), 387–401.

Siegel, D. S., & Zervos, V. (2002). Strategic research partnerships and economic performance: Empirical issues. *Science and Public Policy*, *29*(5), 331–343. doi:10.3152/147154302781780859.

Simon, H. (1945). *Administrative Behavior*. New York: Free Press.

Trucost Plc (2013). Natural capital at risk: The top 100 externatlities of business. Retrieved from www.teebforbusiness.org/js/plugins/filemanager/files/TEEB_Final_Report_v5.pdf.

Unilever Plc (2014). Unilever issues first ever green sustainability bond. *Investor Relations Website*. Retrieved from www.unilever.com/mediacentre/pressreleases/2014/Unileverissuesfirstevergreensustainabilitybond.aspx.

United Nations General Assembly (2010). *Resolution 64/292: The Human Right to Water and Sanitation*. New York: United Nations.

WBCSD (2013). *Sharing Water: Engaging Business* (Vol. 1). Switzerland: World Business Council for Sustainable Development.

Williamson, O. (1975). *Markets and Hierarchies: Analysis and Antitrust Implications: A Study in the Economics of Internal Organization*. New York: Free Press.

Williamson, O. (1979). Markets and hierarchies: Analysis and antitrust implications: A study in the economics of internal organization. *Journal of Law and Economics*, *22*, 233–261.

Williamson, O. (1985). *The Economic Institutions of Capitalism*. New York: Simon & Schuster.

Williamson, O. (1991). Comparative economic organization: The analysis of discrete structural alternatives. *Administrative Science Quarterly*, *36*, 269–296.

Williamson, O. (1999). Strategy research: Governance and competence perspectives. *Strategic Management Journal*, *20*(12), 1087–1108. doi:10.1002/(SICI)1097-0266(199912)20:12<1087::AID-SMJ71>3.0.CO;2-Z.

World Bank (2014). Fact sheet: Infrastructure in sub saharan Africa. Retrieved from http://go.worldbank.org/SWDECPM5S0.

World Economic Forum (2013). Strategic infrastructure. Retrieved from http://www3.weforum.org/docs/AF13/WEF_AF13_Strategic_Infrastructure_Initiative.pdf

World Resources Institute (2010). Aqueduct: Measuring and Mapping Water Risk.

10 Natural resource management and marine protected areas

The importance of balancing environmental sustainability and community support

Tomas Chaigneau

Introduction

Society is increasingly faced with unexpected or uncertain socio-economic and environmental changes and effective natural resource management (NRM) strategies will be required to ensure future sustainable utilisation of environmental resources (Chapin et al., 2010). Nonetheless, many NRM strategies are failing to achieve the ecological or socio-economic objectives placed on them (Chaigneau, 2013). Whilst we may feel confident of broad-scale future environmental changes (such as global mean temperature increases), we still cannot routinely predict changes at local and regional levels (e.g. precipitation patterns: see Millar, Stephenson, & Stephens, 2007). Furthermore, there is growing evidence that humanity is causing widespread, abrupt, and possibly irreversible changes to basic earth-system processes, and driving global environmental change that could threaten development and trigger humanitarian crises across the globe (Griggs et al., 2013; MA, 2005; Steffen et al., 2011).

Facing an unknowable and uncertain future can complicate NRM strategies and lead to a 'paralysis of indecisions' (Chapin et al., 2010; Millar et al., 2007). To counter this, a portfolio of different natural resource management approaches suitable for a range of different contexts are required to respond to uncertain change (Chapin et al., 2010) by 1) preventing, limiting or managing the environmental damage and 2) protecting individuals and populations from the adverse consequences of these changes (MA, 2005).

NRM strategies refer to the sustainable utilisation of major natural resources such as land, water, air, minerals, forests, fisheries, and wild flora and fauna (Castro, 2000). They focus particularly on how management affects the quality of life for both present and future generations and aim simultaneously to tackle a series of objectives such as poverty alleviation, food security, ecological management of productive resources, empowerment of rural communities, and the establishment of supportive policies (See Altieri, 2002).

Sustainable use of natural resources that 'meets the needs of the present without compromising the ability of future generations to meet their own needs' (WCED, 1987) is of utmost importance. Yet, dilemmas remain when considering how to ensure basic needs (such as food, energy, and water security) are met against the background of resource constraints. When there is not enough of what is needed to go around currently, what, if any, duties do we have towards future generations in this respect? (Doyal & Gough, 1991). Greater needs satisfaction in the present is more likely to outweigh our desire for higher levels of need satisfaction in the future (Parfit, 1984, pp. 480–486). These ethical and philosophical arguments have important implications for NRM strategies (especially where there is little enforcement and time lags between NRM implementation and NRM benefits). For example, whilst the poor are often most dependent on natural resources, they may be less likely to manage resources sustainably and comply with management rules for the future if they are focusing on meeting their basic needs in the here and now (Anand & Sen, 2000; Scherr, 2000).

In addition to restoring ecosystem function and biodiversity for the future, NRM strategies also aim to generate substantial socio-economic benefits and contribute to the needs of adjacent communities in the short term. However, very few projects have achieved an optimum balance between these two sets of objectives (Mcshane et al., 2011). The impacts on poverty are frequently thought to be negative (Adams et al., 2004) and NRM strategies are argued to be biological successes and social failures (Christie, 2004). The trade-offs between socio-economic and ecological outcomes of NRM strategies are believed to erode support towards these projects (Chaigneau & Brown, 2016). However, little work has explored support towards NRM strategies and more research is required to investigate what influences community support if the tensions between support and environmental sustainability are to be resolved. This chapter aims to reduce this knowledge gap. It investigates support towards NRM strategies such as Marine Protected Areas (MPAs) and identifies the challenges and opportunities for these strategies, which seek to balance environmental sustainability with community support.

The support of affected communities towards NRM strategies has been shown to be crucial for their success, particularly in the case of protected areas (Infield, 2001; Mehta & Kellert, 1998). This chapter considers community support towards small-scale MPAs in the Philippines. These protected areas are increasingly being implemented (Cabral et al., 2014), yet it is estimated that up to 80 percent in the region are failing to meet the objectives placed on them (Pollnac, Pomeroy, & Harkes, 2001). An understanding of the context-specific factors that influence community support can help inform broader governance principles targeted to improve NRM strategies like MPAs as they seek to maintain the natural resource base from which people meet their food, energy, water, and other needs.

Why MPAs?

The importance in the use of protected areas as a natural resource management tool is well established (Gubbay, 1995). However, whilst there is a rise in the number and coverage of protected areas, there is still a significant mismatch between the larger numbers of terrestrial versus marine protected areas (Agardy et al., 2003). MPAs are defined as 'any area of the intertidal or sub-tidal terrain, together with its overlying water and associated flora, fauna, historical, and cultural features, which has been reserved by law or other effective means to protect part or all of the enclosed environment' (Kelleher & Kenchington, 1992) (See Box 10.1). Despite similar aims and goals, the differences in environmental, ecological, and evolutionary processes between terrestrial and marine ecosystems have made the management of MPAs considerably more challenging (Carr et al., 2003).

Box 10.1 The different types of MPAs

Kelleher and Kenchington's (1992) definition used above is a broad-term definition of MPAs describing the full configuration of protected areas in coastal areas and oceans (Agardy et al., 2003). 'Marine Protected Area' is therefore used as a single, general umbrella term, which can apply to a wide range of different marine habitat protection or resource management tools.

A distinction can however be made between MPAs as 'no take' marine reserves in which all forms of extraction and disturbance by humans are permanently banned and multiple use MPAs with mixed harvest, restricted harvest, and/or complete harvest prohibition areas.

This case study predominantly considers 'no take' community-based MPAs that make up many of the MPAs in the Philippines. This is in part due to the devolution of authority from central to local governments that encouraged their proliferation (White, Courtney, & Salamanca, 2002). In these cases, there is often co-management of MPAs where the community is in charge of monitoring and enforcing the MPA (by spotting encroachers and reporting them to local government officials) with support from municipal governments that help with baseline surveys, delineating boundaries (such as providing buoys), and assistance with enforcement (Chaigneau, 2013).

MPAs aim to maintain or restore marine biodiversity, fish stocks, ecosystem function, and to protect critical habitat. MPAs are also increasingly being designed to address social and economic considerations as well as conservation goals because, without attention to socio-economic issues, MPAs are thought to be less likely to succeed in the long run (Halpern, Lester, & McLeod, 2010). MPAs are therefore considered to contribute positively to the needs of the affected communities (Christie & White, 2007) and are predominantly expected to improve socio-economic conditions by improving food security and increasing revenues from fisheries production in fished areas, through transport of larvae from spawning stocks within the MPA (Gerber, Heppell, Ballantyne, & Sala, 2005) or by a net emigration of adult fish to adjacent fished areas, termed 'spill-over' (Abesamis, Russ, & Alcala, 2006; Russ, Stockwell, & Alcala, 2005). They are often viewed as being a simple and cost-effective way to increase stocks in diverse and complicated tropical fisheries (Roberts & Polunin, 1991; Roberts & Polunin, 1993).

The fact that MPAs often focus on conservation as well as fisheries management objectives is important to acknowledge and may help explain their proliferation as many local communities are dependent on fisheries for food and economic security. For example, it is thought that out of the 120 million people involved in fish production and capture, 95 percent are from developing countries (Allison & Ellis, 2001). Fish play an important positive role in contributing to nutrition (Béné et al., 2016) and in many poor coastal communities such as the Philippines, people depend directly on reef species for most of their protein needs (Gjertsen, 2005). A large decline in fisheries would threaten their livelihoods and food security.

The objective of fisheries management for many years was to maximise the yield taken from a fishery without compromising future catches (Jennings, 2001). However there is still no agreement on the exact levels to which fishing mortality must be reduced and how to reduce by-catch to ensure the sustainability of catches (Browman & Stergiou, 2004). These conventional approaches to fisheries management often require detailed, accurate, and high contrast data to predict the current and future state of fish stocks (Costello et al., 2012; Kelly & Codling, 2006). They are therefore of limited use when data is unreliable or unavailable, which is often the case given the complex nature of marine ecosystems. This lack of data is particularly an issue for developing countries (Worm et al., 2009). Johannes (1998) argues that the costs of scientifically assessing each stock would vastly outweigh the total value of each fishery. Fisheries may even develop and become critically in need of management before any conventional scientific data can be collected (Berkes et al., 2006). It is no surprise therefore that there has been a growing interest in MPAs due in part to their potential benefits and the failures and infeasibility of 'data intensive' conventional forms of management.

MPAs and fisheries in the Philippines

This study considers the case of MPAs in the Philippines where reef fisheries provide livelihoods for more than a million small-scale fishers (White & Vogt, 2000). The coastline is fringed with 25,000 km of coral reefs (Burke, 2011) containing over 2,500 fish species (Froese & Pauly, 2007) and nearly 500 different hard coral species (Licuanan & Capili, 2003). Despite being some of the most biologically diverse reefs in the world, 97 percent of these reefs are at risk from sedimentation, pollution, and fishing pressure (Spalding, 2001).

At the same time this environmental degradation is taking place in the Philippines, a large percentage of the population lacks the resources to meet its basic needs (Gjertsen, 2005). Most live in rural, coastal areas and depend on subsistence farming or fishing for a livelihood (IFAD, 2010). Overfishing in the Philippines has already reduced fish stocks in many areas to well below their maximum sustainable yields (Green, 2003; Tomascik & Mah, 1997) and the Filipino population is expected to double by 2040 (NSCB, 2006, 2013). With a growing Filipino population combined with mounting pressures to export commodities to service government debt, natural resources are extracted at an ever-increasing rate (Central-Intelligence-Agency, 2005). It is imperative to maximise fishery production to meet the needs of the population, but also to ensure that fisheries do not collapse in the longer term.

The concerns outlined previously combined with devolution of central authority to local governments under the Local Government Code initiative of 1991, which encouraged MPA projects through a variety of institutional actors including government, NGOs, fishers' associations, and research institutions (Pomeroy & Carlos, 1997; White et al., 2002). As a result, the Philippines has been at the forefront of establishing community-based MPAs (Christie, White, & Deguit, 2002). Despite this rise in community-based MPAs, most have been rated as ineffective in meeting ecological objectives and there is mixed evidence of the socio-economic benefits they provide to adjacent communities (Chaigneau, 2013; Jones, 2007). These failures are often attributed to poor and inadequate management (Beger, Harborne, Dacles, Solandt, & Ledesma, 2004; Pomeroy, Watson, Parks, & Cid, 2005). In this case, the term 'management' refers primarily to the specific ways in which the MPA is managed (e.g. specific sanctions, law enforcement, MPA maintenance). Management in general is defined and perceived as a technical exercise of employing means to achieve given goals (Chuenpagdee & Jentoft, 2009) in contrast to governance, which is often considered broader than management. This chapter suggests good governance principles by analysing specific facets of the system to be governed.

Why are MPAs failing? The importance of community support

In recent years, much work has been undertaken to determine what factors are linked to successful MPAs and NRM strategies in general (Christie et al., 2009; Milne & Christie, 2005; Ostrom, Janssen, & Anderies, 2007; Pajaro, Mulrennan, Alder, & Vincent, 2010; Pollnac, Crawford, & Gorospe, 2001). Irrespective of whether one is concerned with economic, biological, socio-cultural, or management indicators, an overwhelming number of studies within the common pool resource or natural resource management literature have identified that the support of the community towards MPAs and other conservation strategies is critical for their success (Aldon, Fermin, & Agbayani, 2010; Allegretti, Vaske, & Cottrell, 2012; Beger et al., 2004; Christie, 2005a; Christie & White, 1997; Courtney & White, 2000; Infield & Namara, 2001; Russ & Alcala, 1999; Stoffle & Minnis, 2007; Tissot, Walsh, & Hixon,

2009; Yasué, Kaufman, & Vincent, 2010). Others are more specific and argue that it is the perception of the affected community towards MPAs and their impacts (Dimech, Darmanin, Smith, Kaiser, & Schembri, 2009; Gelcich, Kaiser, Castilla, & Edwards-Jones, 2008; Pomeroy, Pollnac, Katon, & Predo, 1997; Stoffle & Minnis, 2007) or the attitudes of individuals towards the MPA (Allendorf, 2007; Gelcich, Edwards-Jones, & Kaiser, 2005; Lan, 2009; Pita, Pierce, Theodossiou, & Macpherson, 2011) that are of importance. White, Alino, and Meneses (2006) conclude that, in the Filipino context, the success of most small MPAs is thought to lie in the amount of support within the community.

However, despite its apparent importance for MPA success, the term 'community support' remains poorly defined. Most studies suggest that it relates to positive feelings or attitudes, which are supposed to lead to compliance and better enforcement (Ballantine, 1994; Pollnac & Pomeroy, 2005). Indeed, attitude, defined as 'a general and enduring positive or negative feeling about some person, object or issue' (Petty & Cacioppo, 1996, p. 7) is used as a precursor or determinant for most models that attempt to explain pro-environmental behaviour or action (Bamberg & Möser, 2007; Kollmuss & Agyeman, 2002). However, it is now widely believed that there is a large gap between attitude and behaviour and that attitude does not directly determine behaviour (Bamberg & Möser, 2007; Tanner, 1999). Neither can community support purely entail pro-environmental behaviour towards an MPA, as those acting 'pro-environmentally' (e.g. complying with MPA rules) may be doing so despite negative attitudes and this behaviour may undermine longer-term conservation efforts. Consequently, the distinction is made in this case between community members' attitude towards the MPA and MPA-supporting actions, which refer to compliance, monitoring, and enforcement of MPA rules.

Improving MPAs: what influences community support?

The support of individuals in three different communities towards their adjacent MPAs was assessed. The focus of the work was to understand what influenced the attitudes and the actions of individuals towards their MPAs and, as a result, better understand what factors influenced support (Chaigneau, 2013). Focus groups (n=22), in-depth interviews (n=18), and questionnaires (n=166) were used to gain an understanding of community support and to see what is perceived to be important for support by target respondents. The aim was to learn how local communities see, understand, and value the MPA.[1]

Twenty-eight broad factors identified as potentially important through the common pool resource, NRM or MPA literature (a deductive approach), or that arose throughout the qualitative focus groups and in-depth interviews (inductive approach) were found to influence attitude, enforcement, or compliance of individuals towards the MPA (Table 10.1). Furthermore, these factors were argued to influence support through 58 different mechanisms. It is clear therefore that determining what influences community support is a complex issue and can have

Table 10.1 Factors and their mechanisms found to influence attitude, enforcement or compliance of individuals positively, negatively or both, towards the MPA. Grey cells represent no empirical evidence of influence.

	Factors	*Mechanisms*	*Attitude*	*Enforcement*	*Compliance*
MPA Design	**MPA Benefits**				
	Economic	Perceived fishery spillover	**Positive**	**Positive**	**Positive**
		Ease of catch	**Positive**		
		Temptation			**Negative**
	Non Economic	Conservation value	**Positive**		
		For future	**Positive**	**Positive**	**Positive**
		Aesthetic	**Positive**		
		For the community	**Positive**		
	MPA Size	Fishing ground reduction	**Negative**	**Negative**	**Negative**
	MPA Location	Visibility		**Positive**	
		Temptation			**Negative**
		Busy area		**Negative**	
	MPA Network	Jealousy of others' lack of MPA	**Negative**		
		Success stories	**Both**		
	MPA Boundaries	Clear noundaries/ No excuses	**Positive**	**Positive**	**Positive**

important ramifications for marine governance. The important factors influencing community support towards MPAs can be summarised into four broad categories: MPA design, MPA management (both include factors that vary between different systems to be governed), socio-economic characteristics, and emotions (both can vary within a system to be governed). Jentoft, Van Son, and Bjørkan (2007) see marine governance as a relationship between two systems; a governing system (social) and a system to be governed (social and natural). In this case, factors can be organised as those that operate within or between systems. Within system factors operate at an individual level and include changes that occur within the community studied and its adjacent MPA. Between system factors are broader factors that differ between the three sites studied. These context-specific predictors of support can in turn help to identify suitable relevant governance principles that do not undermine the support of affected communities (Figure 10.1).

MPA *design*

The majority of MPAs are designed with ecological and socio-economic expectations in mind (Pomeroy, Watson, Parks, & Cid, 2005). In the case

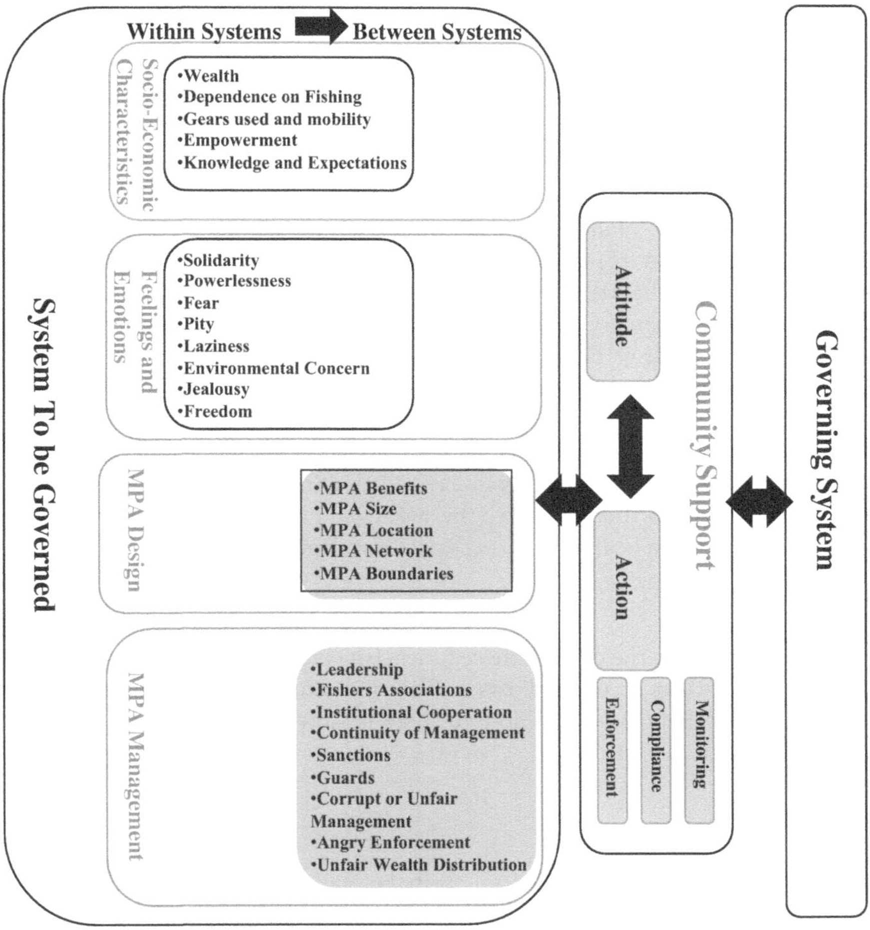

Figure 10.1 Community support as a mediator between the system to be governed and the governing system

of community-based MPAs, if they are designed with consideration for only a subset of people within the affected community, the MPA will likely not persist over time (see Box 10.2). Those that are not supportive of the MPA undermine efforts made to sustain the MPA by actions such as cutting buoys, low levels of compliance with MPA rules, and not reporting encroachers. The establishment of the MPA could lead to rising tensions within the community. It is important therefore to think about all those who will be affected when designing an MPA.

Box 10.2 MPA location as an example of how MPA design can influence support

The location of the MPA was found to be an important factor influencing community support. However unanimous support would be difficult to achieve. For some, an MPA close to the shoreline would lead to increases in positive attitudes, monitoring, and enforcement. Having an MPA that is visible allows communities to be sure the rules are being respected and that there are no free riders. A fisher in Bonbonon illustrates the issues of living far from the MPA: 'We cannot monitor all the time as we are living far from the sanctuary' (fisher bonbonon). This is particularly the case for those fishers that go far out at sea with motor boats: 'For me it is OK [the MPA location] because my husband will go far to fish' (fisher's wife Candaping B).

Others however, would be more negatively affected by a near shore MPA as is described by the fisher's wife below based in Candaping B where the MPA is close to the shore: 'It (the MPA) can affect the fishermen who catch fish on the near area' (fisher's wife Candaping B).

This is especially the case for those fishers (in particular part-time fishers) who do not have access to motor boats and are more likely to remain close to the shoreline. For them, an MPA close to shore is more likely to lead to encroachment and negative attitudes.

MPA *management*

An MPA is not a politically neutral entity but can act as a platform for some to further their own interests (such as aspiring or current government officials, MPA leaders, and fisher association leaders) and have very little do to with concerns about the MPA, its future, and whether it is achieving the expectations placed on it. In turn, community support towards MPAs is also about whether its management is perceived as fair and incorrupt.

Like other NRM tools, MPAs come with restrictions on user behaviour and can influence peoples' ability to sustain themselves. Despite the progression and evolution of coastal management in the Philippines (Christie & White, 1997), the ways that MPAs are managed are not necessarily supported by all within the adjacent community. Certain factors, such as the presence of MPA guards (day and night) and the implementation of graduated sanctions (which progress incrementally based on either the severity or the repetition of violations) were, unsurprisingly, found to influence MPA-related actions positively. These act as deterrents to encroachers and may increase levels of reporting. However, permanent guards are an extravagant expenditure for local governments, and graduated sanctions may not always be enforced and applied to all. Furthermore, it can result in individuals carrying out actions that help fulfil MPA objectives whilst still harbouring a negative attitude towards the MPA itself. Counter-intuitively, it may also lead to a 'laissez-faire' approach and a reduction of pro MPA actions (See Box 10.3).

Box 10.3 MPA guards as an example of how MPA management can influence support

When considering how MPA management can influence support it is important to take into account how management factors can play out differently in different contexts.

The presence of MPA guards, for example, are argued to be important for monitoring, enforcement, and positive attitudes towards MPAs by the scientific community (Jones, 2006; Tissot et al., 2009) as well as respondents at each of the field sites as highlighted by the following extracts:

> 'There should be a guard 24 hours'
>
> (fisher, Bonbonon)

> 'Why do they want to encroach?
> Because there is no permanent guard in the evening'
>
> (fisher, Candaping B)

Counter-intuitively, the presence of guards in other settings can also prevent individuals from reporting as they feel it is 'not their job' to do so. An issue highlighted by a fisher in Suba:

> 'So all of you in the barangay don't help in monitoring?
> No, because there are barangay tanod [guards] and other village officials who monitor the MPA'
>
> (fisher, Suba)

The presence of guards and strict enforcement may lead to perceptions of disempowerment by individuals in affected communities. Platforms through which individuals can help facilitate enforcement need to be encouraged.

Many management approaches such as graduated sanctions or fishermen's associations, may look good on paper; however, these may not work adequately in practice if they cannot be implemented for lack of guards, low membership or infrequent meetings, little enforcement, and unfair wealth distribution.

Socio-economic characteristics

The differences between socio-economic characteristics of individuals can influence the perceived impact of the MPA on their livelihoods and, in turn, their support towards the MPA. Aspects of wealth, dependence on fishing, gear used and mobility, empowerment, and knowledge of the MPA were all found to influence support towards the MPA (see Table 10.1). In particular, livelihood diversity was found to be an important factor (see Box 10.4) and highlights the need for livelihood diversification measures to ensure that the costs harboured by MPA implementation (such as a reduced area available to fishing) are moderated by other income- and food-generating activities.

Box 10.4 Livelihood diversity as an example of how socio-economic characteristics can influence support towards MPAs.

The MPA may be one man's gain by providing some limited fishery spill-over, larval export to other fishing areas, and ensuring the availability of fish for future generations. Those that are wealthy and own boats, or those that fish less frequently, may be able to bear the costs of having to travel further to catch fish whilst reaping the rewards of a larger catch. The same MPA however can also be another man's loss, where he incurs most of the costs from a reduction in fishing grounds, large sanctions, increasing fuel use, or energy to circumvent the MPA whilst receiving limited real-time benefits to offset these costs.

An MPA was significantly more likely to be supported by fishers who had multiple livelihoods (such as working in the tourism sector, as a farmer, or within the local government department) and were not solely dependent on fishing (Chaigneau, 2013). These fishers may benefit from larger and more plentiful fish in the future, and potentially future tourism benefits, whilst turning to other livelihoods for their more immediate needs. Those that are solely dependent on fishing, however, will bear the brunt of the immediate costs of reduced fishing grounds as the following fisher emphasises: 'They [those not supportive] depend on fishing for their food. It is not so much a problem for those who have their own farm for rice' (fisher, Bonbonon).

Feelings and emotions

A host of feelings and emotions arise from the establishment and management of MPAs and can influence support towards them. This was predominantly the result of the inductive approach of the research and in comparison to the socio-economics section does not necessarily assume all individuals simply to be narrow-minded rational actors. This incorporates the importance placed on Filipino culture, values, and customs in understanding support towards MPAs, but also to help in understanding why other factors discussed previously were found to be important predictors of support. Whilst some of these feelings and emotions may be held by oneself, others may arise out of judgement and in relation to others (e.g. pity, fear, jealousy). If one fails to take into account the feelings and emotions, or 'hearts' of the people affected by the MPA, it can have a negative impact on the support of individuals towards the MPA. It is important to consider peoples' value systems and design and manage an MPA in a way that does not contradict such a system.

In the Filipino context, notions of kinship, smooth interpersonal relations, family ties, and shame all are found to influence whether one acts in favour of the MPA or not. Whilst it may initially suggest that individuals will prevent their kin or family from encroaching the MPA as they do not want to be 'shamed' or cause tension within the community, the converse seems to be occurring. Individuals fail to report poachers as they are 'scared' to shame their own family and do not want to cause friction within the kinship system, and would prefer to keep interpersonal relations as calm as possible. Similarly, the valued notion of togetherness could also be encouraged to increase the solidarity value of MPAs by having individuals co-operating with each other and acting in favour of the MPA. This notion of togetherness, or '*kapwa*', refers to the community spirit, not doing things alone, and includes actions such as helping others, participating in community, and preventing conflict (Enriquez, 1986).

There is also a fatalistic attitude that is sometimes expressed in the Philippines when one tends to look at the world and nature as controlled by other beings different from oneself and governed by forces above (Andres, 1981). This leads to a feeling of helplessness and powerlessness as one believes that one's life is shaped by forces beyond their control. This may prevent individuals from co-operating and acting in favour of the MPA, irrespective of their attitudes (See Box 10.5).

Box 10.5 Feelings of powerlessness and agency as an example of feelings and emotions that influence support

Whilst solidarity, jealousy, and concern for the environment may motivate some individuals to co-operate and act in favour of the MPA, some may feel that they themselves cannot make a difference and will be less likely to take action.

When asked about the MPA some individuals would not want to get involved in answering questions and would defer to those 'in charge':

> 'Do you think all of the people here in Bonbonon support the MPA?
> I don't know, maybe the Fisher Association leaders know about that'
> (fisher, Bonbonon)

Others suggested they would not get involved in any action regarding the MPA as it is not up to them: 'That is the project of the government. We can't do anything about that [the MPA]' (fisher, Candaping B). Conversely, some felt like they could make a difference and had what psychologists would call a strong 'locus of control' meaning that individuals feel they can bring about change:

> 'I will make people follow me and I want to be a good example so that they will not abuse'.
>
> (fishers' wife, Candaping B)

> 'Even if I am not a bantay dagat (guard), I will still report encroachers'.
>
> (fisher, Candaping B).

These extracts highlight how support towards the MPA can vary greatly between individuals within communities based on differing personalities and characteristics. It is important to appreciate this diversity when institutionalising an MPA.

Other individual feelings and emotions were best explained by differences in personalities rather than Filipino culture and values. Issues such as jealousy towards other encroachers, laziness to act, a lack of environmental concern or concern about the future of the community, and a desire for freedom were all factors that were also found to negatively influence attitudes and actions towards the MPA (see Table 10.1).

Challenges for sustainable NRM management (and the case for MPAs)

Calls to increase the support of the community towards MPAs are frequently made with a plethora of different suggestions as to how to achieve this in order to increase the 'success' of MPAs (Aldon et al., 2010; Allegretti et al., 2012; Beger et al., 2004; Christie, 2005b; Christie & White, 1997; Courtney & White, 2000; Infield & Namara, 2001; Jentoft, Chuenpagdee, & Pascual-Fernandez, 2011; Russ & Alcala, 1999; Stoffle & Minnis, 2007; Tissot et al., 2009; Yasué et al., 2010). Whilst these are easy suggestions to make, in practice they may be difficult to undertake, as previously highlighted by the complexity of factors influencing MPAs approval and success. Based on the results presented in the earlier section, I suggest there are three issues in particular that complicate the process of understanding and ensuring support towards NRM strategies such as MPAs.

Heterogeneous communities

Not only does one need to consider the different factors that influence support and how they relate to each other, but one also needs to appreciate how these factors can influence certain individuals or groups differently. Whilst it is important to focus at the community level, the interests of a group are not the same as the actions and beliefs of that group and it is important to distinguish between groups and individuals (Agrawal, 2005).

MPAs have been argued to have varied effects on different individuals within a community depending on their wealth (Christie et al., 2003), dependence on the fishery (Gelcich, Edwards-Jones, & Kaiser, 2007), or types of fishing gear used (Dimech et al., 2009). Although these factors have been found to influence support towards MPAs, any intervention that focuses on these factors may only influence the support of certain individuals or groups. Providing alternative livelihoods for fishermen may increase support towards the MPA, but only for those fishermen that were highly dependent on the fishery in the first place. This, however, may do little for those part-time fishers that already have other income-generating activities.

It is therefore important to consider the fact that certain interventions may have different ramifications for different individuals within the community when attempting to develop, implement, or manage an MPA. Similar problems are also found when considering factors pertaining to feelings and emotions as these operate at an individual level. It becomes increasingly clear that one cannot just look at 'the community' as a homogeneous structure (Allison & Horemans, 2006; Coulthard, Johnson, & McGregor, 2011). To investigate what influences support and help explain it, one needs to look at the different individuals within the community and how their social relationships with others in the community and interactions can also influence support.

Temporal uncertainty

So far, I have explained the need to consider the plurality of factors influencing support and the need to explore these factors at both a community and an individual level. A temporal dimension however also needs to be considered when exploring community support. When going through the list of factors identified as influencing support, it becomes apparent that these are not static, but will change over time. This in turn explains why support towards MPAs is variable, and indeed I argue that support for MPAs may wax and wane over time (albeit to different extents) regardless of the community or the MPA considered.

Let us consider factors that are related to the management of the MPA that can influence support. Throughout the villages, there was desire for 'strong' management of the MPA and clear, fair leaders in place. Even if this is achievable, it is unlikely to remain the case over long periods of time. This is highlighted in one of the study sites where a number of individuals have mentioned that the management of the MPA used to be good, but has deteriorated over time. This is not uncommon in the Philippines where village captains get re-elected every three years. Furthermore, corruption and vote-buying is rife in the Philippines, partly due to the importance attributed to interpersonal relationships and reciprocity (Soriano, Dror, Alampay, & Bayugo, 2007) (if a political leader buys you a gift or provides you with a favour, this is normally reciprocated with a vote in this leader's favour). When election time comes, a leader's positive attributes become irrelevant in the face of bribery. The frequent change in leadership has important repercussions on MPAs because leaders can have a very influential impact on their management and support. At one site, a previous village official was very active in managing the MPA and was designated as the 'MPA chairman'. When the leader changed, so did the barangay officials and nobody was given a similar role under the new captain's management. This example highlights how the MPA exists within a social and political setting; the struggle for political control can play a big part in MPA management.

It is not only the way the MPA is managed that is likely to change over time. Socio-economic characteristics are bound to change as individuals' wealth may fluctuate, especially if, as commonly seen, some family members get jobs abroad and send remittances home (Andres, 1981). Individuals' fishing gear may change due to technological improvements or wear and tear of old gear, which will have different impacts on fisheries and associated ecosystems, but also on the amount of benefits individuals will get from the reefs. These changes will affect the ways in which individuals perceive the MPA and the way they interact with it, consequently influencing their support towards the MPA. Even the MPA characteristics can change. In one site, Suba, the size and location of the MPA was altered a few years after its implementation and was almost halved in size, with important consequences for its support. Whilst it initially lead to higher levels of support, this has waned considerably since the changes (Chaigneau, 2013). Importantly, in early years many community members may have high expectations as to the benefits the MPA can produce, but these expectations may shatter if the benefits are not forthcoming over time.

Uncontrollable factors

Actively influencing support towards the MPA may also prove to be difficult because some factors just cannot be controlled. Some MPA design factors can be changed, such as the size of the MPA and its boundaries. However, the presence of MPA benefits such as fishery spill-over is one that is difficult to control as it involves a wide range of factors. The nature and type of MPA benefits such as fishery spill-over are still contested in the MPA literature as they are dependent on a large array of factors. Current increases in the Filipino population (Salas, Abrigo, & Racelis, 2012), offshore trawling, illegal fishing from other countries (Palma, Tsamenyi, & Secretariat, 2008), and other environmental impacts, such as climate change, could mask, prevent, or limit any fishery benefits occurring from the MPA, irrespective of what happens at the community level.

So it is important to note that although certain factors may influence support, they may not be susceptible to change through interventions and may prove to be uncontrollable. Whilst many fail to report encroachers who may be friends, family, or neighbours, it may be difficult or unethical to promote enforcement at the expense of important values attributed to interpersonal relations and could lead to dire consequences for community solidarity. Broadly, most community-level factors such as MPA design and MPA management could be amenable to change with often direct consequences to the MPA. When considering differences between individuals such as personality traits and socio-economic differences, these factors may prove more difficult or impossible to change. Nevertheless, the MPA can be designed and managed in ways that appreciate these important factors that influence support.

Implications for sustainable natural resource management

The acknowledgement of the three challenges identified previously will be a necessary but not sufficient step to improve MPA management. Deciding on how best to boost support towards MPAs or NRM strategies through interventions is a daunting task when faced with the complex set of factors that are found to influence support.

The potential costs of intervening however may exceed the benefits of heightened support and, whilst it is unlikely that an MPA will be fully supported by the community, there are a few lessons learned that could be applied to most community-based MPAs in the Philippines, which could have important repercussions for the governance of natural resources in general.

Three opportunities for sustainable future NRM strategies that are in line with broader good governance principles (Lockwood, Davidson, Curtis, Stratford, & Griffith, 2010) and also consider community support are identified from this research and highlighted in the text that follows.

Realism and honesty with objectives

Firstly, it is important to explore the potential repercussions of MPA implementation and management on different individuals within the community. This would require village meetings that are representative of most individuals affected (not solely those more empowered or more involved) and an honest discussion about the potential negative and positive impacts of the proposed MPA or MPA changes. Whilst it is difficult to incorporate multiple and marginalised voices into decision making (Adger, Barnett, Brown, Marshall, & O'brien, 2012), it is important to ensure support and community buy-in to MPAs. Mental models approaches, for example, which explore peoples' thought processes, can be an effective way to understand knowledge systems and integrate plural perspectives to support design of management strategies (Adger et al., 2012). There is an increasing amount of literature on participatory processes that aim to incorporate multiple perspectives for decision making (for good practice in participation for decision making see Dietz & Stern, 2008).

Furthermore, differences in priorities are likely to escalate unless these different voices are considered (Hicks, Graham, & Cinner, 2013). It is important to understand how MPA priorities and goals came to fruition and whether they take into account concerns and interests of different stakeholders (Jentoft et al., 2011). It is thought to be particularly constructive to engage those affected in a process that explores their priorities. It ensures that respect can be built towards alternative views, it improves trust between those involved, and it highlights areas of agreement (Hicks et al., 2013). An honest representation and deliberation of the potential benefits and costs of an MPA (or various proposed changes) for various stakeholders should be clearly articulated by those in charge of the NRM strategy. This is an argument that is also believed to apply to all protected areas, not solely MPAs (Brockington & Wilkie, 2015). Adaptive co-management that focuses on shared 'learning-by-doing' and concentrates on the relationships, requirements, and capacity of managers, is one such approach that involves the sharing of information and participation of affected communities (Armitage, Berkes, & Doubleday, 2010). Only by these honest representations of costs and benefits can we explore the potential trade-offs, reach compromises, and ensure that most people are supportive of NRM strategies.

Consideration of the different costs and benefits

Secondly, it important to realise that support towards MPAs is not solely based on monetary or economic incentives. Whilst it is clear that the prospect of fishery spill-over and increased fish catch or tourism is an attractive prospect for many, and would undoubtedly boost support of individuals towards the MPA, there are a whole range of other factors that can influence support. As Table 10.1 highlights, other non-economic benefits of MPAs are important (Also see Coulthard et al., 2011). Furthermore, even if an MPA could lead to economic benefits to all within the community, inequalities between how much different individuals received (e.g. income obtained through tourism versus income obtained through increases in fish catch) would still lead to tensions through issues of fairness, jealousy, and equity. The local consequences of protected areas in general are highly contrary and are thought to distribute fortune and misfortune unevenly (Brockington & Wilkie, 2015). It is essential therefore not solely to describe MPAs as 'win–win' scenarios for all, but as a precautionary tool that can potentially reduce negative impacts of overfishing, climate change, and ensure that *future* generations may be better off than if no MPAs were implemented. A shift in focus from MPA benefits, to a reduction of costs arising from MPAs may help prevent issues of jealousy and of shattered expectations and could help boost support in the long run.

Take cultural values into account

Finally, building support towards MPAs may not necessarily involve the MPA or issues surrounding its management. Various cultural values were also found to be important factors influencing support. It is indeed more broadly argued that governance challenges could be reduced if these values of stakeholders are made explicit and taken into account when managing natural resources (Song, Chuenpagdee, & Jentoft, 2013; Sowman, 2015). The notion of togetherness in Filipino society, for example, highlights the importance of respect for mutual help, but with increasing globalisation, some of these values are becoming eroded. There may be other ways to boost support by promoting cooperation and trust between individuals and improving solidarity between individuals within the community, which could have little to do directly with the MPA and issues surrounding it. For example, this notion of 'team spirit' or community cohesion is felt during village celebrations, but not in day-to-day activities. Promoting community activities such as fiestas, beach clean-ups, and mangrove planting for example could promote cooperation, solidarity, and a feeling of team spirit that are important to strengthen support towards MPAs. Different cultural contexts would have different types of community values, processes etc. that could be fostered in order to achieve a more united perspective of NRM practices for managing common resources.

Conclusion

Many MPAs are failing to reach the various objectives placed on them due in large part to a lack of community support, a fate that is shared by other natural resource management strategies, especially those that are community based with little extended funding for monitoring and enforcement. However, looking at the predictors of community support, it is a complex endeavour with no clear 'one size fits all' solution. Nevertheless, we now have a clearer idea as to how the design of the MPA, the way it is managed, and its impacts can influence attitudes and behaviours of the community. Furthermore, different socioeconomic characteristics of individuals can influence support and alter the way they perceive the MPA and are impacted by it. Finally, not only cultural values, but also the feelings and emotions of individuals need to be considered in how MPAs are designed, implemented and managed. These sometimes supersede the thoughts and actions of the assumed *rational* self-interested individual. Indeed the rational-actor model is often deemed to be inadequate for explaining attitudes and behaviours and richer models are required that look to sociology to obtain a more accurate picture of social influences on human behaviour (Ellickson, 1989). Although it may be impossible to have an MPA that is fully supported by the whole community over time, striving to develop and manage MPAs or other NRM strategies to increase support may be better than not acting at all.

A detailed look at what influences support has helped identify challenges (the need to consider the heterogeneity of communities, temporal uncertainties, and uncontrollable factors), but also opportunities (to focus on a wider variety of benefits that NRM strategies can provide, to be more honest and realistic with NRM objectives, and to take context-specific cultural values into account) for NRM strategies. These important lessons learned for improving support towards

NRM strategies also coincide with broader governance principles of fairness, transparency, and inclusiveness, which are acknowledged as essential for successful management of natural resources (Lockwood et al., 2010).

These suggestions imply that for supported, successful NRM strategies that work in the long term, one needs to consider the wide variety of costs and benefits these strategies have on different individuals within communities and to be honest and realistic as to how these impacts will play out in both the short and long term. A move away from solely focusing on the economic benefits (that are sometimes untenable and are unlikely to benefit all within the community equally) towards the cost-reduction benefits that MPAs can provide in the longer term may bolster support towards NRM strategies, such as MPAs, regardless of whether people perceive direct benefits to themselves.

Note

1 In terms of attitude, they were asked to discuss how they felt about the MPA, what they like or dislike about it, and more specifically in the questionnaire, what effect they thought the MPA had on local fisheries. They were also asked: how effective the MPA was at preserving the reef; how well it was managed; whether they would like to keep or remove the MPA; whether their attitude had changed since implementation; and how they thought the MPA had affected the community. In terms of enforcement and compliance they were asked about what they do or would do if they spotted an encroacher and what keeps them from reporting encroachers or from poaching; what was preventing them from acting in favour of the MPA; and what were the incentives for monitoring and reporting encroachers.

References

Abesamis, R. A., Russ, G. R., & Alcala, A. C. (2006). Gradients of abundance of fish across no-take marine reserve boundaries: Evidence from Philippine coral reefs. *Aquatic Conservation-Marine and Freshwater Ecosystems*, *16*(4), 349–371. doi:10.1002/aqc.730.

Adams, W. M., Aveling, R., Brockington, D., Dickson, B., Elliott, J., Hutton, J., . . . Wolmer, W. (2004). Biodiversity conservation and the eradication of poverty. *Science*, *306*(5699), 1146–1149.

Adger, W. N., Barnett, J., Brown, K., Marshall, N., & O'Brien, K. (2012). Cultural dimensions of climate change impacts and adaptation. *Nature Climate Change*, *3*(2), 112–117.

Agardy, T., Bridgewater, P., Crosby, M. P., Day, J., Dayton, P. K., Kenchington, R., . . . Peau, L. (2003). Dangerous targets? Unresolved issues and ideological clashes around marine protected areas. *Aquatic Conservation-Marine and Freshwater Ecosystems*, *13*(4), 353–367. doi:10.1002/aqc.583.

Agrawal, A., Gupta, A., Hathaway, M., Narotzky, S., Raffles, H., Skaria, A., Sundar, N., 2005. Environmentality: Community, intimate government, and the making of environmental subjects in Kumaon, India. *Current Anthropology* 46, 161–190.

Aldon, M. E. T., Fermin, A. C., & Agbayani, R. F. (2010). Sociocultural context of fishers' participation in coastal resources management in Anini-y; Antique in west central Philippines. *Fisheries Researhc*, 107, 112–121.

Allegretti, A. M., Vaske, J. J., & Cottrell, S. (2012). Predicting stakeholder support for fishery policies. *Coastal Management*, *40*(1), 20–32.

Allendorf, T. D. (2007). Residents' attitudes toward three protected areas in southwestern Nepal. *Biodiversity and Conservation*, *16*(7), 2087–2102.

Allison, E. H., & Ellis, F. (2001). The livelihoods approach and management of small-scale fisheries. *Marine Policy*, *25*(5), 377–388.

Allison, E. H., & Horemans, B. (2006). Putting the principles of the sustainable livelihoods approach into fisheries development policy and practice. *Marine Policy*, *30*(6), 757–766. doi:10.1016/j.marpol.2006.02.001.

Altieri, M. A. (2002). Agroecology: The science of natural resource management for poor farmers in marginal environments. *Agriculture, Ecosystems & Environment*, *93*(1), 1–24.

Anand, S., & Sen, A. (2000). Human development and economic sustainability. *World Development*, *28*(12), 2029–2049.

Andres, T. Q. D. (1981). *Understanding Filipino Values: A Management Approach*. Quezon City. New Day Publishers.

Armitage, D., Berkes, F., & Doubleday, N. (2010). *Adaptive Co-Management: Collaboration, Learning, and Multi-Level Governance*. Vancouver, Toronto. UBC Press.

Ballantine, W. J. (1994). Networks of 'no-take' marine reserves are practical and necessary. In N. Shackell & J. Willison (Eds.), *Marine Protected Areas and Sustainable Fisheries* (pp. 13–20). Wolfville, NS, Canada: Science and Management of Protected Areas Association.

Bamberg, S., & Möser, G. (2007). Twenty years after Hines, Hungerford, and Tomera: A new meta-analysis of psycho-social determinants of pro-environmental behaviour. *Journal of Environmental Psychology*, *27*(1), 14–25.

Beger, M., Harborne, A. R., Dacles, T. P., Solandt, J. L., & Ledesma, G. L. (2004). A framework of lessons learned from community-based marine reserves and its effectiveness in guiding a new coastal management initiative in the Philippines. *Environmental Management*, *34*(6), 786–801. doi:10.1007/s00267-004-0149-z.

Béné, C., Arthur, R., Norbury, H., Allison, E. H., Beveridge, M., Bush, S., . . . Squires, D. (2016). Contribution of fisheries and aquaculture to food security and poverty reduction: Assessing the current evidence. *World Development*, *79*, 177–196.

Berkes, F., Hughes, T. P., Steneck, R. S., Wilson, J. A., Bellwood, D. R., Crona, B., . . . Worm, B. (2006). Ecology: Globalization, roving bandits, and marine resources. *Science*, *311*(5767), 1557–1558. doi:10.1126/science.1122804.

Brockington, D., & Wilkie, D. (2015). Protected areas and poverty. *Phil. Trans. R. Soc. B*, *370*(1681), 20140271.

Browman, H. I., & Stergiou, K. I. (2004). Perspectives on ecosystem-based approaches to the management of marine resources. *Marine Ecology-Progress Series*, *274*, 269–270.

Burke, L. R., K. Spalding, M, & Perry, A. (2011). Reefs at Risk Revisited. Executive Summary. World Resources Institute, Washington, DC, USA, 10 pp.

Cabral, R. B., Aliño, P. M., Balingit, A. C. M., Alis, C. M., Arceo, H. O., Nañola Jr, C. L., . . . Partners, M. (2014). The Philippine marine protected area (MPA) database. *Philippine Science Letters*, *7*(2), 300–308.

Carr, M. H., Neigel, J. E., Estes, J. A., Andelman, S., Warner, R. R., & Largier, J. L. (2003). Comparing marine and terrestrial ecosystems: Implications for the design of coastal marine reserves. *Ecological Applications*, *13*(sp1), 90–107.

Castro, G. (2000). Natural resource management. *A Draft Discussion Paper Submitted to the World Bank.*

Central Intelligence Agency (2005). The World Factbook. Retrieved from www.cia.gov/library/publications/the-world-factbook/geos/rp.html.

Chaigneau, T. (2013). *Understanding community support towards three marine protected areas in the visayas region of the Philippines*. (PhD), University of East Anglia, Norwich.

Chaigneau, T., & Brown, K. (2016). Challenging the win–win discourse on conservation and development: Analyzing support for marine protected areas. *Ecology and Society, 21*(1), 36.

Chapin, F. S., Carpenter, S. R., Kofinas, G. P., Folke, C., Abel, N., Clark, W. C., . . . Young, O. R. (2010). Ecosystem stewardship: Sustainability strategies for a rapidly changing planet. *Trends in Ecology & Evolution, 25*(4), 241–249.

Christie, P. (2004). Marine protected areas as biological successes and social failures in southeast Asia. *Aquatic Protected Areas as Fisheries Management Tools, 42*, 155–164.

Christie, P. (2005a). Is integrated coastal management sustainable? *Ocean & Coastal Management, 48*(3–6), 208–232. doi:10.1016/j.ocecoaman.2005.04.002.

Christie, P. (2005b). Observed and perceived environmental impacts of marine protected areas in two Southeast Asia sites. *Ocean & Coastal Management, 48*(3–6), 252–270. doi:10.1016/j.ocecoaman.2005.04.012.

Christie, P., McCay, B. J., Miller, M. L., Lowe, C., White, A. T., Stoffle, R., . . . Pollnac, R. B. (2003). Toward developing a complete understanding: A social science research agenda for marine protected areas. *Fisheries, 28*(12), 22–26.

Christie, P., Pollnac, R. B., Oracion, E. G., Sabonsolin, A., Diaz, R., & Pietri, D. (2009). Back to basics: An empirical study demonstrating the importance of local-level dynamics for the success of tropical marine ecosystem-based management. *Coastal Management, 37*(3–4), 349–373. doi:10.1080/08920750902851740.

Christie, P., & White, A. T. (1997). Trends in development of coastal area management in tropical countries: From central to community orientation. *Coastal Management, 25*(2), 155–181.

Christie, P., & White, A. T. (2007). Best practices for improved governance of coral reef marine protected areas. *Coral Reefs, 26*(4), 1047–1056. doi:10.1007/s00338-007-0235-9.

Christie, P., White, A. T., & Deguit, E. (2002). Starting point or solution? Community-based marine protected areas in the Philippines. *Journal of Environmental Management, 66*(4), 441–454. doi:10.1006/jema.2002.0595.

Chuenpagdee, R., & Jentoft, S. (2009). Governability assessment for fisheries and coastal systems: A reality check. *Human Ecology, 37*(1), 109–120.

Costello, C., Ovando, D., Hilborn, R., Gaines, S. D., Deschenes, O., & Lester, S. E. (2012). Status and Solutions for the World's Unassessed Fisheries. *Science*, 338(6106), 517-520.

Coulthard, S., Johnson, D., & McGregor, J. A. (2011). Poverty, sustainability and human wellbeing: A social wellbeing approach to the global fisheries crisis. *Global Environmental Change-Human and Policy Dimensions, 21*(2), 453–463. doi:10.1016/j.gloenvcha.2011.01.003.

Courtney, C. A., & White, A. T. (2000). Integrated coastal management in the Philippines: Testing new paradigms. *Coastal Management, 28*(1), 39–53.

Dietz, T., & Stern, P. C. (2008). *Public Participation in Environmental Assessment and Decision Making*. National Academies Press.

Dimech, M., Darmanin, M., Smith, I. P., Kaiser, M. J., & Schembri, P. J. (2009). Fishers' perception of a 35-year old exclusive Fisheries Management Zone. *Biological Conservation, 142*(11), 2691–2702. doi:10.1016/j.biocon.2009.06.019.

Doyal, L., & Gough, I. (1991). *A Theory of Human Need*. London: Palgrave Macmillan.

Ellickson, R. C. (1989). Bringing culture and human frailty to rational actors: A critique of classical law and economics. *Chicago-Kent Law Review*, 65, 23.

Enriquez, V. G. (1986). Kapwa: A Core Concept in Filipino Social Psychology. In A. Aganon & S. M. Assumpta David (Eds.). *Sikolohiyang Pilipino*. Quezon City, The Philippines: National Book Store.

Froese, R., & Pauly, D. (2007). FishBase. *FishBase*, www.fishbase.org.

Gelcich, S., Edwards-Jones, G., & Kaiser, M. J. (2005). Importance of attitudinal differences among artisanal fishers toward co-management and conservation of Marine Resources. *Conservation Biology*, 19(3), 865–875.

Gelcich, S., Edwards-Jones, G., & Kaiser, M. J. (2007). Heterogeneity in fishers' harvesting decisions under a marine territorial user rights policy. *Ecological Economics*, 61(2–3), 246–254. doi:10.1016/j.ecolecom.2006.02.017.

Gelcich, S., Kaiser, M. J., Castilla, J. C., & Edwards-Jones, G. (2008). Engagement in co-management of marine benthic resources influences environmental perceptions of artisanal fishers. *Environmental Conservation*, 35(1), 36–45.

Gerber, L. R., Heppell, S. S., Ballantyne, F., & Sala, E. (2005). The role of dispersal and demography in determining the efficacy of marine reserves. *Canadian Journal of Fisheries and Aquatic Sciences*, 62(4), 863–871.

Gjertsen, H. (2005). Can habitat protection lead to improvements in human well-being? Evidence from marine protected areas in the Philippines. *World Development*, 33(2), 199–217. doi:10.1016/j.worlddev.2004.07.009.

Green, S. J. (2003). *Philippine fisheries in crisis: A framework for management*: Coastal Resource Management Project of the Department of Environment and Natural Resources, 2003, 77pp.

Griggs, D., Stafford-Smith, M., Gaffney, O., Rockström, J., Öhman, M. C., Shyamsundar, P., . . . Noble, I. (2013). Policy: Sustainable development goals for people and planet. *Nature*, 495(7441), 305–307.

Gubbay, S. (1995). *Marine Protected Areas: Past, Present and Future*. Chapman & Hall, London, p. 232.

Halpern, B. S., Lester, S. E., & McLeod, K. L. (2010). Placing marine protected areas onto the ecosystem-based management seascape. *Proceedings of the National Academy of Sciences*.

Hicks, C. C., Graham, N. A., & Cinner, J. E. (2013). Synergies and tradeoffs in how managers, scientists, and fishers value coral reef ecosystem services. *Global Environmental Change* 23.6 (2013): 1444-1453.

IFAD (2010). Enabling poor rural people to overcome poverty in the Philippines. Retrieved from www.ruralpovertyportal.org/web/guest/country/home/tags/philippines.

Infield, M. (2001). Cultural values: A forgotten strategy for building community support for protected areas in Africa. *Conservation Biology*, 15(3), 800–802.

Infield, M., & Namara, A. (2001). Community attitudes and behaviour towards conservation: An assessment of a community conservation programme around Lake Mburo National Park, Uganda. *Oryx*, 35(1), 48–60.

Jennings, S., M. J. Kaiser, & J. D. Reynolds. (2001). *Marine Fisheries Ecology*. Blackwell. Science, Oxford.

Jentoft, S., Chuenpagdee, R., & Pascual-Fernandez, J. J. (2011). What are MPAs for: On goal formation and displacement. *Ocean and Coastal Management*, 54(1), 75–83.

Jentoft, S., van Son, T. C., & Bjørkan, M. (2007). Marine protected areas: A governance system analysis. *Human Ecology*, 35(5), 611–622.

Johannes, R. E. (1998). The case for data-less marine resource management: Examples from tropical nearshore finfisheries. *Trends in Ecology & Evolution*, 13(6), 243–246.

Jones, P. J. S. (2006). Collective action problems posed by no-take zones. *Marine Policy, 30*(2), 143–156. doi:10.1016/j.marpol.2004.10.003.
Jones, P. J. S. (2007). Point-of-View: Arguments for conventional fisheries management and against no-take marine protected areas: Only half of the story? *Reviews in Fish Biology and Fisheries, 17*(1), 31–43.
Kelleher, G., & Kenchington, R. (1992). *Guidelines for Establishing Marine Protected Areas. A Marine Conservation and Development Report.* Gland, Switzerland: IUCN.
Kelly, C. J., & Codling, E. A. (2006). 'Cheap and dirty' fisheries science and management in the North Atlantic. *Fisheries Research, 79*(3), 233–238. doi:10.1016/j.fishres.2006.03.007.
Kollmuss, A., & Agyeman, J. (2002). Mind the gap: Why do people act environmentally and what are the barriers to pro-environmental behavior? *Environmental Education Research*, 8(3), 239–260.
Lan, D. T. K. (2009). *The Attitudes and Perceptions of Resource Users and Managers Towards the Nha Trang Bay Marine Protected Area management.* Vietnam. MS thesis. Universitetet i Tromsø.
Licuanan, W. Y., & Capili, E. B. (2003). Range extension of Japanese Scleractinia to the Philippines. *Journal of the Japanese Coral Reef Society* 2003.5 (2003): 63–67.
Lockwood, M., Davidson, J., Curtis, A., Stratford, E., & Griffith, R. (2010). Governance principles for natural resource management. *Society and Natural Resources, 23*(10), 986–1001.
MA (2005). *Ecosystems and human well-being.* (Vol. 5). Washington, DC: Island Press.
Mcshane, T. O., Hirsch, P. D., Trung, T. C., Songorwa, A. N., Kinzig, A., Monteferri, B., . . . Pulgar-Vidal, M. (2011). Hard choices: Making trade-offs between biodiversity conservation and human well-being. *Biological Conservation, 144*(3), 966–972.
Mehta, J. N., & Kellert, S. R. (1998). Local attitudes toward community-based conservation policy and programmes in Nepal: A case study in the Makalu-Barun Conservation Area. *Environmental Conservation, 25*(4), 320–333.
Millar, C. I., Stephenson, N. L., & Stephens, S. L. (2007). Climate change and forests of the future: Managing in the face of uncertainty. *Ecological Applications, 17*(8), 2145–2151.
Milne, N., & Christie, P. (2005). Financing integrated coastal management: Experiences in Mabini and Tingloy, Batangas, Philippines. *Ocean & Coastal Management, 48*(3–6), 427–449. doi:10.1016/j.ocecoaman.2005.04.004.
NSCB (2006). Annual report. Retrieved from www.nscb.gov.ph/download/NSCB2006AnnualReport.pdf.
NSCB (2013). Population projections. Retrieved from www.nscb.gov.ph/secstat/d_popnProj.asp.
Ostrom, E., Janssen, M. A., & Anderies, J. M. (2007). A diagnostic approach for going beyond panaceas. *Proceedings of the National Academy of Sciences of the United States of America, 104*, 15176–15178. doi:10.1073/pnas.0701886104.
Pajaro, M. G., Mulrennan, M. E., Alder, J., & Vincent, A. C. J. (2010). Developing MPA effectiveness indicators: Comparison within and across stakeholder groups and communities. *Coastal Management*, 38(2), 122–143. doi:10.1080/08920751003633094.
Palma, M. A., Tsamenyi, M., & Secretariat, A. P. E. C. (2008). *Case Study on the Impacts of Illegal, Unreported and Unregulated (IUU) Fishing in the Sulawesi Sea.* APEC Fisheries Working Group. Australian National Center for Ocean Resources and Security.
Parfit, D. (1984). *Reasons and Persons.* Oxford: Oxford University Press.

Petty, R. E., & Cacioppo, J. T. (1996). *Attitudes and Persuasion: Classic and Contemporary Approaches*. Boulder, Colorado. Westview Press.

Pita, C., Pierce, G. J., Theodossiou, I., & Macpherson, K. (2011). An overview of commercial fishers' attitudes towards marine protected areas. *Hydrobiologia, 670*(1), 289–306.

Pollnac, R. B., Crawford, B. R., & Gorospe, M. L. G. (2001). Discovering factors that influence the success of community-based marine protected areas in the Visayas, Philippines. *Ocean & Coastal Management, 44*(11–12), 683–710.

Pollnac, R. B., & Pomeroy, R. S. (2005). Factors influencing the sustainability of integrated coastal management projects in the Philippines and Indonesia. *Ocean & Coastal Management*, 48(3–6), 233–251. doi:10.1016/j.ocecoaman.2005.04.003.

Pollnac, R. B., Pomeroy, R. S., & Harkes, I. H. (2001). Fishery policy and job satisfaction in three Southeast Asian fisheries. *Ocean & Coastal Management, 44*(7), 531–544.

Pomeroy, R. S., & Carlos, M. B. (1997). Community-based coastal resource management in the Philippines: A review and evaluation of programs and projects, 1984–1994. *Marine Policy, 21*(5), 445–464.

Pomeroy, R. S., Pollnac, R. B., Katon, B. M., & Predo, C. D. (1997). Evaluating factors contributing to the success of community-based coastal resource management: The central Visayas regional project-1, Philippines. *Ocean & Coastal Management*, 36(1–3), 97–120.

Pomeroy, R. S., Watson, L. M., Parks, J. E., & Cid, G. A. (2005). How is your MPA doing? A methodology for evaluating the management effectiveness of marine protected areas. *Ocean & Coastal Management*, 48(7–8), 485–502. doi:10.1016/j.ocecoaman.2005.05.004.

Roberts, C. M., & Polunin, N. V. C. (1991). Are marine reserves effective in management of reef fisheries? *Reviews in Fish Biology and Fisheries, 1*(1), 65–91.

Roberts, C. M., & Polunin, N. V. C. (1993). Marine Reserves: Simple solutions to managing complex fisheries. *Ambio, 22*(6), 363–368.

Russ, G. R., & Alcala, A. C. (1999). Management histories of Sumilon and Apo Marine Reserves, Philippines, and their influence on national marine resource policy. *Coral Reefs, 18*(4), 307–319.

Russ, G. R., Stockwell, B., & Alcala, A. C. (2005). Inferring versus measuring rates of recovery in no-take marine reserves. *Marine Ecology-Progress Series, 292*, 1–12.

Salas, J., Abrigo, M. R. M., & Racelis, R. H. (2012). *Implications of Projected Philippine Population Growth, Age Structure Change, and Aging: Using National Transfer Accounts Results* (No. 2012-30). PIDS Discussion Paper Series.

Scherr, S. J. (2000). A downward spiral? Research evidence on the relationship between poverty and natural resource degradation. *Food Policy, 25*(4), 479–498.

Song, A. M., Chuenpagdee, R., & Jentoft, S. (2013). Values, images, and principles: What they represent and how they may improve fisheries governance. *Marine Policy, 40*, 167–175.

Soriano, E., Dror, D., Alampay, E., & Bayugo, Y. V. (2007). Attitudes toward solidarity, risk, and insurance in the rural Philippines. In D. M. Dror and A. S. Preker (Eds.), *Social Reinsurance: A New Approach to Sustainable Community Health Financing* (pp. 377–394). Washington: World Bank/ ILO.

Sowman, M. (2015). Governability challenges facing small-scale fishers living adjacent to Marine protected areas in South Africa. In Jentoft, S., Chuenpagdee, R. (Eds.). *Interactive Governance for Small-Scale Fisheries* (pp. 359–377). Springer.

Spalding, M. D., Ravilious, C, & Green, E. P. (2001). *World Atlas of Coral Reefs: Prepared at the UNEP World Conservation Monitoring Centre*. Berkeley, CA: University of California Press.

Steffen, W., Persson, Å., Deutsch, L., Zalasiewicz, J., Williams, M., Richardson, K., . . . Gordon, L. (2011). The anthropocene: From global change to planetary stewardship. *Ambio*, *40*(7), 739–761.

Stoffle, R., & Minnis, J. (2007). Marine protected areas and the coral reefs of traditional settlements in the Exumas, Bahamas. *Coral Reefs*, *26*(4), 1023–1032. doi:10.1007/s00338-007-0264-4,

Tanner, C. (1999). Constraints on environmental behaviour. *Journal of Environmental Psychology*, *19*(2), 145–157.

Tissot, B. N., Walsh, W. J., & Hixon, M. A. (2009). Hawaiian islands Marine ecosystem case study: Ecosystem- and community-based management in Hawaii. *Coastal Management*, *37*(3–4), 255–273. doi:10.1080/08920750902851096.

Tomascik, T., & Mah, A. J. (1997). *The Ecology of the Indonesian Seas*. Oxford: Oxford University Press, 829–906.

WCED, U. (1987). Our common future. *World Commission on Environment and Development*. Oxford: Oxford University Press.

White, A. T., Alino, P. M., & Meneses, A. T. (2006). Creating and managing marine protected areas in the Philippines. *Fisheries Improved for Sustainable Harvest Project, Coastal Conservation and Education Foundation, Inc. and University of the Philippines Marine Science Institute, Cebu City, Philippines*, 83, 00–03.

White, A. T., Courtney, C. A., & Salamanca, A. (2002). Experience with marine protected area planning and management in the Philippines. *Coastal Management*, *30*(1), 1–26.

White, A. T., & Vogt, H. P. (2000). Philippine coral reefs under threat: Lessons learned after 25 years of community-based reef conservation. *Marine Pollution Bulletin*, *40*(6), 537–550.

Worm, B., Hilborn, R., Baum, J. K., Branch, T. A., Collie, J. S., Costello, C., . . . Zeller, D. (2009). Rebuilding global fisheries. *Science*, *325*(5940), 578–585. doi:10.1126/science.1173146.

Yasué, M., Kaufman, L., & Vincent, A. (2010). Assessing ecological changes in and around marine reserves using community perceptions and biological surveys. *Aquatic Conservation: Marine and Freshwater Ecosystems*, *20*(4), 407–418.

11 An ontology of development in the geopolitical North

Resource extraction in the Canadian Northwest Territories and the shift in Indigenous experiences of nature

Brice Perombelon

Introduction

Arctic societies are now faced with the systemic impacts of melting sea-ice (Meier et al., 2014). Through a sleight of hand that transformed an ecological and human rights issue into a geopolitical one, circumpolar States have interpreted these impacts in terms of future potentialities for economic growth (Sejersen, 2015a). In Canada, over a period of nine years of Conservative majority,[1] the Federal government implemented the Northern Strategy: an approach to Arctic issues that related climate change to opportunities for social modernisation and political change, particularly for Indigenous peoples (GOC, 2009).

As the following quotation shows, Prime Minister Stephen Harper's Northern Strategy aimed to assert Canada's sovereignty in its northernmost regions by ensuring their sustainable development through an increase in the number of resource extraction projects (GOC, 2009, pp. 1, 8, 16):

> The Northern tradition of respect for the land and the environment is paramount and the principles of responsible and sustainable development anchor all decision-making and action [. . .]. Few countries are more directly affected by changes in the Arctic climate – or have as much at stake – as Canada. We have an important role to play in the ongoing stewardship of the Canadian Arctic, its vast resources and its potential. [. . .]. Canada will continue to support the sustainable development of these strategic resource endowments.

This quotation also shows that, in the Northern Strategy, the sustainable preservation of the circumpolar region's pristine ecosystems and the exploitation of its natural resources, opposed as these may be, sustain, rather than contradict, one another (GOC, 2009). It also confines the question of sustainable development in the North to immaterial ideals (of sovereignty, stabilisation, progress) by conveniently excluding the everyday lives of Indigenous peoples, which are still embedded in the material experience of inequality, disparity, and exclusion (through the commodification of traditional lands). Indeed, for many

Indigenous peoples living in these areas, oil and gas development have often meant environmental destruction, economic and social deprivation, as well as political and cultural dispossession (Wilson, 2009; Duhaime & Caron, 2006). In this context, the Northern Strategy's sustainable discourse also helped to further co-opt local Indigenous elites into believing a grand geopolitical narrative whose final aims are the extraction of hydrocarbon resources and the assertion of the State's authority and capitalist ideology in the barren lands of Canada's High North. After all, most rights over the NWT's sub-surface remain under Federal control.

In this chapter I ask what a utilitarian version of sustainable development, as promoted by Harper's government, has meant for Indigenous peoples. And how has it been integrated within the politically decolonised context of the post-land claims settlement era, particularly when it comes to the exploitation of natural resources?

I argue that one explanation can be found in the experience of colonisation of the Canadian North (Coates, 1985). As I will demonstrate, colonisation and post-colonisation,[2] essentially through an historical process of commensuration of the North's natural capital, is the structural force that has transformed the ways that Northern Indigenous peoples live, act, and behave since the time of first contact (Milloy, 2003; Perley, 1993; Fumoleau, 2007; Krech III, 1990). I have called this process of refashioning a society's ontological norms, or its sets of validated and legitimised ways of being, an *ontological shift*.[3] I contend that it is through the everyday experiences and understandings of dispossession and empowerment that one can understand the implications of this shift for Northern Indigenous peoples. I will show that in the Indigenous North, it is indeed at the ontological level (where individual ways of being are negotiated) that narratives pertaining to natural resources are translated into traditional or EuroCanadian[4] practices (which underpins the continuous exploitation of nature, even if at a reduced pace).

In the following sections, I will use the example of Tulita[5] to analyse the nature of the model of economic development that is being pursued in the Sahtu region, an area predominantly populated by Indigenous groups and which is located in Canada's Northwest Territories. I begin by presenting the historical and socio-political context of this community. I focus in particular on the impacts of past colonial policies on experiences of land dispossession and imposed capitalist forms of economic development. I then review the community's main decolonial[6] response to this phenomenon: the implementation of a land claims settlement agreement. I go on to suggest that the effects of this agreement, when applied to the sustainable management of natural resources, have been socially and politically divisive because they have embedded a non-traditional ontology based on the domination and exploitation of nature within local traditional practices and belief systems. I conclude by suggesting that resource development and sustainable development governance strategies need to be channelled through the empowerment of individuals in order for them to be successful catalysts of positive existential, social, cultural, and economic change.

History of resource exploitation in the Sahtu region

Tulita is one of the five communities of the Sahtu region (See Figure 11.1). It is predominantly composed of Sahtu Dene and Metis[7] peoples and its development as a permanent settlement for semi-nomadic Indigenous groups living in this remote area of the Canadian Sub-arctic is directly linked to resource extraction activities (Piper, 2009). This process began in the 19th century with the fur trade industry, which introduced commercial exchanges and a debt-based economy into Sahtu Dene society (Krech III, 1990; Ray, 1990; Yerbury, 1986). Then, following the discovery of oil 72 kilometres north of Tulita in 1920, Treaty 11 gave the Canadian State full control over land surface and sub-surface rights of access and exploitation (Fumoleau, 2007). Over the course of the 20th century, two other major resource development projects took place in the Sahtu area. The first was the extraction of pitchblende, uranium, and silver at Port Radium, on the eastern shore of Great Bear Lake, from 1932 to 1982. The second project was the building of the Canol pipeline by the US Army between 1942 and 1944 (Van Wyck, 2010; Lackenbauer & Farish, 2007, p. 925). By the

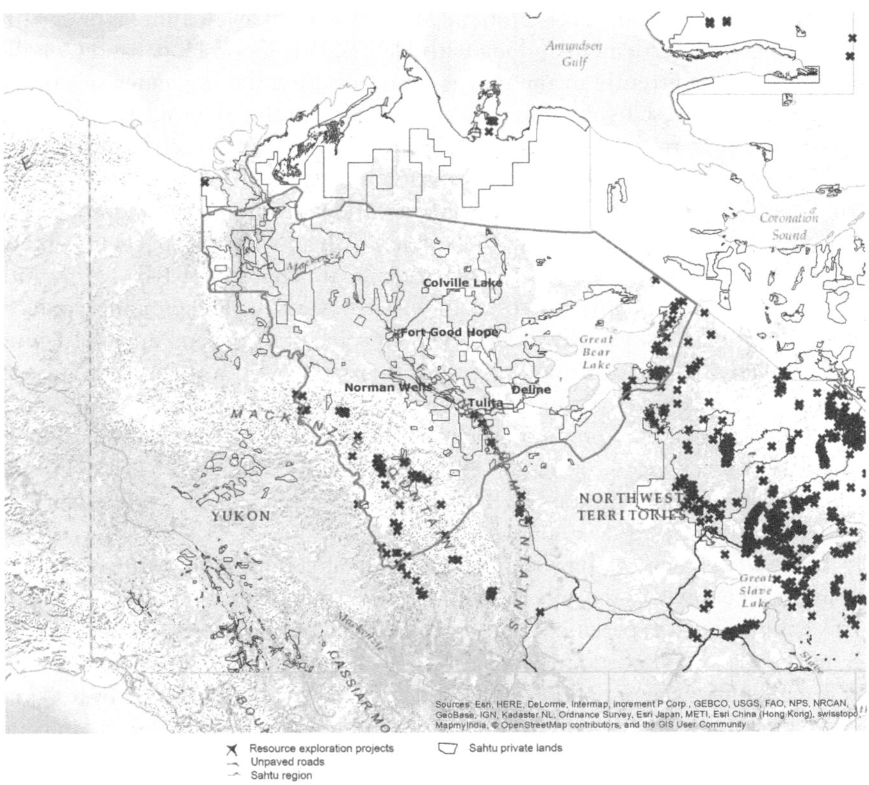

Figure 11.1 Map of the Sahtu region with an emphasis on resource exploitation activities

end of the 1960s, State-funded social, health, housing, and education programs achieved the transformation of Tulita into a year-round place of habitation for four Sahtu Dene groups: the hare, willow lake, mountain, and bear lake peoples (Sabin, 1995). All three groups shared a common history of environmental rights violation and physical violence that culminated in the dispossession of their traditional lands by the colonial State (Fumoleau, 2007; Berger, 1999; see also Crosby, 2015). This ensured the pursuit of an exploitative and structurally unequal relationship throughout the majority of the 20th century (Helm, 2000; Fumoleau, 2007).

In the 1970s, however, when faced with the prospect of yet another resource development project, the Mackenzie Valley Pipeline, proposed by a group of multinational oil companies, a group of 16 Dene leaders filed a legal caveat with the Federal government (Dickerson, 1992). It insisted that the Dene still had a legal title to 450,000 square miles of land and opened the way to the formal recognition by the Canadian State that Dene rights and title to land had not been extinguished. This led to the signing of a Sahtu Dene and Metis Comprehensive Land Claims Settlement Agreement in 1993, following five years of negotiation (Kulchyski, 2005). This agreement is considered a landmark achievement in the process of recognition of Indigenous rights to politico-environmental and cultural self-governance in Canada (Dalton, 2006). At the local level, Tulita is one of the four Indigenous settlements (along with Deline, Fort Good Hope, and Colville Lake) that are currently in the process of negotiating the implementation of a community-based self-government, a right secured in the SDLMCA.

As can be inferred from Figure 11.1, Tulita's geographic and economic proximity to most, if not all, current resource exploitation activities in the central Mackenzie valley, makes it the only predominantly Indigenous settlement of the Sahtu region where residents continue to experience the combined effects of capital-intensive resource development activities, climate change (SRRB, 2014), and, more recently, sovereignty-assertion strategies by way of a close proximity to Norman Wells (CBC News, 2012). However, for some of my informants, these phenomena only represented the most recent forms taken by the Canadian State to exert colonial control. For example, during an interview I conducted in March 2015 in Yellowknife, a prominent youth Dene leader, Denezeh Nakehko interpreted climate change as one composing part of a neo-colonial continuum linking resource development in the NWT to global capitalism and to Stephen Harper's Northern Strategy.

Tulita's everyday life has been significantly affected by the intensity of the combined cumulative impacts wrought by these phenomena. For many residents this has also been experienced through a colonial prism. For instance, the community's social and economic cycles are now highly dependent on the business contracts, employment opportunities, and financial transfers provided by resource companies. Economic and industrial activities start in January–February and end in April–May, when large supply and equipment trucks can travel to this area of the High North. Life in Tulita has thus become determined by the particular temporality of global petro-capital. In this sense, this structure of depency reproduces that of the colonial economy. The Northern Strategy relied on and made strategic use of this situation. For instance, whilst this political narrative officially aimed to assert the country's sovereignty over its Arctic territories (CBC News, 2010; GOC, 2009), it in fact presented Canada's northern Indigenous peoples as lacking enough social or human capital to make a profitable use of the land on their own (GOC, 2009). It thus strengthened the narrative according to which the social and economic development of Indigenous communities can only be ensured through a capitalist-industrial mode of production and development.

Indeed, the Northern Strategy interpreted improvements in northern governance structures as necessary only in so far as they responded to the resource industry's demands for regulatory rationalisation and simplification and with only limited input from Indigenous peoples (Abele, 2011, p. 224). In this sense, the overall impacts of this policy and of other previous similar initiatives amounted to what I have referred to as an *ontological shift* for local Indigenous residents, that is a shift in modes of being and inter alia, in modes of thinking and knowing. For example, in Tulita, waged labour is now perceived as being the only viable relationship with the outside world. Similarly, sharing, which is a central principle in traditional Dene epistemology (McMillan & Parlee, 2013), is now being refashioned into hybrid commodified/traditional systems of exchanges. The Dene traditional resource ownership system, which is informed and maintained by egalitarian rules relying on ecological knowledge, is also slowly being eroded (McMillan & Parlee, 2013, p. 437; Savishinsky, 1994). The people of Tulita are now socially and economically divided along wealth, age, and mastery of the Canadian State's political and cultural values.

Critical natural capital and ontological shift

The critical natural capital (CNC) approach can help to further delineate the ontological nature of this transformation. A society's CNC is made of the environmental processes and systems that are essential to its very existence and its wellbeing (Pellenc & Ballet, 2015, p.36; see also Ekins, 2003; Ekins, Simon, Deutsch, Folke, & De Groot, 2003). It is, in other words, the ecological factors and systems that enable a society to exist along with the cultural, political, and social rules and beliefs that give them an existential importance (Dedeurwaerdere, 2014). Natural resources are thus more than factors of production; they operate at an ontological level, determining the type of relationships that arise between an individual and the social community in which she/he lives. Similarly, in Dene epistemology, nature and human societies are interdependent. This phenomenon can be summarised through the Dene concept of 'Denneneh', which means 'the land by which all creatures live and exist'. Here, the term 'land' takes a holistic meaning and natural resources involve the following:

> Preserving the land and water which produce a rich complex of traditional foods that provide, nutritionally and otherwise, the spiritual, cultural, and

> economic well-being of the Dene. When the land is not healthy, the people are not healthy, and when the people are not healthy the land will also be unhealthy.
>
> (Paci & Villebrun, 2005, p. 75)

As done in the CNC approach, Dene traditions therefore directly associate the spatial integrity of Denneneh to societal and individual well-being (Legat, 2012; Blondin, 1997). Furthermore, according to Dene traditional epistemology, the process of exploitation of natural resources entails a form of kin-based reciprocity and respect that ensures their sustainable use. This takes place within a particular socio-ecosystem and with the aim of preserving a reciprocal relationship between society and nature based on the idea of a long-term balance (Helm, 1965; Hara, 1980; Blondin, 1997). This is a holistic view, which does not separate the use of the land, water, or ice from the living beings that dwell on or in it. In contrast to this, the Federal and Territorial governments approach to the sustainable use of natural resources entails a disjuncture between their exploitation and the eco-societal well-being of human societies since nature in this perspective is mediated by its economic transformation into man-made capital (Government of the Northwest Territories, 2005; GOC, 2009).

When viewed from this perspective, it can be argued that Tulita's critical natural capital is being substituted into financial capital through resource development activities. The metamorphosis of traditional lands into a kaleidoscopic series of extraction/exploration sites (e.g.: drilling of wells, clearing of areas for access roads) can in this sense be interpreted as an *ontological shift*, since for the Tulita Dene, it is their very being, their geopolitical identity (i.e.: an existential–spatial connection to the land), their understanding of existence, that are altered through these changes. The next section will examine the resource governance framework set up by the SDLMCA to manage the Sahtu region's natural resources and ask whether this can also be considered part of this ontological shift.

The SDLMCA, the failure of decolonisation and the shift towards capitalism

The Dene Nation and the SDLMCA

The SDLMCA was signed in 1993 and progressively implemented from 1994. It is a direct outcome of the activism of the Dene Nation (called the Indian Brotherhood of the NWT until 1975), an NWT-wide Indigenous and Metis decolonial movement. The Dene Nation sought full emancipation from Canada. The following quotations details what that entailed:

> We, the Dene, are not seeking sovereignty. [. . .] What we envisage is a new political entity, a Dene territory governed provincially by a Dene government and federally by the Canadian government.
>
> (Dene Nation, 1977, p. 6–7)

> The assumption of racial and cultural superiority led to the imposition of non-Dene political institutions which completely ignored our tradition of decision-making by consensus and government by agreement [. . .] The result is the creation of a small neo-colonial elite who benefit from a continuing colonial relationship [. . .] The colonial development of non-renewable resources has made it extremely difficult for us to maintain our traditional land-based activities [. . .] Clearly, we must develop our own economy [. . .] which would not only encourage continued renewable resource activities, such as hunting, fishing, and trapping but would include community-scale activities designed to meet our needs in a more self-reliant fashion [. . .] In accordance with our emphasis on sharing, Dene development will not permit a few to gain at the expense of the whole community [. . .] As such, we believe the conditions that govern the development of individuals will determine the conditions for the development of the whole Dene community.
>
> (Dene Nation, 1977, p. 27–29)

As is evident from these quotations, in 1977, the Dene envisioned their decolonisation as a single, united group, which included all Indigenous and Metis peoples in the NWT. Their objectives comprised a rejection of the capitalist ethos in favour of a traditional egalitarian socio-political system, the re-valorisation of renewable resources as their core developmental base, full political control over the entire NWT territory within the Canadian confederacy and sovereignty based on a form of endogenously designed land-based economic self-reliance.

Almost 40 years have passed since these declarations were made. The process that the Dene Nation initiated gave birth to a governance system based on political–economic and social compromises whose effects can only be seen as running opposite to the decolonial aims the Dene Nation had initially set. In fact, the signature of three different land claims settlement agreements (Gwich'in, Inuvialuit, Sahtu Dene and Metis), one land claims and self-government agreement (Tłįchǫ), one self-government agreement (Deline), and one treaty settlement agreement (Salt River First Nation) between 1984 and 2015 in the NWT can arguably be explained in part by the Indigenous failure to act as a single political entity, a nation.

These agreements created a territorial division based on the geographical compartmentalisation of each group, according to colonial categories of residence (in settlements/communities rather than nomadism), percentage of indian-ness (establishing the division of rights between Metis, Indian, and Inuit individuals), and history of non-renewable resource development (the cause for the current demographic composition of the North). Furthermore, and most relevant for this chapter, by prioritising the settlement of Indigenous rights over land and natural resources (rather than, say, rights to an autonomous governing entity), it made their commodification the cornerstone of any future development, whatever form it took.

This compartmentalisation created an imbalanced polity as the nature and effects of each agreement ended up reflecting the political particularities and tribal composition of the Indigenous groups they involved. For instance, the Inuvialuit signed a final agreement in as early as 1984 as a single five settlements-strong entity. This agreement emphasised a tribally owned resource corporation model of development with equal shares given to each Inuvialuit individual (Indian Affairs and Northern Development Canada, 1984). However, the land/resource claims negotiation process is still on-going. In some cases, the reason for this delay is ideological and involves the rejection of the SDLMCA's land corporation model (Dehcho and Akaitcho Dene) or an attempt to secure full control (surface and sub-surface) over the entire portion of tribal lands through the creation of a regional government (following the Tłįchǫ model) (Irlbacher-Fox, 2009; IANDC, 2006; Dehcho First Nations, 2015). For others, it is the result of a political strategy whose aim is to obtain as much local autonomy as possible (Acho Dene Koe/Fort Liard Metis and K'atlodeeche First Nation) or simply to gain the recognition of a particular historical context (NWTMN) (IANDC, 2007a; 'DAAIR, 2014a; Department of Aboriginal Affairs and Intergovernmental Relations, 2014a; Wohlberg, 2014; Tambour, 2010).

Similarly, for those areas where a land and resource claim agreement has already been implemented and where negotiations for the creation of self-governments have to be undertaken to complete the process, the situation is highly contrasted. For instance, the Tłįchǫ Dene have secured the setting up of a tribal self-government that has jurisdiction over the entire Tłįchǫ territory as part of a final agreement signed in 2003 (IANDC, 2003). In parallel, the Gwich'in Dene are still negotiating the creation of a similar entity. However, for the Sahtu Dene and Metis, the situation is more complex as the region's tribal composition is less homogenous. For communities like Tulita and Fort Good Hope whose demographic composition is made of different cultural–linguistic tribes, any decision-making process became particularly long and difficult. By contrast, Deline, whose population is predominantly composed of members of the same tribe (Bear Lake Dene), became the first and only Sahtu community to have had a self-government put in place in 2015. Tom Ranin, a 45-year old acquaintance who owns small construction businesses in Tulita once explained to me that he viewed Deline as having chosen a different path of development, one that is more traditional. He believed that political disagreements were more rife in Tulita but the reason for this was that Tulita had chosen a more conventional developmental path. A similar assessment was provided to me by Douglas Johns, one of Tulita's business leaders, whose portrait I develop in section IV of this paper:

> There is a lack of coordination between the different groups that live in Tulita. The Metis, White and Dene don't get on well. There's too much politics involved in each and every decision. There's no common goal [. . .] These differences that we have here were created during the negotiations. It would have been a lot better if all these groups could have worked together at the time. If we had negotiated as one single group, we

> would have obtained a much better deal and gained more control over our lands.
>
> (Interview, Tulita, January 2015)

Hence, the NWT's 'divide and administer' system of government is as much the consequence of disagreements between Indigenous groups than a result of Ottawa's choices. However, each cycle of negotiations has been clearly advantageous to the Federal government whose legal procedures and objectives have structured the process since it began.

The SDLMCA is nevertheless a clear success from Ottawa's perspective as only 15 percent of the Sahtu original tribal (settlement) lands' surface and 0.64 percent of their sub-surface remained under Indigenous control after its signature (Sahtu Land Use Planning Board, 2010). In fact, through a mirroring process, capitalist rent-seeking expectations with regard to the manner in which these future payments would be shared (rather than regarding a way of retaining control over settlement lands) became the central cause for dispute among Indigenous groups.

The shift towards capitalism embedded in the principles and aims of the SDLMCA

It is in this particular normative framing that one can see a shift towards ontologies that fit the capitalist fetishism for commensuration, when the commodified value of an entity becomes more important for a society's continuity than the manner in which it is used (Marx, 1993). For instance, by making the *model of the settlement corporation* the main apparatus of its strategy of development, the SDLMCA entrenched the idea that the profit-oriented exploitation of non-renewable resources would lead to the political empowerment of Sahtu communities' as well as to their economic self-sufficiency.

In fact, the SDLMCA made principles of economic development central to the current resource-governance architecture. It indeed confirmed the constitutional protection of the Sahtu Dene and Metis right to self-governing economic and political institutions, particularly as regards the right to be consulted on resource exploitation projects (Dalton, 2006; Northwest Territories Board Forum 2015; DAAIR, 1993). It thus facilitated the extraction of natural resources by clarifying the rules and regulations pertaining to projects taking place on or near Indigenous lands (Anderson et al., 2005; Dokis, 2010b). As such, it made petro-capitalist entrepreneurial ways of being a condition for the continuation of the traditional subsistence economy. In doing so, it paved the way to a future riven with accrued social and power disparities by strengthening the economic and symbolic capital of individuals already involved in the resource economy. For those individuals unable or unwilling to adapt to the rules of this new system, the situation only became worse (Nadasdy, 2003a; 2003b).

The SDLMCA's governance structure and its effects: Tulita's decolonial capitalism

As noted in the previous section, the SDLMCA generated the conditions for the capitalist mind-set to become the norm among the Sahtu Dene and Metis of Tulita. It also made the paradigm of economic development as progress their only viable decolonial strategy (DAAIR, 1993). These processes gave birth to what I would describe as a form of *decolonial capitalism*, a political ideology that functions through capitalist economic infrastructures and networks, but that seeks to follow decolonial aims. For instance, at the regional level, the agreement created a business-like organisation, the Sahtu Secretariat Incorporated, to serve as the key entry-point for the administration of the access to Sahtu lands. Several Indigenous land and financial corporations were also set up to manage these responsibilities at settlement level (See Figure 11.2 for a detailed representation of the Sahtu region's resource governance framework). These included three District Land Corporations and a dozen member sub-corporations, each representing one particular tribal component of the agreement (McArthur, 2009, p. 201). Their responsibilities include the administration of money transfers from the Federal and Territorial governments, as well as compensations received in exchange for rights of access and/or use of land parcels given to resource exploitation companies (the Access/Impacts and Benefits Agreements).

In Tulita, there are two land corporations: the Tulita Land/ Financial Corporation (LFC) and the Fort Norman Metis Land Corporation. Each LFC has a legal

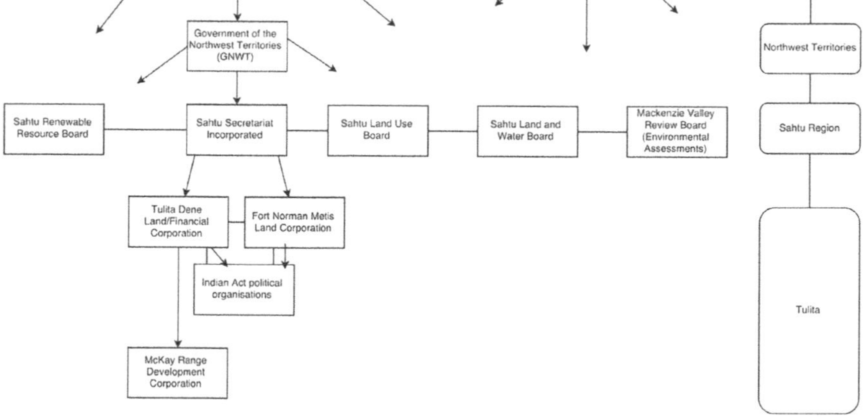

Figure 11.2 Resource governance framework – Sahtu region and Tulita. (Please note that lines indicate working relationships and arrowheads indicate directions of capital flows)

jurisdiction over the rights of access and ownership of a section of Tulita's share of settlement lands (as agreed in the SDLMCA). They are also responsible for transferring to Dene/Metis beneficiaries, the community's share of financial payments formerly made by the Federal Government directly to them. A third organisation, the Mackay Range Development Corporation (MCR) was created as a community-owned subsidiary to take advantage of the opportunities generated by the investments made in the community by external financial capital holders (Mackay Range Development Corporation, 2015). It is, in other words, the capitalist/corporatist arm of this hybrid institutional structure, which differs from the pre-colonial, family-based not-for-profit economic structure of Dene society.

However, the MCR is a hybrid structure since it follows egalitarian ideals (it acts on behalf of and for the benefit of the entire community), whilst having capitalist for-profit objectives. The MCR's ethos indeed remains intrinsically rooted in Dene traditional views of nature, since the organisation seeks to 'protect and conserve the wildlife and environment of the settlement area for present and future generations' (MCR, 2015). It also supports traditional subsistence practices in innovative ways (Southcott and Walker, 2009). Indeed, communal and individual hunts, which are critical to the Dene subsistence economy and culture, are supported by the MCR and through this structure, by the land corporations, and thus indirectly by the financial revenues provided by resource multinationals. The MCR also capitalises on the traditional characteristics of Dene society such as strong social cohesion and the existence of extended kin-based sharing networks in order to enable a higher proportion of individuals to benefit from resource development projects even if only a few are actually employed. Furthermore, its effectiveness relies on Dene traditional ontological connections to nature in the sense that they constitute an added motivation for the monitoring and reporting of environmental crimes (Parlee, 2015, p. 78).

However, it is precisely because of hybrid structures like the MCR that the shift towards a capitalist type of resource management takes place *without the Dene realising it*. For example, in the case of resource development projects, the MCR and the LFCs are the only points of entry for outside businesses seeking to work on Sahtu lands located within Tulita's surroundings. The Access/Impacts and Benefits Agreements (A/IBA) signed between these entities constitute the main sources of revenue for Tulita's land corporations in addition to governmental transfers. As such, these agreements create a dependent relationship between the community and the financial capital provided by the resource industry (Irlbacher-Fox, 2009; Kulchyski & Bernauer, 2014). Furthermore, the resource industry's social/environmental responsibility guidelines are devised from the perspective of rationalised (and as such non-Indigenous) epistemologies since they proceed through standardised procedures (Mikkelsen, Camp II, & Anderson, 2008; Bridge, 1998; Andrews, Klassen, & Searle, 2014). These A/IBAs can thus be viewed as one of the vehicles through which the transformation of the community's political and social structure takes place. This happens also because the A/IBAs are negotiated in secret, away from the traditional guidance and leadership of the community's elders (Abel, 2005) and without the informed consent of all residents. This favours the input of those individuals that are already involved in the administration of the land corporations and thus strengthens their power.

The capitalist nature of the SDLMCA architecture of governance also lies in the organisational structure of Tulita's land corporations, which loosely follows the Canadian States as member-shareholders elect rather than select their respective corporations' president and board members. Furthermore, board meetings are held in secrecy and their members are often chosen among the business and political elite since they possess the necessary skills and are willing to exert these responsibilities (Dokis, 2010a). This intrinsically limits the ability of other residents of the community to participate actively in the management of natural resources. As such, the practices and procedures of the resource industry further entrench already existing power disparities in communities like Tulita. They exacerbate the factors that lead to the dispossession of some and the enrichment of others. At a macro-structural level, this leads to a process of unsteady, unequal growth (Harvey, 2006; Smith, 2008; see also Sejersen, 2015b), which limits from a local perspective, incentives for economic diversification (Taylor, Severson-Baker, Woynillowicz, & Winfield, 2004). In Indigenous communities like Tulita, resource-based economic development indeed reduces the possibilities for non-skilled individuals to access reliable, long-term forms of employment and as such increases their vulnerability to social and economic stresses (Dana, Anderson, & Meis-Mason, 2009, pp. 105–110; Perombelon, 2013).

The SDLMCA, sustainable development in Tulita, and their dependency on oil

By placing the extraction of non-renewable resources at the cornerstone of its model of development, the SDLMCA has created an economy of expectations, an underlying systemic dependency based on the constant anticipation of future revenues/opportunities (i.e. in the form of economic growth or social or political empowerment). Sustainable development, as conceptualised under Canada's Northern Strategy, participates in this process via a type of reverse colonialism whereby the community envisions itself as a positive force that enables the physical transformation of Denenneh for its own future benefit whilst allowing, almost unintentionally, the same exploitative strategies to continue. Northern sustainable development or 'balanced development', as reformulated in the GNWT's Land Use and Sustainability Framework (GNWT, 2014a), then becomes paradoxically sustained through the extraction of non-renewable resources.

Examples of this paradox can be found in the processes of consultation and knowledge-sharing that occurs between the industry and the community. I have participated in two different community consultation meetings organised by resource extraction companies whilst residing in Tulita. These events take advantage of the community's economic, political, and geographic isolation, as residents are generally desperate to attract and retain these companies.

However, the exact conditions of their installation often become the subject of heightened social and political conflicts and creates power disparities within the community. For instance, during those meetings, non-elite members were unable to express their concerns as the A/IBAs, in both cases, had already been agreed upon, prior to the consultation itself. Furthermore, the terms used by the industry representatives were very technical. This intrinsically limited the ability of local residents, some of whom are illiterate or simply not used to complex jargon, to provide any meaningful input. Lastly, industry representatives are aware of the vague nature of what is entailed by the duty to consult. They capitalise on this vagueness by equating attendance to these meetings with acceptance of their proposed project. It is then easy to transform a community feast, where an abundance of free food and beverages is provided to a population that remains predominantly poor, into a successful consultation exercise. Through this, habituation takes place and a certain experience of the process of resource development, sustained by the resource industry's practices, becomes embedded within the community's lifeworld.

As can be inferred from the following quotation, it is also via this process of habituation that Tulita's leaders end up equating the exploitation of natural resources to future economic self-sufficiency:

> Of course, fracking is a very important issue, with all the water that will be needed. Water sustains animals. But we want to do this, and the industry wants to work with us, so we need a balanced approach [. . .]. I think that this is going to become economically sound over the next few years. [. . .] Certain sections of the land claims have remained under the control of the government [. . .] But what can happen is that a share goes to the industry and we get the price of the license [*the A/BA*].
>
> (Interview with LK, president of one of Tulita's land corporation, March 2015)

However, not all Dene leaders see this 'balanced approach' as sustainable in a traditional sense. As Denezeh Nakehko said:

> Balanced approach? What is a 'balanced approach'? I mean English is the language of commerce. No, nature, in the way it is, is balanced already so there's no such thing as a balanced approach [. . .] Sustainable development is about going out there and do things differently.
>
> (Recorded, structured interview, Yellowknife, March 2015)

As I will demonstrate in the next section, it is this difference in understandings of what development entails that informs diverging views of what sustainability means in the Dene context. In Tulita, this difference is, again, informed by socio-economic and power disparities.

Northern sustainable resource development as geo-colonialism

The concept of sustainable development as formulated in the Brundtland report sustains a geopolitical order based on transnational liberalism, the control of multinational capitalist actors on the epistemological appropriation of nature, and the domination of the West on the ontological conditions of this appropriation (Agnew & Corbridge, 1995, pp. 205–206; Escobar, 1995, p. 195; Rios Osorio, Ortiz Lobato, & Alvarez Del Castillo, 2005). It informs and strengthens the dominant occidental discourse on sustainable development, which retains the exploitation of nature as a necessary condition for human progress (Moore, 2015; Malm, 2016). When used under the umbrella of the Northern Strategy, the concept of sustainable development thus participates in the same geocolonial logic, since it seeks to reconcile Canada's resource-based economic model (Easterbrook & Watkins, 2003; Berland, 2009; Abele, 2009), with Indigenous holistic interpretations of nature. Similarly, as part of the NWT's land claims settlement political system, sustainable development implies that the social and environmental impacts of resource extraction activities should not be a cause of concern for Indigenous communities since they will eventually lead to their political emancipation. For some Dene decolonial scholars, however, this governance architecture acts as a hidden form of political domination 'that works through the medium of state recognition and accommodation' with the purpose of ensuring the continuation of the exploitation of natural resources (Coulthard, 2014, p. 25; see also Freeland Ballantyne, 2014; Keeling & Sandlos, 2009; Kulchyski & Bernauer, 2014; Irlbacher-Fox, 2009; 2014). This was clearly analysed by Denezeh Nakehko:

> '"NWT"? That's not a name; it's a direction from Ottawa. They don't care about us so much so that they'd allow us to decide on our own name' (Interview, Yellowknife, March 2015).

In effect, this geocolonial domination is experienced via two concomitant phenomena that have been historically constitutive of the colonial project in Canada's High North. The first is a pathological habituation to the psychology, practices, language, and belief-system of the Canadian State (Coulthard, 2014; Fanon, 1967). The second is the dehumanisation of Indigenous traditional ontologies via the dispossession of their lands (Alfred, 2009; Memmi, 2013).

In Tulita, these phenomena are deeply entrenched in the lifeworlds of the community's inhabitants. They reside within a material–semiotic structure of embodied habits, practices, beliefs, and discourses, which act to constrain their existence within a limited set of horizons (i.e.: the possibilities of realisation of personal and common subjectivities) (Haraway, 1988; Lazzarato, 2014, p. 220). For example, an acquaintance, John Gorda, also a member of Tulita's leadership, explained to me that he believed that the development of resource extraction activities in and around the settlement would only generate more inequalities. He clearly expected that such an economy would create haves and have-nots and that what is needed is better preparation, planification, training and coordination. This suggests that resource development processes embed the dispossession of land in the everyday lives of residents since they generate disparities in wealth that are visible and affect the physical integrity of Denenneh (Coulthard, 2014; Freeland Ballantyne, 2014).

However, John Gorda's point of view is not widespread in the community and the result is that, as Eric (another informant) explained to me, the line separating those in favour of oil development and those against it, is now also the one that lies between the wealthy/educated and vulnerable fractions of Tulita's residents.

Social disparities and competing visions of sustainable development

The exploitation of oil generates a process of social differentiation, which disrupts Dene traditional values of solidarity and intergenerational support, and increases the dependency of the poorest upon the influx of global capital (via the seasonal income provided by oil activities) (Elias, 1995, p. 21). Elite individuals, who are able to respond to the demands of the industry, monopolise the business opportunities these companies provide (Mitchell, 1996; Elias, 1995; Dokis, 2010a). In this sense, the SDMLCA has rendered the local political system less transparent and more prone to capture and rent-seeking behaviours (Dokis, 2010a; Loxley, 2010, p. 82):

> The families that are already in power will stay in power. If one family is bigger than another, the person it has chosen to run for a position will win. It can also find allies in another big family in exchange for certain types of favour. Before, the chief was chosen because he was the most experienced and knowledgeable about the land.
>
> (Interview, Tulita, November 2014)

Similarly, John Gorda explained to me that the lack of coordination between businesses and management institutions in Tulita created the conditions for the enrichment of some to the detriment of others.

These issues also stem from an opposition, among the community's residents, between two competing visions of sustainable development: with the first lying somewhere between traditional practices and capitalist objectives and the second firmly rooted in the subsistence economy (see Dokis, 2010a). It would be wrong however to see a static dichotomy in this opposition as decolonial objectives dynamically drive both viewpoints. The latter gives rise to economic practices that are rooted in traditional values, away from the model of the capitalist corporation, and based on individual experiences of ecological scarcity. One can see this model exemplified in the selling of craft labour on an as-needed basis when access to subsistence activities or seasonal employment with large-scale construction, government, or resource projects is limited. On the other hand, for those who share the former position, resource development offers an opportunity to 'complete the process of decolonisation' started in the 1970s, as stated by former Indigenous NWT premier, and former leader of the Dene Nation, Stephen Kakfwi in 2000 (Sarkadi, 2000, p. 25; Abele, 2005). Many among Tulita's political-economic elite also share this point of view. The reason is that this group emerged with the SDLMCA governance architecture and was subsequently nurtured by its hybrid decolonial capitalist model of development. It is arguably the incarnation of the colonised elite described by Fanon (1967) that works, often without realising it, against the full emancipation of their subaltern counterparts. In Tulita, 21 years after the implementation of the agreement, this group has come to be responsible for this architecture. They are thus perceived by the most traditional and poorer fractions of the community's population to be the new products and agents of colonialism.

As the following quotation from Douglas Johns, one of the most prominent member of this elite group demonstrates, what seems to drive these individuals is an association of decolonial objectives with a creative–destructive mind-set. That mind-set is shaped by a belief that the sustainable preservation of Denenneh can be achieved through personal enrichment. For this elite, individual empowerment can only serve the good of the entire community. Douglas Johns is one of the most prominent Metis representatives of the settlement's business community. His entrepreneurial success epitomises the paradoxical traditional–capitalistic attachment to land described previously. He is also an important member of the settlement's Canadian Ranger unit as well as, at the time of writing, the president of the Fort Norman Métis Lands and Financial Corporation. These factors arguably explain that his primary concern, when it came to the future of the community, was the global petro-economy:

> When the price of oil changes, it affects all of us here. A lot of people will loose their jobs. So what we need is to be able to control the production of oil from here. We should set up our own refinery so we can sell our oil.
>
> (Interview, Tulita, January 2015)

Sustainable development is not sustainable in a traditional sense

Hence, Tulita's sustainable development strategies, when applied to the governance of natural resources, whether they result from Canada's Northern Strategy or from the SDLMCA's governance architecture, are not sustainable from a traditional knowledge perspective. Rather, they embed industrial modernity's transformation of nature into factors of production within a complex geopolitical nexus of transnational capitalist exchanges, national policies of colonial dispossession, and local decolonial struggles for empowerment, all centred on the (re) appropriation of land. What seems nevertheless critical to the actors evolving within this nexus is their personal history of past and present deprivation. However, as I was told by Julia,[8] a member of a youth network involved in discussions on the improvement of social and economic life in Tulita, it is a shared willingness to remain in the community and a common understanding that dependency (to the State, to the resource industry, or to a complex capitalist mixture of these entities) creates the conditions for the community's lack of opportunity, that appears to motivate the younger generation to propose alternative views of their future. If for some, like Eric, this envisioned future is one made of radical forms of Indigenous traditionalism, for others it remains firmly anchored in a perspective that sees the capitalistic resource-based economy as a necessity for survival and modernisation.

Conclusion

The social experiment that started with the SDLMCA has been an engine for endogenous economic innovation. This model has been successful in that it has generated the necessary impetus for the paradigm of development as progress to be sought and accepted as a viable strategy of emancipation, giving birth to a Dene version of Indigenous capitalism (Bunten, 2011), characterised by a collective will to support key traditional practices and knowledge and a sense of economic empowerment. However, the resource governance system created by the agreement, mirrored in the Canadian State's own ways of functioning (Nadasdy, 2009), is also experienced as a modern expression of the colonial relationship that has existed with the Canadian State since the period of first contact. In fact, whilst the dependency on governmental financial transfers has not disappeared, the community and its inhabitants now rely extensively on the capital provided by multinational resource corporations. A mixture of local (e.g. the SDMLCA and its land corporations), national (e.g. the Northern Strategy, the resource industry's social and environmental responsibility guidelines), and global (fluctuations of commodity prices) factors interacts to generate a divisive and disempowering experience. Whilst a few individuals have been able to use these processes to their advantage, a significant number of Tulita's residents perceive the post land claims era as more environmentally and socially disruptive and less politically empowering (Perombelon, 2013). The fact that the non-renewable industry accounts for 38 percent of the NWT's GDP growth over the last 16 years confirms this perception (GNWT, 2014b). It also gives further ground to the idea that the Northern Strategy was never meant to be sustainable in a traditional Northern Indigenous sense but rather that its primary purpose was geopolitical. After all, the Federal government has retained jurisdiction over the rights of access for most of the Territories' sub-surface (Bankes, 2010).

The experientially-situated character of these phenomena calls for solutions that focus on individuals and their personal experience (Wilson, 1996; McLaughlin & Davidson, 1994). This is especially relevant if one were to suggest a strategy of development for Indigenous communities that would conform to their holistic views of nature whilst enabling their socio-economic self-determination at both a personal and community level (Kuokkanen, 2011). The 'grassroots business model' arguably offers the most adapted approach to achieve these objectives (Wilson, 1996). This framework, which is linked to the concept of social entrepreneurship, can be defined as the setting up of businesses by individuals, within a socially closely inter-connected community, according to their economic needs and social, political, and cultural values (Wilson, 1996, p. 617; Anderson et al., 2006; Doherty, Haugh, & Lyon, 2014). It seeks, in other words, to dissociate work from a notion of experienced alienation and to transform it into a process of existential liberation within particular ontologies and broader social, economic, and ecological systems (Gorz, 1985, 1999, 2010).

The notion of Sustainable Social Environmental Enterprise (SSEE), developed by Thornton and Wanasuk, offers a convincing example of how this vision can apply to Indigenous communities (2016). For them, an SSEE 'combines social, economic and environmental sustainability with values, principles and strategies that seek to radically improve sociocultural and environmental benefits to a community' (Thornton & Wanasuk, 2016, pp. 44–45). It is based on 'four pillars of sustainability' (environmental, economic, cultural, and political) (Thornton & Wanasuk, 2016, p. 45), which can, in the case of Tulita, be synthesised through a geopolitical understanding of the concept of critical natural capital. Here, an SSEE would serve as a decolonial weapon for individual and community *empowerment* within a *familiar* territory, in this case, Denneneh. It aims, as such, to preserve not only the local natural capital which enables it to function, but also the sets of ontological relationships that mediate the existence of this natural capital vis-a-vis the Dene, and thus the continuation of Dene society itself. The Deline Social Economy Project offers an interesting example of how such an approach could take shape. This community-based research, drawn from the long-term embedded collection of information from local Indigenous stakeholders, showed that traditional Dene land-based concepts could also be put at the centre of an Indigenous community's model of economic development (Simmons et al., 2015, pp. 253–257).

Beyond these suggestions, this chapter has also sought to show that in order for sustainability to emerge successfully as a viable economic development paradigm in these small Indigenous communities, it needs to be informed by personal experiences of empowerment in the face of dispossessing phenomena, such as resource exploitation activities, climate change, or an increased dependency on global capital. It has thus also pointed to the need to undertake more research to develop sustainable resource development strategies that are grounded in the particular ontologies of individuals, away from the often path-dependent character of complex political (e.g.: States, Multinational Corporations), socio-technical (e.g.: transportation), or economic (e.g.: capitalism) systems (Yin & Herfel, 2011, p. 406).

Notes

1 'Conservative' refers to Canada's Conservative party.
2 'Post-colonial' is taken to mean the chronological period that immediately follows the end of formal colonisation (Sharp, 2009). In this chapter, formal colonisation is assumed to have ended with the signature of a comprehensive land claims settlement agreement in 1993 (Fumoleau, 2007).
3 I call this ability to change the conditions of life via changes in a population's environment, geopower (Clare, 2013; Yusoff, Grosz, Clark, Saldanha, & Nash, 2012).
4 I use EuroCanadian, Western and Occidental interchangeably to mean a mostly capitalist, rationalist way of understanding and exploiting nature
5 This chapter is based on ethnographic data (interviews and participant-observation) collected during a period of fieldwork undertaken in the Sahtu community of Tulita in the Northwest Territories of Canada, and in the administrative capital of this region, Yellowknife, in 2014–2015. For ethical reasons, all my informants' names have been changed.
6 In the geographical and historical context of Tulita, decolonisation can be defined as an organised attempt by the Dene to access a form of sovereignty based on the recognition of political, social, and cultural rights over a territory that constituted their traditional hunting, fishing, and trapping grounds (Dene Nation, 1977). At the experiential–individual level, as I observed and was told during my fieldwork, it is also a rejection of practices and values that are associated with the post-colonial world (for example, the use of motorised vehicles or the notion of individual enrichment).
7 The distinction between Dene and Metis groups emerged as a political byproduct during the negotiation process that led to the signature of the Comprehensive Sahtu Dene and Metis Land Claim Settlement Agreement (SDMLCA). As such, for the purposes of this chapter, I will not use this distinction as it risks undermining the fact that both groups share a common history of colonisation, a central point of my argument.
8 This interview was conducted in February 2015 in Tulita.

References

Abel, K. (2005). *Drum Songs: Glimpse of Dene History*. Montréal: McGill-Queen'sUniversity Press.

Abele, F. (2005). The smartest steward?: Indigenous people and petroleum-based economic development in Canada's North. In G. B. Doern (Ed.), *Canadian Energy Policy and the Struggle for Sustainable Development* (pp. 223–245). Toronto, Buffalo, London: University of Toronto Press.

Abele, F. (2009). Northern development: Past, present and future. In F. Abele, T. J. Courchene, F. L. Seidle, & F. St-Hilaire (Eds.), *Northern Exposure: Peoples, Powers and Prospects in Canada's North* (pp. 19–65). Montréal: Institute for Research on Public Policy.

Abele, F. (2011). Use it or lose it? The conservatives' Northern Strategy. In C. Stoney & G. B. Doern (Eds.), *How Ottawa Spends 2011–2012: Trimming Fat or Slicing Pork?* (pp. 218–242). Montréal: McGill University Press.

Agnew, J., & Corbridge, S. (1995). *Mastering Space: Hegemony, Territory and International Political Economy*. London and New York: Routledge.

Alfred, T. (2009). *Peace, Power, Righteousness: An Indigenous Manifesto* (2nd ed.). New York: Oxford University Press.

Anderson, R. B., Camp II, R. D., Dana, L. P., Honig, B., Nkongolo-Bakenda, J. M., & Peredo, A. M. (2005). Indigenous land rights in Canada: The foundation for development? *International Journal of Entrepreneurship and Small Business*, *2*(2), 104–133.

Andrews, W., Klassen, C., & Searle, H. (2014). Aboriginal relations guiding principles and guidelines (Imperial Oil Limited). *Proceedings from Society of Petroleum Engineers International Conference on Health, Safety, and Environment*, 17–19 March. Long Beach, CA, USA.

Bankes, N. (2010). Oil and gas and mining development in the Arctic: Legal issues. In N. Loukacheva (Ed.), *Polar Law Text Book* (pp. 101–122). Copenhagen: Nordic Council of Ministers.

Berger, T. R. (1999). *A Long and Terrible Shadow: White Values, Native Rights in the Americas since 1492*. Washington: University of Washington Press.

Berland, J. (2009). *North of Empire: Essays on the Cultural Technologies of Space*. Durham and London: Duke University Press.

Blondin, G. (1997). *Yamoria, the Lawmaker: Stories of the Dene*. Edmonton: NeWest Press.

Bridge, G. (1998). Excavating nature: Environmental narratives and discursive regulation in the mining industry. In A. Herod, G. O' Tuathail, & S. M. Roberts (Eds.), *Negotiating Unruly Problematics. Unruly World?: Globalization, Governance and Geography* (pp. 219–243). London, New York: Routledge.

Bunten, A. C. (2011). A call for attention to indigenous capitalisms. *New Proposals: Journal of Marxism and Interdisciplinary Inquiry*, *5*(1), 60–71.

CBC News (2010). Arctic sovereignty a priority: Harper. *CBC News Online*. Retrieved from www.cbc.ca/news/politics/arctic-sovereignty-a-priority-harper-1.951536.

CBC News (2012). Harper leaves room for mining near North's new national park. *CBC News Online*. Retrieved from www.cbc.ca/news/politics/harper-leaves-room-for-mining-near-north-s-new-national-park-1.1157611.

Clare, S. (2013). Geopower: The politics of life and land in Frantz Fanon's writing. *Diacritics*, *41*(4), 60–80.

Coates, K. (1985). *Canada's Colonies: A History of the Yukon and Northwest Territories*. Toronto: James Lorimer and Company Publishers.

Coulthard, S. G. (2014). *Red Skin, White Masks: Rejecting the Colonial Politics of Recognition*. Minneapolis, London: University of Minnesota Press.

Crosby, A. W. (2015). *Ecological Imperialism* (2nd ed.). Cambridge, U.K.: Cambridge University Press.

DAAIR, Department of Aboriginal Affairs and Intergovernmental Relations (1993). *Sahtu Dene and Metis Comprehensive Land Claim Agreement: Volume I (SDLMCA)*. Minister of Indian and Northern Affairs Canada. Ottawa: Canada. Retrieved from www.daair.gov.nt.ca/_live/pages/wpPages/SahtuLandClaim.aspx.

DAAIR, Department of Aboriginal Affairs and Intergovernmental Relations (2014a). *Acho Dene Koe First Nation Fort Liard Métis: Agreement in Principle*. Yellowknife: Department of Aboriginal Affairs and Intergovernmental Relations. Retrieved from www.daair.gov.nt.ca/sites/www.daair.gov.nt.ca/files/adkfn_agreement_in_principle_0.pdf

Dalton, J. E. (2006). Aboriginal self-determination in Canada: Protection afforded by the judiciary and government. *Canadian Journal of Law and Society*, *21*(1), 11–37.

Dana, L. P., Anderson, R. B., & Meis-Mason, A. (2009). A study of the impact of oil and gas development on the Dene First Nations of the Sahtu (Great Bear Lake) Region of the Canadian Northwest Territories (NWT). *Journal of Enterprise Communities: People and Places in the Global Economy*, *3*(1), 94–117.

Dedeurwaerdere, T. (2014). *Sustainability Science for Strong Sustainability*. Northampton: Edward Elgar.

Dene Nation (1977). *Dene Nation: The Struggle of Canada's Internal Colony for Self-Determination*. Geneva: World Council of Churches.

DFN, Dehcho First Nations (2015). *Backgrounder Dehcho Process: MLA Briefing Note*. Fort Simpson: Dehcho First Nations. Retrieved from www.dehcho.org/documents/deh_cho_process/ddehcho_process_updated_november%20_2015.pdf.

Dickerson, O. (1992). *Whose North?: Political Change, Political Development, and Self-Government in the Northwest Territories*. Vancouver: University of British Columbia Press.

Doherty, R., Haugh, H., & Lyon, F. (2014). Social enterprises as hybrid organizations: A review and research agenda. *International Journal of Management Reviews, 16*(4), 417–436.

Dokis, C. A. (2010a). *People, land, and pipelines: Perspectives on resource decision-making processes in the Sahtu region*. Northwest Territories (PhD thesis). University of Alberta.

Dokis, C. A. (2010b). Modern day treaties: 'Development', politics, and the corporatization of land in the Sahtu Dene and Métis comprehensive land claim agreement. *Geography Research Forum, 30*, 32–49.

Duhaime, G., & Caron, A. (2006). Economic and social conditions of Arctic regions. In S. Glomsrød & I. Aslaksen (Eds.), *The Economy of the North 2008* (pp. 11–21). Oslo-Kongsvinger: Statistics Norway.

Easterbrook, W. T., & Watkins, M. (2003). *Approaches to Canadian Economic History: A Selection of Essays*. Montréal: McGill-Queen's University Press.

Ekins, P., Simon, S., Deutsch, L., Folke, C., & De Groot, R. (2003). A framework for the practical application of the concepts of critical natural capital and strong sustainability. *Ecological Economics, 44*(2–3), 165–185.

Ekins, P. (2003). Identifying critical natural capital: Conclusions about critical natural capital. *Ecological Economics, 44*(2–3), 277–292.

Elias, P. D. (1995). *Northern Aboriginal Communities: Economies and Development*. Concord: Captus University Publications.

Escobar, A. (1995). *Encountering Development: The Making and Unmaking of the Third World*. Princeton: Princeton University Press.

Fanon, F. (1967). *Black Skin, White Masks*. New York: Grove Press.

Freeland Ballantyne, E. (2014). Dechinta Bush University: Mobilizing a knowledge economy of reciprocity, resurgence and decolonisation. *Decolonisation: Indigeneity, Education & Society, 3*(3), 67–85.

Fumoleau, R. (2007). *A History of Treaty II, 1870–1939: As Long as this Land shall last*. Calgary: University of Calgary Press.

Gorz, A. (1985). *Paths to Paradise: On the Liberation from Work*. London: Pluto Press.

Gorz, A. (1999). *Reclaiming Work: Beyond the Wage-Based Society*. Cambridge and Malden: Polity press.

Gorz, A. (2010). *Ecologica*. Seagull Books: Kolkata.

GOC, Government of Canada, (2009). Canada's Northern Strategy: our North, our heritage, our Future. Minister of Indian Affairs and Northern Development and Federal Interlocutor for Métis and Non-Status Indians. Ottawa: Minister of Public Works and Government Services, Canada.

Government of the Northwest Territories (GNWT). (2005). *Statement of Policy: Sustainable Development*. Yellowknife, Canada: Northwest Territories Policy (Publication No. 53.02).

Government of the Northwest Territories (GNWT). (2014a). *Northern Lands, Northern Leadership: The GNWT Land Use and Sustainability Framework*. Yellowknife, Canada: Department of Lands, GNWT.

Government of the Northwest Territories (GNWT). (2014b). *NWT GDP by Industry, 1999 to 2014*. Yellowknife, Canada: Northwest Territories Bureau of Statistics, GNWT.

Hara, H. J. (1980). *The Hare Indians and Their World*. Ottawa: Canadian Ethnology Service, National Museums of Canada.

Haraway, D. (1988). Situated knowledges: The science question in feminism and the privilege partial perspective. *Feminist Studies*, *14*(3), 575–599.

Harvey, D. (2006). *Spaces of Global Capitalism: A Theory of Uneven Geographical Development*. London and New York: Verso.

Helm, J. (1965). Bilaterality in the socio-territorial organization of the Arctic Drainage Dene. *Ethnology*, *4*(4), 361–385.

Helm, J. (2000). *The People of Denendeh: Ethnohistory of the Indians of Canada's Northwest Territories*. Montréal & Kingston, London, Ithaca, NY: McGill-Queen's University Press.

IANDC, Minister of Indian Affairs and Northern Development Canada (1984). *The Western Arctic Claim: The Inuvialuit Final Agreement*. Ottawa, Canada: Minister of Public Works and Government Services. Retrieved from www.collectionscanada.gc.ca/webarchives/20071125181720/www.ainc-inac.gc.ca/pr/agr/inu/wesar_e.pdf.

IANDC, Minister of Indian Affairs and Northern Development Canada (2003). *Land Claims and Self-Government Agreement Among the Tłįchǫ and the Government of the Northwest Territories and the Government of Canada*. Ottawa, Canada: Minister of Public Works and Government Services. Retrieved from www.aadnc-aandc.gc.ca/DAM/DAM-INTER-HQ/STAGING/texte-text/ccl_fagr_nwts_tliagr_tliagr_1302089608774_eng.pdf.

IANDC, Minister of Indian Affairs and Northern Development Canada (2006). *Backgrounder: Akaitcho Process Negotiations*. Ottawa, Canada: Minister of Public Works and Government Services. Retrieved from www.aadncaandc.gc.ca/DAM/DAMINTERNWT/STAGING/textetext/ntr_pubs_abe_1331059977676_eng.pdf.

IANDC, Minister of Indian Affairs and Northern Development Canada (2007a). *NWT Plain Facts on Land and Self-Government: Northwest Territory Métis Nation Process Negotiations*. Ottawa, Canada: Minister of Public Works and Government Services. Retrieved from www.aadnc-aandc.gc.ca/DAM/DAM-INTER-NWT/STAGING/texte-text/mvrma_1328635567694_eng.pdf.

Irlbacher-Fox, S. (2009). *Finding Dashaa: Self-Government, Social Suffering, and Aboriginal Policy in Canada*. Vancouver: University of British Columbia Press.

Irlbacher-Fox, S. (2014). Traditional knowledge, co-existence and co-resistance. *Decolonisation: Indigeneity, Education & Society*, *3*(3), 145–158.

Keeling, A., & Sandlos, J. (2009). Environmental justice goes underground? Historical notes from Canada's Northern mining frontier. *Environmental Justice*, *2*(3), 117–125.

Krech III, S. (1990). *The Subarctic Fur Trade: Native and Social Economic Adaptations*. Vancouver: The University of Columbia Press.

Kulchyski, P. (2005). Sahtu Land Claims Agreement. In M. Nuttall (Ed.), *Encyclopedia of the Arctic: Volumes 1, 2 and 3, A-Z* (pp. 1822–1823). New York, Abingdon: Routledge.

Kulchyski, P., & Bernauer, W. (2014). Modern treaties, extraction, and imperialism in Canada's indigenous north: Two case studies. *Studies in Political Economy: SPE, a Socialist Review*, 93, 3–24.

Kuokkanen, R. (2011). Indigenous economies, theories of subsistence, and women: Exploring the social economy model for indigenous governance. *American Indian Quaterly*, *35*(2), 215–241.

Lackenbauer, P. W., & Farish, M. (2007, October). The cold war on Canadian soil: Militarizing a Northern environment. *Environmental History*, *12*, 920–950.

Lazzarato, M. (2014). *Signs and Machines: Capitalism and the Production of Subjectivity*. Los Angeles: MIT Press.

Legat, A. (2012). *Walking the Land, Feeding the Fire: Knowledge and Stewardship among the Tlicho Dene*. Tucson, AZ: The University of Arizona Press.
Loxley, J. (2010). *Aboriginal, Northern, and Community Economic Development: Papers and retrospective*. Winnipeg: Arbeiter Ring Publishing.
Malm, A. (2016). *Fossil Capital: The Rise of Steam Power and the Roots of Global Warming*. London: Verso Books.
Marx, K. (1993). *Grundrisse: Foundations of the Critique of Political Economy*. London, New York, Victoria, Toronto, Auckland: Penguin Books.
McArthur, D. (2009). The changing architecture of governance in Yukon and the Northwest Territories. In F. Abele, T. J. Courchene, F. L. Seidle, & F. St-Hilaire (Eds.), *Northern Exposure: Peoples, Powers and Prospects in Canada's North* (pp. 187–232). Montréal: Institute for Research on Public Policy.
McKay Range Development Corporation (MCR). (2015). Retrieved from www.mackayrange.com/.
McLaughlin, C., & Davidson, G. (1994). *Spiritual Politics: Changing the World from the Inside Out*. New York: Ballantine.
McMillan, R., & Parlee, B. (2013). Dene hunting organization in fort good hope, Northwest Territories: Ways we help each other and share what we can. *Arctic*, 66(4), 435–447.
Memmi, A. (2013). *The coloniser and the colonised*. Lexington, MA: Ebook, Plunkett Lake Press.
Meier, W. N., Hovelsrud, G. K., Van Oort, B. E. H., Key, J. R., Kovacs, K., Michel, C., Haas, C., Granskog, M. A., Gerland, S., Perovich, D. K., Makshtas, A., & Reist, J. D. (2014). Arctic sea ice in transformation: A review of recent observed changes and impacts on biology and human activity. *Review of Geophysics*, 52(3), 185–217.
Mikkelsen, A., Camp II, R. D., & Anderson, R. E. (2008). Managerial implications. In A. Mikkelsen & O. Langhelle (Eds.), *Arctic Oil and Gas: Sustainability at risk?* (pp. 350–380). London and New York: Routledge.
Milloy, J. S. (2003). *A 'National Crime': The Canadian Government and the Residential School System, 1879 to 1986*. Winnipeg: The University of Manitoba Press.
Mitchell, M. (1996). *From Talking Chiefs to a Native Corporate Elite: The Birth of Class and Nationalism among Canadian Inuit*. Montréal: McGill-Queen's University Press.
Moore, J. W. (2015). *Capitalism in the Web of Life: Ecology and the Accumulation of Capital*. London: Verso Books.
Nadasdy, P. (2003a). *Hunters and Bureaucrats: Power, Knowledge, and Aboriginal–State Relations in the Southwest Yukon*. Vancouver, Toronto: University of British Columbia Press.
Nadasdy, P. (2003b). Reevaluating the co-management success story. *Arctic*, 56(4), 367–380.
Nadasdy, P. (2009). The antithesis of restitution? A note on the dynamics of land negotiations in the Yukon, Canada. In D. Fay & D. James (Eds.), *The Rights and Wrongs of Land Restitution: 'Restoring What Was Ours'* (pp. 85–98). Abingdon, New York: Routledge-Cavendish.
Northwest Territories Board Forum (NWT Board Forum). (2015). *Land and Resource Management in the NWT*. Yellowknife: Ministry of Aboriginal Affairs and Northern Development Canada.
Paci, C., & Villebrun, N. (2005). Mining Denendeh: A Dene nation perspective on community health impacts of mining. *Pimatisiwin: A Journal of Aboriginal and Indigenous Community Health*, 3(1), 71–86.
Parlee, B. (2015). The social economy and resource development in Canada. In C. Southcott (Ed.), *Northern Communities Working Together: The Social Economy of Canada's North* (pp. 70–87). Toronto, Buffalo, London: University of Toronto Press.

Pellenc, J., & Ballet, J. (2015, April). Strong sustainability, critical natural capital and the capability approach. *Ecological Economics*, *112*, 36–44.

Perley, D. G. (1993). Aboriginal education in Canada as internal colonialism. *Canadian Journal of Native Education*, *20*(1), 118–128.

Perombelon, B. (2013). *Arctic resources, Arctic rights. Perceptions of oil activities in the Sahtu region, Northern Canada: Curse or blessing?* Master's dissertation. London School of Economics and Political Science.

Piper, L. (2009). *The Industrial Transformation of Canada*. Vancouver, Toronto: UBC Press.

Ray, A. J. (1990). Periodic shortages, native welfare, and the Hudson's Bay Company 1670–1930. In S. Krech III (Ed.), *The Subarctic Fur Trade: Native and Social Economic Adaptations* (pp. 1–20). Vancouver: The University of Columbia Press.

Rios Osorio, L. A., Ortiz Lobato, M., & Alvarez Del Castillo, X. (2005). Debates on sustainable development: Towards a holistic view of reality. *Environment, Development and Sustainability*, *7*(4), 501–518.

Sabin, P. (1995). Voices from the hydrocarbon frontier: Canada's mackenzie valley pipeline inquiry (1974–1977). *Environmental History Review*, *19*(1), 17–48.

Sarkadi, L. (2000). What Goes Around . . . Comes Around. *Far North Oil and Gas Review* (Winter), p. 23.

Savishinsky, J. S. (1994). *The Trail of the Hare: Environment and Stress in a Sub-Arctic Community* (2nd ed.). Langhorne, PA: Gordon and Breach.

Sejersen, F. (2015a). Future-makers in the Arctic: A critical view on globalisation, urbanisation and change: Keynote, Aalborg, August 12–15, 2015. *Post graduate summer school 'Change and Continuation in the Arctic'*. University of Aalborg.

Sejersen, F. (2015b). *Rethinking Greenland and the Arctic in the Era of Climate Change: New Northern Horizons*. New York and Abingdon: Earthscan Routledge.

Sharp, J. P. (2009). *Geographies of Postcolonialism: Spaces of Power and Representation*. New London, Thousand Oaks, Delhi and Singapore: Sage Publications.

Simmons, D., Bayha, W., Fink, I., Gordon, S., Rice, K., & Taneton, D. (2015). Gúlú Agot' T'á Kə Gotsúha Gha (Learning about Changes): Rethinking Indigenous Social Economy in Délı̨nę, Northwest Territories. In C. Southcott (Ed.), *Northern Communities Working Together: The Social Economy of Canada's North* (pp. 240–265). Toronto, Buffalo, London: University of Toronto Press.

SLUPB, Sahtu Land Use Planning Board (2010). *Sahtu Land Use Plan Background Report: The Sahtu Settlement Area*. Fort Good Hope: Sahtu Land Use Planning Board.

Smith, N. (2008). *Uneven Development, Nature, Development, and the Production of Space*. Athens: University of Georgia Press.

Southcott, C., & Walker, V. (2009). A portrait of the social economy in Northern Canada. *The Northern Review*, *30*(Spring), 13–36.

SRRB, Sahtú Renewable Resources Board (2014). *Staying Strong: Sahtú Youth and Elders Building Healthy Communities in the Face of Climate Change, Volume I: Final Report*. Tulita: Sahtú Renewable Resources Board.

Tambour, H. (2010). *K'atlodeeche First Nation Framework Agreement: Newsletter to Members of K'atlodeeche First Nation*. Hay River: K'atl'odeeche First Nation.

Taylor, A., Severson-Baker, C., Woynillowicz, D., & Winfield, M. S. (2004). *When the Government is the Landlord: Economic rent, Non-Renewable Permanent Funds, and Environmental Impacts Related to Oil and Gas Development in Canada*. Toronto: Pembina Institute.

Thornton, T., & Wanasuk, P. (2016). Indigenous tourism as a sustainable social-environmental enterprise: The political ecology of tourism in Southeast Alaska. In S. Nepal & J. Saarinen (Eds.), *Political Ecology and Tourism*. Abingdon and New York: Routledge.

USEIA, U.S. Energy Information Administration (2013). *State Energy Data 2013: Consumption, Glossary*. Washington: U.S. Energy Information Administration.

Van Wyck, P. C. (2010). *The Highway of the Atom*. Montréal, Kingston, London, Ithaca, NY: McGill-Queen's University Press.

Wilson, J. (2009). *Poverty Reduction Policies and Programs: Northwest Territories*. (Social Development Report Series). Kanata: Canadian Council on Social Development, Canadian Council on Social Development.

Wilson, P. A. (1996). Empowerment: Community economic development from the inside out. *Urban Studies*, *33*(4–5), 617–630.

Wohlberg, M. (2014). *Fort Liard One Step Closer to Settling Land Claim*. Yellowknife: Northern Journal.

Yerbury, J. C. (1986). *The Subarctic Indians and the fur trade 1680–1860*. Vancouver: The University of British Columbia Press.

Yin, G., & Herfel, W. (2011). *Philosophy of Complex Systems*. Oxford and Amsterdam: Elsevier.

Yusoff, K., Grosz, E., Clark, N., Saldanha, N., & Nash, C. (2012). Geopower: A panel on Elizabeth Grosz's Chaos, Territory, Art: Deleuze and the framing of the Earth. *Environment and Planning D: Society and Space*, *30*(6), 971–988.

12 Bridging ICTs with governance capabilities for food–energy–water sustainability

Timothy Karpouzoglou, Laura M. Pereira and Samir Doshi

Introduction

In this chapter, we critically explore the important role that Information and Communication Technologies (ICTs) play in food, energy, and water (FEW) governance. In particular, we perceive an important role for ICTs in bridging the knowledge gap associated with FEW development goals and for addressing the many synergies and trade-offs between these goals and existing governance arrangements (Mol, 2006; UN, 2014; WWAP, 2015). FEW governance presents us with a 'wicked' problem in that it is unstructured, complex, and contains multiple and interconnected subsets of problems (Weber & Khademanian, 2009). A wicked problem can be understood as a problem that is typically ill-defined and involves many uncertainties and contrary views of how to address it (Dewulf & Termeer, 2015). In a similar way in which climate change adaptation is a wicked problem, FEW governance cannot be precisely formulated or solved due to diverging understandings of FEW development priorities as well as inherent complexities associated with their monitoring and evaluation (Murthy et al., 2013). Yet, FEW governance is becoming a key political priority as the adoption of the Sustainable Development Goals (SDGs) places huge importance on FEW systems and the need for putting in place a comprehensive monitoring framework to track progress towards global sustainability (EAT Initiative, 2015). It is in this context that we envisage ICTs will play a substantial role in creating better conditions for FEW governance and thus aid in achieving the SDGs over the coming years.

Theoretical and practical engagement with ICTs in the field of development and environmental sustainability is growing. This expansion has created new opportunities for governing through 'small infrastructures' that support information gathering and exchange, which in turn are changing relationships between different governance actors (namely, civil society, state, and business actors) in FEW-related sectors (Zulkafli et al., 2017). Many of these small infrastructures have developed with strong public–private partnerships as well as direct involvement of citizens in ICT initiatives. ICTs have been deployed in early warning systems and disaster management – such as the response efforts to the Ebola Viral Disease in West Africa in 2014 and the 7.8 scale earthquake in

Nepal in April 2015 – and have been recognised as important tools for monitoring, integration, and sharing of data on climate change mitigation and adaptation (Pettit et al., 2013).

Research suggests that access to mobile phones further plays an increasingly important role in economic development (Donner & Escobari, 2010). In agriculture, mobile phones are key tools in monetary transactions of small-holder farmers and micro credit schemes (Chowdhury, 2006; Qiang, Kuek, Dymond, & Esselaar, 2012). More broadly, a recent study in Science has shown that M-PESA, a mobile money service used by 96 percent of Kenyan households, has directly lifted 2 percent of the population out of extreme poverty by allowing money to move more easily between people (Suri & Jack, 2016; See Dawson, 2017). There has also been increasing focus on the promise of ICTs as a medium for improving knowledge management in African agricultural and rural development policies as long as investments in ICT infrastructure, technology, and expertise are sustained (Nyirenda-Jere & Kazembe, 2014).

In complex governance arenas, several new forms of governance capabilities need to be cultivated (Dewulf & Termeer, 2015; Termeer, Dewulf, Breeman, & Stiller, 2013). Drawing upon scholarship on governance capabilities, the aim is to understand better how FEW governance can be improved with the increase in ICTs that we are witnessing globally. A governance capability can be broadly defined as the ability of governance actors to act wisely when facing wicked problems, and the ability of the governance system to enable such acting (Termeer & Bruinsma, 2016). As an approach to understanding governance, it departs from the usual criteria of governance evaluation, such as good or bad, effective or ineffective, and takes into account a greater variety of governance qualities, while being more sensitive to institutional complexity. A governance capability further relates to ways of observing and understanding a problem, different strategies to cope with it, and the institutional conditions that enable these observations and strategies (Dewulf & Termeer, 2015).

Public policy and administration scholars have expanded the notion of a governance capability into an operative framework that consists of five main governance capabilities that are: *reflexivity*, *resilience*, *responsiveness*, *revitalisation*, and *rescaling* (Dewulf & Termeer, 2015). We focus on these five capabilities using a literature survey that has helped us identify key examples where ICTs and governance capabilities interact. As has been noted in previous studies, a natural fit between ICT expansion and demands for information collection, processing, storage, and dissemination is often celebrated (Nyirenda-Jere & Kazembe, 2014). Here we take a more cautious approach, recognising that, despite the opportunities, a tendency to idealise ICTs without placing appropriate attention on how ICTs enter different governance areas or how they might be part of both enabling or preventing sustainable outcomes for FEW governance may be problematic (Soma, Termeer, & Opdam, 2016). Drawing on the governance capabilities framework helps to confront this potential paradox, but also helps to make more sense of the underpinning governance qualities that can be strengthened through association with ICTs.

This chapter is organised according to the five 'R' governance capabilities: reflexivity, resilience, responsiveness, revitalisation, and rescaling. In each section, we describe the governance capability and further illustrate how it relates to the use of ICTs in the FEW governance sphere. As ICTs are both rapidly evolving and highly dynamic, we do not claim to have covered the entire spectrum, instead we use key examples. Some of the technologies include remote and automated technologies, which can provide independent and objective gathering of information and are increasingly important in water monitoring. We are also interested in ICTs that imply citizen involvement – i.e. relying on people to gather information, to report on, and to monitor systems. In the ICT examples discussed, we would argue that there exist new possibilities emerging from cross-level dynamism offered by these technologies as they span different scales from the local to global and involve multiple actors. After we have discussed each of the governance capabilities separately, we reflect in the conclusion upon the synergies and conflicts that emerge between the various governance capabilities as well as the long-term possibilities and challenges of ICTs for FEW governance.

Few governance and ICTs: mapping five domains of interaction

Reflexivity

As a capability, reflexivity underlines how there can be disagreement and contestation in a governance setting about the nature, associated meanings, and actions of the problem being addressed (Termeer, Dewulf, Karlsson-Vinkhuyzen, Vink, & Vliet, 2016). Governance actors including scientists and policy makers frame issues differently depending on their position, or the particular interactional setting in which they operate (Dewulf & Termeer, 2015). Decision making in FEW governance can sometimes be overtly focused on one type of problem framing. Take the example of food security, which has for a long time been framed solely as an issue of raising farm productivity and elevating agricultural outputs – a perspective that guided much of post-World War II global agriculture policy (Friedmann & McMichael, 1989; Karpouzoglou & Barron, 2014). ICTs can help create conditions where different types of users can contribute to the production of information, but can also allow for greater diversity in terms of problem framing, allowing certain frameworks to be challenged, or to bring about a different set of viewpoints and perspectives.

The greater involvement of citizens in scientific practice through forms of citizen science is one way in which this type of reflexivity can be leveraged and where ICTs have a central role to play. While citizen science initiatives have been around for over a century (e.g. citizens monitoring seasonal migration patterns of birds), developments in sensing technology, data visualisation, and wider accessibility of low cost and low maintenance sensing equipment is expanding the ways in which citizens interact with science (see Sullivan et al., 2009 for an example of a citizen based bird observation network, eBird). This new paradigm in data collection is also creating conditions for more locally grounded approaches to FEW governance (Buytaert et al., 2014). In the context of water resources management, for example, citizen science has the potential to support new hydrological knowledge as the data retrieved from citizen-driven data collection processes may often complement or provide entirely new insights, sometimes even contrasting with the information received by traditional water-monitoring stations (Buytaert et al., 2014). At the same time, using ICTs in a way that involves communities in generating their own data further aids more systemic learning and provides citizens with tools for actively engaging in governance and taking greater control over how problem-framing choices are constructed and acted upon (Landström et al., 2011).

In a study focusing on local tailoring of ICT tools for environmental decision support systems in Peru, it was found that water users in upland Andean environments were far more inclined to receive environmental information that links directly to their everyday activities and needs (for example, 1–3 days of temperature data) as opposed to model-driven forecasts projecting several months into the future (Zulkafli et al., 2017). This is an area that may require more systematic reflection at the technology pre-design and experimentation phase and it needs to be recognised that not all these benefits occur in all cases (Karpouzoglou et al., 2016). More emphasis can be placed for instance on the tacit knowledge of users by using methods such as cognitive mapping, focus groups, games and experiments, and interactive group exercises (Elsawah, Guillaume, Filatova, Rook, & Jakeman, 2015).

Looking at ICTs from the perspective of reinforcing the interaction between users and sources of information is therefore critical for strengthening reflexivity. The web-based nature of some ICTs brings about new possibilities for fostering user interaction, but also to bring disagreements to the forefront of public debate. Social media platforms can sometimes bring to the surface deep contestation and entrenched ideologies. These platforms also have the potential to unite different actors and create coalitions that construct alternative framings of FEW governance in line with new modes of collective action and sustainability. For an example in the energy sphere, Twitter provides a powerful tool for connecting social movements and civil society in opposing fossil fuel-based economies through their own counter framing and narrative strategies of decarbonisation that often link intersectionally with other movements, (e.g. the 'Occupy' movement that arose in opposition to the role of capitalism in the financial crisis of 2008). In terms of governance, this can have potentially transformative implications given that new social networks arise around shared ideas (Kelly Garrett, 2006; Segerberg & Bennett, 2011). At the same time the global nature of social media networks means that a local issue can swiftly become global news and therefore attract widespread attention (Stevens, Aarts, Termeer, & Dewulf, 2016).

Recent developments in visualisation are also an important catalyst for reflexivity in FEW governance since they are key for enabling data to be presented in more visually appealing ways whilst also creating space for visualisation design that is better tailored for different users (McInerny et al., 2014). The dashboard-style web application designed for small-holder farmers in Kenya to access online real-time data such as weather forecasts and crop market data,[1] is one such example. It is based on novel channels of collaboration between the visualisation community, scientists, governments, and end-users and offers a catalyst for making complex datasets more accessible to farmers, but also provides scope for farmers to interact with each other and to learn from the types of farming practices adopted.

The voluntary and non-profit sector has played an important role in shaping this particular pathway of ICTs into the development arena. For example, Digital Green is a non-profit international development organisation that uses digital services for community engagement and knowledge transfer around issues such as agricultural extension in parts of India, Ethiopia, Ghana, Niger, and Tanzania.[2] Using video and mobile phones, it has created new channels for farmers to share with each other successful practices in neighbouring villages that might not have spread without the technology. The technology is essentially meant to supplement (and not substitute) community dialogue. Making this form of knowledge exchange interactive allows other types of stakeholders, such as development practitioners and donor organisations, to interact with this knowledge too. Reflexivity therefore relates to making visible the different frames associated with FEW, but also facilitates interaction between actors with different perspectives. ICT provides new forms of opportunities for enabling effective FEW governance.

Resilience

Resilience is the capability of governance to adapt flexibly to uncertainties and constantly changing conditions. Resilience theory as applied to the governance context highlights the crucial role of information and knowledge co-production processes in developing adaptive capacity in governance and for developing long-term resilience (Folke et al., 2005). Resilience, as a governance capability, relates to using real-time data to better understand vulnerabilities in the governance system, to adapt flexibly to disturbance (i.e. disease, floods, cyclones), and also to cultivate monitoring and learning (Dewulf & Termeer, 2015).

Policy makers are tasked with reconciling the need to reduce communities' current vulnerability to disruptions in their access to FEW resources with the need to invest in development projects that offer long-term resilience to communities facing an uncertain future. By improving access to current information on the state of FEW needs in communities and then mapping these onto larger vulnerability maps (e.g. given future scenarios of climate, economy, etc.), FEW resources could become targeted in a way that improves both current and future access (Murthy et al., 2013). This is why the public sector is increasingly engaged with bridging ICTs with FEW-related issues, for instance in the area of improving seasonal and long-term forecasts and prediction capacity relevant to climate services (Street, Jacob, Parry, Runge, & Scott, 2015). Research programs like Euporias[3] develop methods and tools to translate data into climate services in important sectors such as water, energy, health, and food security that match the needs of specific users.

In light of variable and changing climatic conditions, these issues need to be further interwoven with the role of information for developing essential capabilities for decision making under high uncertainty. The call for integration of multiple sustainable development sectors, as in the SDGs, increases the scope of interdependent dynamics in the system where an intervention is implemented. This subsequently increases the complexity of the intervention (Ramalingam, 2013). In complex environments that are characteristic of high uncertainty and non-linear dynamics, a higher frequency of information is necessary in order to understand how an intervention impacts communities and governance systems over time. ICTs can help address this information deficit in real-time. However, as ICTs can have a profound impact on communities where there is little existing technology, any new technological system that is implemented would need to be closely monitored. Creating an enabling environment in which to tighten feedbacks between information–decision cycles and to adapt to on-going dynamics for increasing the positive impact of the intervention is important.

There are caveats to how applicable ICTs data can be for FEW decision making. The rapid rise in data collection and mapping through ICT has not necessarily improved our predictive ability, not only because it takes a long time to collect, process and analyse data properly, but also because the complex, non-linear characteristics of FEW systems make predictions virtually impossible (See Murthy et al., 2013). For example, vulnerability to food insecurity can be next to impossible predict because it is not simply about food production, but is also about access to markets, dietary habits, household income, nutrition, and cultural traditions (See Lang & Barling, 2012). Furthermore, we have not yet studied social systems and their non-linear feedbacks as well as we have done in the physical environment, largely due to a lack of large-scale data. An example is that, in retrospect, a hike in food prices may be an accurate indicator of an impending food crisis, but given the numerous other variables at play, it may not be a consistent forecaster for all situations (Murthy et al., 2013). A step in the right direction would be for more real-time data to be measured and analysed so as to be able to recognise trends that could lead towards a 'tipping point' or crisis situation. However, it must be emphasized that an increase in information does not make a complex system more predictable, it just makes navigating uncertainty more reliable. Use of ICTs for real-time data collection can strengthen resilience by means of increase understanding of how complex FEW systems respond to disturbances, and provide a basis for fostering adaptive capacity in FEW governance.

Responsiveness

Responsiveness is the governance capability to respond appropriately to continuously changing demands. In the absence of responsiveness, decision makers may risk neglecting citizens' interests and losing legitimacy (Termeer, Dewulf, Breeman, & Stiller, 2013). FEW-related decision making is often fragmented among different actors (government, industry, multilateral institutions) and influenced by various institutional interests. This can hamper responsiveness and distance citizens from the decision-making process. For instance, municipal delivery of water and sanitation services is impacted not only by the way water is allocated and managed as a resource, but also by laws that govern pollution, urban planning, and land tenure (Subbaraman & Murthy, 2015). In other instances, development goals in one sector can create conflicts with sustainability goals in another. A policy that seeks to intensify agricultural development in a region could conflict with a desire to improve water security and sustainability; or a policy designed to promote energy access through biofuels could inadvertently result in an increase in food prices or unsustainable agricultural practices (Murthy et al., 2013, Chapter 1). Understanding these types of trade-offs requires access to different types of information and keeping track of developments taking place across FEW sectors. ICT services, such as access to the internet, create new ways for exploring pieces of related data and information coming from different sources in real time (i.e. information systems and databases), which can help create more responsive governance arrangements in the long term.

In a world where there is an increased pressure from donors and funding organisations to provide evidence of project successes, the governance capability of responsiveness can also be very elusive. Following the Millennium Development Goals (MDGs), and now with the SDGs, development projects are increasingly more 'outcomes'-oriented with a particular emphasis on metrics such as the number of water taps built; the number of children fed in a school program; or the number of households with access to electricity. This can lead to high expectations around the role of ICT applications in achieving the sustainable development goals and the extent to which they can play a central role in making governments more responsive is greatly emphasised (Termeer & Bruinsma, 2016). ICTs can further run the risk of creating an artificial sense of responsiveness while masking certain risks and an inability to appreciate trends and issues of accountability, security, and privacy (Murthy et al., 2013).

These risks include questionable accuracy of data, the lack of context for some of the data generated, and the inability to control the 'sample' of those providing crowd-sourced data. For these issues to be overcome, the role of cultivating trust as part of developing more responsive governance arrangements is central. There are two indirect contributions that ICTs can make towards this. By providing more real-time, 'objective' data, ICTs can be used to build trust between sometimes disparate stakeholders. They can also aid in understanding what decisions are helped by more information and where uncertainty needs to be handled through more flexible governance. These feedback systems can create a virtual conversation, which allows for multiple voices to be heard and reciprocated.

With the ubiquitous nature of ICTs and the Internet of Things (IoT) deploying a massive amount of sensors and data-collection devices for big data throughout the world, there is an obvious risk to privacy and security of individuals through personably identifiable information (PII). As outlined in the Principles for Digital Development – a set of best practices and considerations for designing and implementing ICT for development (ICT4D) programs – the integration of ICTs with governance programs requires new principles, policies, and practices in regards to data privacy and protection of individuals (Waugaman, 2016). The Principles for Digital Development have been signed on by over 60 organisations, including the leading donors and practitioner organisations working in the ICT4D field.[4] These privacy, security, and ethical measures are critical in developing trust between users and decision makers when ICTs are used as a communication and information gathering medium. In recognition of the need for increased privacy, security, and ethics with ICTs and data collection, several organisations have developed privacy policies and responsible data practices, such as Oxfam International, Amnesty International, and USAID. The Responsible Data Forum is a collaborative of donors and practitioners that are developing policies and tools for ethical use of data in evidence-based policy and decision making.[5]

The more trust there is in the system, the more information is likely to be shared and the more likely it is that responsiveness is embedded in the overall governance process, although this is not always the case.[6] ICTs can be used for building trust through the transparency when different participants come together to learn and discuss the information provided by ICTs in order to build common understandings of problems and solutions. The internet has helped generate the new 'carbon economy' that is based on new ways of integrating greenhouse gas emissions data and combining them with social networking tools. This allows for information exchange among regulators, the regulated communities, and energy companies worldwide (Gilbert, 2009; Soma et al., 2016). In this respect, ICTs gradually help build knowledge that can ultimately help decision makers prioritise investments based upon what concerns citizens (public legitimacy) and where problems are occurring (real-time data gathering). The capability of responsiveness can therefore be seen as a strong incentive for politicians to engage with ICTs in a FEW governance context.

Revitalisation

Revitalisation takes responsiveness one step further since it is the capability that relates to unblocking stagnation. Stagnation in governance processes can emerge when actors are no longer able to change the course of action, even in cases in which 'more of the same' no longer works or becomes counterproductive (Dewulf & Termeer, 2015). In a complex FEW governance context, revitalisation relates to the way in which actors may try to change established patterns of behaviour in order to enhance new ways of thinking and acting (Termeer et al., 2016). This may not be straightforward, particularly since stagnation or path dependency can also be the outcome of entrenched power configurations between the actors involved.

In the context of developing the revitalisation capability for FEW governance, ICTs can provide important tools for facilitating a more forward-looking perspective in governance contexts with a history of tense relations. This challenge can be found in the case of transboundary water basins where regional cooperation on water remains contentious because of deep mistrust between governance actors involved or due to chronic political tensions (Pedregal, 2015). The Nile Basin Initiative (NBI) is an inter-governmental group that is trying to develop an appropriate management plan for the Nile basin that can overcome some of these challenges. The Nile has huge potential for improved use and development, but there is an enormous challenge in trying to get Nile countries to see the benefit of cooperation. For example, there have been contestations as to the amount of water that could be saved due to evapotranspiration if a dam were to be built in Ethiopia instead of Egypt. Historically, each country would each use their own country expert to provide evidence because of lack of trust between the countries (Whittington, Waterbury, & Jeuland, 2014). As part of the NBI, scientists now use automated sensors to provide real-time flow data and this is creating a different type of dynamic in the governance process. By talking about the data and making it more readily accessible, scientists are slowly gaining the interest and trust of policy makers and this more transparent environment will hopefully lead to more evidence-based decision making in the future (Whittington et al., 2014).

In the context of transboundary water cooperation in the South Asian context, there is a reluctance of state authorities to share water-related information in spite of a number of transboundary water treaties in place (Prasai & Surie, 2015). As a result, regional cooperation between South Asian countries can often lead to stagnation masked by problems of mistrust and political tension. This provides an emerging opportunity context for governance revitalisation on transboundary cooperation associated with ICTs. Drawing on the examples of telemetry, remote-sensing, and modelling techniques such as Water Accounting +, new technologies are rapidly changing the ways in which data are collected, stored, and used by government agencies (Prasai & Surie, 2015). These developments challenge governments' current inclination towards data secrecy through encouraging easier and more reliable access to information. At the same time, these tools help governments communicate with constituencies within and outside their borders. This in turn may make it easier for governments to navigate away from stagnating situations and to move towards making transparent political choices in complex situations. It is important to note that these ICT-driven processes can take time, but ultimately hold promise towards changing the norms of transboundary cooperation, by lowering nationalistic tendencies around water, promoting more open political contestation in agenda-framing and scope, and deliberating alternative political choices (Pedregal, 2015; Prasai & Surie, 2015).

Rescaling

Rescaling is the final governance capability. It relates to the interconnectivity inherent in FEW governance. Rescaling necessitates observing and engaging with scale mismatches[7] and re-organising connections across levels and scales so that interconnectivity is reflected in governance arrangements (Dewulf & Termeer, 2015). The possibilities for fostering interconnectivity are significantly enhanced with ICTs as a result of acquiring and disseminating more accurate spatially distributed data (Buytaert, Dewulf, De Bièvre, Clark, & Hannah, 2016). This can help understanding of system dynamics across scales and how FEW systems interact with each other.

One way in which ICTs contribute to rescaling is through linking local level data and knowledge with regional to global data sets and models. Rescaling can boost responsiveness since it is related to the way different sources of evidence come together, but it is distinct in helping think through the polycentric arrangement of information gathering and dissemination organised at multiple governance levels (Buytaert et al., 2016; Karpouzoglou et al., 2016). In the case of citizen science initiatives where non-expert involvement is strong, new forms of interconnectivity are being created that are changing the relationship between monitoring and decision making. Citizens act as human 'sensors'. With mobile phones, like sensors recording temperature or river flow, citizens can report issues to municipal agencies regarding public service failures (Georgiadou et al., 2011). Open-sourced geo-mapping applications, such as the OpenStreetMap (OSM) wiki app are also increasingly as part of strategies for fostering citizen-orientated involvement around key FEW priorities (Pedregal et al. 2015).

In another example, UNICEF created a user-centred citizen engagement tool based on SMS messages in 2010, polling opinions on issues related to food, nutrition, water, sanitation, health, and other development issues. The platform, which spread to over 1 million users across 17 countries, relies on volunteer community members serving as 'U-Reporters' to provide information that affects their own communities.[8] Some organisations that initially focussed on crowdsourcing information have shifted to involve trusted information sources (such as utility valve men in the case of NextDrop and health workers in the case of mWater) to provide a uni-directional flow of important information to users (Sharmila Murthy, pers. comm.). In order to take full advantage of crowd-sourced datasets and analysis, common data architectures and standards are needed to enable interoperable data systems that can aggregate diverse sources in order to obtain a macro level view of the issue. With so many privatised platforms being developed and a large number of new systems that are not open source, this becomes more difficult. Appropriate legal policies are needed not only to ensure privacy of data, but also effective use (Sharmin, Faith, Prieto Martin, & Ramalingam, 2017).

The use of human sensors can have broad-ranging applications for rescaling governance strategies as new information, connected across different levels and scales, becomes available. In Bangladesh where disaster floods and droughts are major natural disasters that occur regularly, communities using the radio or mobile phones can quickly broadcast warnings to volunteers, media, and rural centres. Information can reach decision makers and first responders more quickly (Karim, 2004). New sensor technologies are also used to create micro-scale climate maps in rural regions where large-scale climate data are unavailable. At the same time, much more technologically sophisticated sensors, such as drones and micro-satellites (through innovation start-ups such as Planet Labs[9]), are gaining traction and used for deforestation mapping and land conservation projects, tied closely to agricultural land tenure issues. In the energy sector, Smartgrids enable utilities to monitor electricity demands from their customers. They provide financial incentives for customers to use energy more efficiently, through the use of smart meters in houses that connect to smart appliances, which in turn respond to information on energy usage from the utility – lowering their energy usage during peak times.[10]

It is important to emphasise that rescaling can also be highly politicised and whilst it can challenge power positions, there are always consequences for who is included and who is excluded from the decision making process. Those providing information may not always stand to benefit from the information collected. An anecdotal story involves a community in the Amazon that agreed to participate in a project where they would send information about malaria prevalence; yet, the project never benefitted the local people (Murthy et al., 2013). Whilst the community was supposed to provide 'sweat equity' to install the satellite tower and then send in reported cases of malaria, the organisation coordinating the project could not send medicine or other help once malaria was actually reported (Murthy et al., 2013:8).

The case highlights that it is important to work with communities of different users in a polycentric information flow arrangement and to develop methods of communication that work for them. As Avgerou (2010) highlights, it is critical to understand the socio-political context in which ICT takes place, to identify the power brokers of the information flow, and what the outcomes of the data collection are, i.e. what is the response and who is expected to respond once the data have been collected? This caveat emphasises the concept of an 'end-to-end' project design and talks specifically to projects in rural areas where ICT allows for the flow of information more easily than ever before, but does not necessarily improve the ability to respond to the issues that these data flag.

Conclusion: ICT as part of a new approach to FEW governance?

Recent advancements in ICTs, some of which have been captured in this chapter, highlight how ICTs have the potential to support FEW governance systems. The ability of ICTs to provide data in a variety of forms and through diverse processes has resulted in a concomitant push to make use of these data, for everything from advocacy to decision making, large-scale modelling to local service delivery. However, there is currently little to be found in the literature in terms of a guiding theory for the role of ICTs in FEW governance. This is where we find that a discussion that attempts to go beyond a purely technical discussion of ICTs opportunities and that links up to the less understood assemblages of ICTs is both timely and important.

Drawing insights from the capabilities framework, we have described concrete entry points where FEW governance capabilities can become better supported by ICTs (see also Table 12.1). For example, social media tools such as Twitter can create new possibilities for governance reflexivity by rendering visible different problem frameworks. A virtual space of interaction can also create new opportunities for actors that hold different 'problem' frameworks

Table 12.1 Food, energy, and water (FEW) governance capabilities and ICTs

Governance capability	*Key features*	*ICT opportunity space*	*Challenges*
Reflexivity	Rendering visible different problem frames Facilitating interaction between actors with different perspectives	Social media tools to connect governance actors in opposing dominant problem frames or constructing new frames	Internet accessibility and familiarity with social media networking tools
Resilience	Fostering adaptive capacity in governance Flexibility in dealing with uncertainty and fast-changing conditions	Use of web-services and mobile phone technology for 'real-time' monitoring of FEW-related data	Ensuring robustness of predictions in light of complex non-linear FEW characteristics Limits in usability of scientific data on climate by non-expert actors
Responsiveness	Communicating sensitively in light of multiple demands and concerns	New ICT capacity to examine large datasets across governance sectors	Questionable data accuracy and weak trust in data generated Skewed emphasis on metrics disconnected from local contexts
Revitalisation	Unblocking stagnations in governance and policy deadlocks	Remote sensing technologies and modelling tools that encourage multi-stakeholder dialogue	Challenging political contexts Institutional barriers for data transparency and access to information
Rescaling	Enabling connections across different scales Fostering interaction between citizens, policy makers, and scientists	Citizen involvement in local level FEW data collection using mobile phones, radio, and web services	Information power-brokers Weak coordination of information flows between actors at different scales

(adapted from Dewulf & Termeer, 2015)

and who may not have an opportunity to interact directly in policy. They may instead use social networking tools and the internet. ICTs do not necessarily improve our ability to predict new events or to be more prescriptive about them. However, they do help improve real-time monitoring, which in turn can help strengthen resilience in governance. For example, a range of new web-related information services and mobile phone applications have been shown to create new capacities for real-time monitoring of FEW-related issues. This can help decision makers become more quickly aware of crucial information to deal with uncertainty and fast-changing conditions and hopefully to plan better for the future.

In the discussion on governance responsiveness, it was described how new ICTs capabilities for dealing with large datasets can also play a central role in helping governance actors become more responsive and better able to navigate changing demands and cross-sectoral complexities. Furthermore, remote sensing technologies and modelling tools that encourage multi-stakeholder dialogue can create governance revitalisation and help unblock policy stagnation. Policy stagnations are in fact highly common in FEW governance, particularly in challenging contexts for effective political dialogue such as in the case of transboundary water management or decarbonised energy systems. Lastly, we have highlighted the potential role of citizen-orientated involvement with ICTs as part of data collection and dissemination, and how this type of involvement can make a difference in terms of strengthening governance rescaling and for enabling connections across scales to become more visible.

In conclusion, we have used the governance capabilities framework as a conceptual lens to consider the broad spectrum of FEW issues and domains. However, there are still many gaps to be addressed and to strengthen this line of enquiry. We envisage that insights generated from this work could also be useful and can easily tie in with other FEW related interventions discussed in this book. For example, by helping the development donor community to design interventions that bridge more effectively the ICTs opportunity space with governance complexity. It is also important not to limit the governance capabilities to ICTs that are typically associated with 'big data' gathering interventions but instead to explore how the capabilities may form a basis for smarter forms of public dialogue around ICTs that involve stronger participation of concerned citizens. This approach can render visible unforeseen or surprising directions for achieving broader societal goals (such as the SDGs which relate to FEW) while empowering actors that are not always part of the policy debate.

Notes

1 See: Thought for food – We are bringing satellite data to smallholders, http://blog.vizzuality.com/post/114042422481/thought-for-food.
2 See: www.digitalgreen.org/media/docs/publications/rural_video_based_social_network.pdf.
3 See: www.euporias.eu/objectives.
4 For list of sponsoring organisations and other resources, see http://digitalprinciples.org/.
5 See https://responsibledata.io/ for a list of resources on Responsible Data.
6 See a series of blog posts by Daraja and Reboot: http://blog.daraja.org/2012/02/why-did-maji-matone-fail-1-low-tech.html; https://reboot.org/2016/10/12/longreads-moving-real-time-data-real-time-programs/.

7 A scale mismatch occurs when there is a mismatch between the scale of a problem and the scale at which it is governed, for example when a city municipality tries to govern its water resources that come from a watershed in a different country upstream.
8 See: www.unicefstories.org/2015/08/31/u-report-nigeria-reaches-more-than-1-million-responders/.
9 See: www.planet.com/.
10 See: www.smartgrid.gov/the_smart_grid/index.html.

References

Avgerou, C. (2010). Discourses on ICT and development. Information Technologies & International Development, 6(3), pp–1.

Buytaert, W., Dewulf, A., De Bièvre, B., Clark, J., & Hannah, D. M. (2016). Citizen science for water resources management: Toward polycentric monitoring and governance? *Journal of Water Resources Planning and Management*, *142*(4), 1816002. Retrieved from http://doi.org/10.1061/(ASCE)WR.1943-5452.0000641.

Buytaert, W., Zulkafli, Z., Grainger, S., Acosta, L., Bastiaensen, J., De Bièvre, B., . . . Zhumanova, M. (2014). Citizen science in hydrology and water resources: Opportunities for knowledge generation, ecosystem service management, and sustainable development. *Frontiers in Earth Science*, *2*, 1–21.

Chowdhury, S. K. (2006). Investments in ICT-capital and economic performance of small and medium scale enterprises in East Africa. *Journal of International Development*, *18*(4), 533–552. Retrieved from http://doi.org/10.1002/jid.1250.

Dawson, S. (2017, January 18). *Why does M-PESA Lift Kenyans Out of Poverty?* Washington, D.C.: Consultative Groups to Assist the Poor. Retrieved from www.cgap.org/blog/why-does-m-pesa-lift-kenyans-out-poverty.

Dewulf, A., & Termeer, C. J. A. M. (2015). Governing the future? The potential of adaptive delta management to contribute to governance capabilities for dealing with the wicked problem of climate change adaptation. *Journal of Water and Climate Change*, 6(4), 759–771. Retrieved from http://doi.org/10.2166/wcc.2015.117.

Donner, J., & Escobari, M. X. (2010). A review of evidence on mobile use by micro and small enterprises in developing countries. *Journal of International Development*, *22*(5), 641–658. Retrieved from http://doi.org/10.1002/jid.1717.

EAT Initiative (2015). Integrated Indicators for Sustainable Food Systems and Healthy Diets in the Post-2015 Development Agenda_17 Sept Final Statement. https://drive.google.com/file/d/0B7TR9GImQpPUMGItWktvOFgtUnc/view

Elsawah, S., Guillaume, J. H. a, Filatova, T., Rook, J., & Jakeman, A. J. (2015). A methodology for eliciting, representing, and analysing stakeholder knowledge for decision making on complex socio-ecological systems: From cognitive maps to agent-based models. *Journal of Environmental Management*, *151C*, 500–516. Retrieved from http://doi.org/10.1016/j.jenvman.2014.11.028.

Folke, C., Hahn, T., Olsson, P., & Norberg, J. (2005). Adaptive governance of social-ecological systems. Annu. Rev. Environ. Resour., 30, 441-473.

Friedmann, H., & McMichael, P. (1989). Agriculture and the State System: The rise and decline of national agricultures, 1870 to the present. *Sociologia Ruralis*, 29(2), 93–117. Retrieved from http://doi.org/10.1111/j.1467-9523.1989.tb00360.x.

Georgiadou, Y., Bana, B., Becht, R., Hoppe, R., Ikingura, J., Kraak, M. J., . . . Verplanke, J. (2011). Sensors, empowerment, and accountability: A digital earth view from East Africa. *International Journal of Digital Earth*, *4*(4), 285–304. Retrieved from http://doi.org/10.1080/17538947.2011.585184.

Gilbert, J. (2009). Climate change and GHG legislation invigorate the EH&S software market. Air & Waste Management Association, december: 34–36. Retrieved from http://pubs.awma.org/gsearch/em/2009/12/itinsight.pdf.

Karim, N. (2004). Options for floods and drought preparedness in Bangladesh. In *Proceedings of the Second International Conference on Post-disaster reconstruction: Planning for Reconstruction*.

Karpouzoglou, T., & Barron, J. (2014). A global and regional perspective of rainwater harvesting in sub-Saharan Africa's rainfed farming systems. *Journal of Physics and Chemistry of the Ear*, *72–75*, 43–53. Retrieved from http://doi.org/10.1016/j.pce.2014.09.009.

Karpouzoglou, T., Zulkafli, Z., Grainger, S., Dewulf, A., Buytaert, W., & Hannah, D. M. (2016). Environmental Virtual Observatories (EVOs): Prospects for knowledge co-creation and resilience in the information age. *Current Opinion in Environmental Sustainability*, *18*, 40–48. Retrieved from http://doi.org/10.1016/j.cosust.2015.07.015.

Kelly Garrett, R. (2006). Protest in an information society: A review of literature on social movements and new ICTs. *Information, Communication & Society*, 9(2), 202–224. Retrieved from http://doi.org/10.1080/13691180600630773.

Landström, C., Whatmore, S. J., Lane, S. N., Odoni, N. a., Ward, N., & Bradley, S. (2011). Coproducing flood risk knowledge: Redistributing expertise in critical 'participatory modelling.' *Environment and Planning* A, *43*(2006), 1617–1633. Retrieved from http://doi.org/10.1068/a43482.

Lang, T., & Barling, D. (2012). Food security and food sustainability: Reformulating the debate. *The Geographical Journal*, *178*(4), 313–326.

McInerny, G. J., Chen, M., Freeman, R., Gavaghan, D., Meyer, M., Rowland, F., . . . Hortal, J. (2014). Information visualisation for science and policy: Engaging users and avoiding bias. *Trends in Ecology and Evolution*, *29*(3), 148–157. Retrieved from http://doi.org/10.1016/j.tree.2014.01.003.

Mol, A. P. J. (2006). Environmental governance in the information age: The emergence of informational governance. *Environment and Planning* C*: Government and Policy*, *24*(4), 497–514. Retrieved from http://doi.org/10.1068/c0508j.

Murthy, S., Pereira, L., Harley, A., Shemie, D., Lee, E., Guardabassi, P., . & Moore, S. (2013). The Role of Information and Communications Technology (ICT) in Helping Decision-Makers Meet Food, Energy and Water (FEW) Needs. Harvard Kennedy School.

Nyirenda-Jere, T., & Kazembe, J. (2014). Improving Policymaking for Agricultural and Rural Development in Africa The role of ICTs and Knowledge Management. London: International Institute for Environment and Development (IIED) http://pubs. iied. org/pdfs/14636IIED. pdf.

Pedregal Mateos, B., Cabello Villarejo, V., Hernández Mora, N., Limones Rodríguez, N., & Moral Ituarte, L. D. (2015). Information and Knowledge for Water Governance in the Networked Society. Water Alternatives, 8(2), 1–19.

Pettit, C., Williams, S., Bishop, I., Aurambout, J.-P., Russel, A. B. M., Michael, A., . . . Borda, A. (2013). Building an ecoinformatics platform to support climate change adaptation in Victoria. *Future Generation Computer Systems*, *29*(2), 624–640. Retrieved from http://doi.org/10.1016/j.future.2011.07.004.

Prasai, S., & Surie, M. D. (2015). Water and climate data in the ganges basin: Assessing access to information regimes and implications for cooperation on transboundary rivers. *Water Alternatives*, 8(2), 20–35.

Qiang, C. Z., Kuek, S. C., Dymond, A., Esselaar, S., & Unit, I. S. (2011). Mobile applications for agriculture and rural development. World Bank, Washington, DC.

Ramalingam, B. (2013). *Aid on the Edge of Chaos: Rethinking International Cooperation in a Complex World.* Oxford: Oxford University Press.

Segerberg, A., & Bennett, W. L. (2011). Social media and the organization of collective action: Using twitter to explore the ecologies of two climate change protests. *The Communication Review*, *14*(3), 197–215. Retrieved from http://doi.org/10.1080/10714421.2011.597250.

Sharmin, S., Faith, B., Prieto Martín, P., & Ramalingam, B. (2017). The Contribution of Digital Technologies to Service Delivery: An Evidence Review (No. IDS Evidence Report; 221). IDS.

Soma, K., Termeer, K., & Opdam, P. (2016). Informational Governance: A systematic literature review of governance for sustainability in the information age. *Environmental Science & Policy*, 56, 89–99. Retrieved from http://doi.org/10.1016/j.envsci.2015.11.006.

Stevens, T., Aarts, N., Termeer, C. J. A. M., & Dewulf, A. (2016). Social media as a new playing field for the governance of agro-food sustainability. *Current Opinion in Environmental Sustainability*, *18*, 99–106. Retrieved from http://doi.org/10.1016/j.cosust.2015.11.010.

Street, R., Jacob, D., Parry, M., Runge, T., & Scott, J. (2015). *A European Research and Innovation Roadmap for Climate Services*. European Commission. Brussels.

Subbaraman, R., & Murthy, S. L. (2015). The right to water in the slums of Mumbai, India. *Bulletin of the World Health Organization*, 93(11), 815–816. Retrieved from http://doi.org/10.2471/BLT.15.155473.

Sullivan, B. L., Wood, C. L., Iliff, M. J., Bonney, R. E., Fink, D., & Kelling, S. (2009). eBird: A citizen-based bird observation network in the biological sciences. *Biological Conservation*, *142*(10), 2282–2292. Retrieved from http://doi.org/10.1016/j.biocon.2009.05.006.

Suri, T., & Jack, W. (2016). The long-run poverty and gender impacts of mobile money. *Science*, *354*(6317), 1288–1292.

Termeer, C. J. A. M., & Bruinsma, A. (2016). ICT-enabled boundary spanning arrangements in collaborative sustainability governance. *Current Opinion in Environmental Sustainability*, *18*, 91–98. Retrieved from http://doi.org/10.1016/j.cosust.2015.11.008.

Termeer, C., Dewulf, A., Breeman, G., & Stiller, S. J. (2013). Governance capabilities for dealing wisely with wicked problems. *Administration & Society*, *47*(6), 680–710. Retrieved from http://doi.org/10.1177/0095399712469195.

Termeer, C. J. A. M., Dewulf, A., Karlsson-Vinkhuyzen, S. I., Vink, M., & Vliet, M. va. (2016). Coping with the wicked problem of climate adaptation across scales: The five r governance capabilities. *Landscape and Urban Planning*, *154*, 11–19. Retrieved from http://doi.org/10.1016/j.landurbplan.2016.01.007.

UN (2014). A world that counts: Mobilising the data revolution for sustainable development. UN.

Waugaman, A. (2016). From principle to practice: implementing the principles for digital development. Proceedings of the Principles for Digital Development Working Group. http://digitalprinciples.org/wp-content/uploads/2016/03/From_Principle_to_Practice_v5.pdf

Weber, E. P., & Khademanian, A. M. (2009). Wicked problems, knowledge challenges, and collaborative capacity builders in network settings. *Public Administration Review*, 68, 334–339.

Whittington, D., Waterbury, J., & Jeuland, M. (2014). The Grand Renaissance Dam and prospects for cooperation on the Eastern Nile. Water Policy, 16(4), 595-608.

WWAP. (2015). *The United Nations World Water Development Report 2015: Water for a Sustainable World*. UNESCO Publishing. Paris.

Zulkafli, Z., Perez, K., Vitolo, C., Buytaert, W., Karpouzoglou, T., Dewulf, A., . . . Shaheed, S. (2017). User-driven design of decision support systems for polycentric environmental resources management. *Environmental Modelling & Software*, 88, 58–73. Retrieved from http://doi.org/10.1016/j.envsoft.2016.10.012.

13 Out of sight, out of mind?

Bringing the governance of mining and water risk into focus

Caitlin A. McElroy

Introduction

The management of water and mining is local, technical, and tied to human rights for sustainable livelihoods that are contingent on food and energy as well. This results in a significant governance challenge. This chapter argues that this challenge is now exacerbated by the increasingly remote or precarious locations of new mines and because of the increasing effects of climate change. Declining ore grades globally is leading to the development of new mines in locations that are remote. Metal and mineral deposits in more accessible locations are less abundant and, in seeking new resources, mines are pushing into terrains that were previously less appealing due to the additional difficulty of extracting there. The remote and precarious nature of such locations often not only means that they are distant from the infrastructure of large population centres, but also that they are often unique as environmental and cultural places thus far cushioned by this same remoteness. Many such places are also on the front lines of the effects of climate change. Such effects are increasing aridity and reducing the quantity of water reliably available in some parts of the world. In other places, climate change is increasing the intensity of storms and the volume of water added to water systems in short time frames. As large-scale mines are increasing, stakeholders in these regions are also becoming subject to the consequences of limited water and managing contamination in times of excess water (ICMM, 2015). Lack of water can stop mining operations quickly and an inability to manage contamination after an environmental event, like a flood or earthquake, can close mines long term. The on-going challenges of the political economy of water and mining impacts are therefore now joined by immediate corporate risk.

Governing water use and mining is also the co-governance of energy because it is used to power water production. Similarly, food production relies upon the quantity and the quality of water available and can also be in competition with a mine's access to water or affected by mine-water contamination. Water governance, independent of mining interactions, is also an active field of governance innovations from which more should be shared in managing the relationship of mining with water systems (Pahl-Wostl, Holtz, Kastens, & Knieper, 2010).

Mechanisms of further collaboration and new governance tools are now necessary to address these inherent resource-use conflicts.

This chapter suggests two positive approaches to improve the governance of water and mining in remote locations. First, that new digital network technologies can better connect affected communities to both each other and advocacy networks, and that new spatial imaging technologies can improve the monitoring and governance of remote locations. Second, the chapter presents the potential opportunity to address the effects of climate change in these locations through the water and energy infrastructure that mines must build in order to operate in these locations. This chapter focuses on large-scale mines managed by industrial actors rather than artisanal or small-scale mining (ASM). While there are many similarities between these mining processes, the governance approaches are deemed different enough to merit a separate discussion and so this chapter focuses on large-scale industrial mining alone.

The chapter will start with an overview of the impacts of mining on water and an historical overview of approaches to mine governance. This section aims to provide context for the evolution of simultaneously changing mining technology, global climate, and mining and water governance. In doing so it will highlight why the mining industry needs water, how this need has changed, and why specific conflicts arise as a result. From this background, the chapter then discusses the emerging positive approaches in digital technologies and environmental infrastructure. The final section of the chapter discusses the implications for these approaches on our collective reconceptualizations of intertwined resource governance and the urgency of new approaches as our complex environmental systems are experiencing rapid change.

Mining and water governance

The ways in which mining has used and needed water have changed considerably in even the past 50 years (Franks, 2015). This section has two parts: first, an introduction to the technical uses of water by the mining industry; and, second, the changing governance approaches to mining and environmental and social impacts.

Mining and water

The mining process not only engages with water because it cuts into the earth, but because the use of this water is part of the production of the minerals being extracted. Here we provide some examples of: 1) how mines disrupt water systems by their physical presence; 2) their potential to contaminate water; and 3) their 'production' of water needed for mine operations. In the first instance, the disruption of water systems from mining can be related to changing the hydrological system of the area. Natural surface and groundwater flows can be altered by the way a mine disrupts the topography of an area, including where water flows and collects above and below ground. This can have many effects, but for an example, picture what happens when a hole is dug in the wet sand on a beach and water begins to seep from all edges and collect in the newly formed space. A similar effect occurs with many mines and they must engage in a process of de-watering. This is the removal of water that has seeped or flowed into the excavated spaces to allow for mining to continue (Morton & Mekerk, 1993). Subsequently, the disposal of this removed water must go to a new location, which can create further disruptions in the water system. Climate change also effects this process; for example, storms of greater intensity have flooded mine-pits in Queensland, Australia and left few places to then release this water (Loechel, 2013). Through this process the hydrology of the site of extraction is altered.

A second well-known effect from the earth disruption of mining is a process called acid mine drainage or AMD. The breaking-up and grinding-up of rock to separate and access the desired ore exposes much more surface area of these materials. When they are disposed of in tailings facilities, these exposed rocks interact with air and water to release acids that further release other toxins. These toxins can then travel in water resources and spread this contamination across great distances for many decades or more (Bebbington & Williams, 2008; Akcil & Koldas, 2006). Gold and coal mines are particularly at risk of AMD because of the presence of particular related minerals in these deposits. Recently, the management of AMD in South Africa has become a very public environmental governance issue as the consequences of previous generations of mining are becoming better known due to recent impacts on urban areas, as well as areas of archeological importance (McCarthy, 2011).

Depending on where a mine is located, there are of course different consequences for the water system and different political processes that ascribe value and risk to these consequences. Take for example two recent controversies over mining in remote and environmentally sensitive areas of South America. In Peru, a highly contentious aspect of the revised Water Law of 2009 concerned permitting mining in the Andean highland headwaters of river systems (Budds & Hinojosa, 2012; Bebbington & Williams, 2008). Many argued that interfering with the unique nature of these areas would have significant and far-reaching effects on the water systems downstream from either altered hydrology or water contamination or both of these effects (Budds & Hinojosa, 2012). However, after a long political process the new legislation did not specifically prohibit mining in headwaters.

In contrast, in another area of controversy, there was a proposed mining project on the border of Chile and Argentina that would have significantly impacted, if not destroyed, an already climate-stressed glacier in order to access the ore body (Urkidi, 2010). In this instance, again after a long political process, the mine has not been developed. Urkidi finds that the difference in this situation was the influence of internationalizing conflict – or it becoming 'glocal' (2010). The debate in Peru stayed mostly within the national government, with the mining industry as the strongest international actor. Whereas in the Chile/Argentina mining case, the proposed project was put into many international news forums over years of sustained activism. Environmentalists and anti-mining advocates vilified the companies and regulators involved in this project and pursued various legal actions (Li, 2016). It quickly becomes clear that although all mines disrupt the natural landscape, and some in locations with potential for greater environmental impact, that the allowance of impact or degree of impact is socially not environmentally determined.

Relatedly, other impacts on water systems from mining are more diffuse than these examples and ascribing attribution for changes is a discursive as much as a scientific process. An example of such a situation is ongoing in the South Gobi Desert of Mongolia. There the nomadic herders are experiencing a range of changes to the quality and quantity of their water supply. The impacted area is expansive and it is very difficult even for scientific experts to unpick the causes for different changes between the impact of the development of several large mines, desertification due to climate change, and changing livelihood practices related to both of the former (Jackson, 2015, 2014). As can be expected, each stakeholder in the process has a different view of the situation about both the causes for the changes and the best ways forward to manage their entangled effects. These views here in Mongolia are often complicatedly intertwined, as they are in other resource-rich countries, with further different expectations of benefits and imaginaries of the impacts of mining on individual livelihoods and national development (Weszkalnys, 2008).

Second, beyond the intrusion of mining into a natural landscape, the process of extracting and processing ores requires water as an input and thus it also becomes an output. Some of the most infamous controversies regarding mines and the environment are related to the problems of this output and especially the potential for it to contaminate water. Water can be used in a myriad of ways, from keeping dust down on-site to serving as the key ingredient in separating out the desired mineral from the extracted ore. Many of these uses involve mixing water with other minerals or chemicals that are difficult to remove after. If released back into the environment, contaminated water can harm humans and wildlife. The waste rock and water from mining is therefore often stored in purpose-built tailings storage facilities (TSF). Depending on the inputs, these facilities may require decades of maintenance to keep the contents secure. If they break or leak, incidents such as at the Samarco dam collapse in Brazil in November of 2015 can occur.[1] Around the world TSF failures are the leading cause of mine disasters and, although technically complex, Schoenberger (2016) analyzed three-major TSF failures and argued that nonetheless the majority of these failures are more political than technical failures. Further, she emphasizes the importance of social scientists and those governing these facilities engaging with their technical engineers in order to make better decisions (Schoenberger, 2016). Again, we can witness another way in which this resource-intensive process is socio-technically contingent.

Third, in many locations the operational demand for water to run a mine exceeds the immediately available capacity. Mining operations often then must 'produce' water for the operations (Budds & Hinojosa, 2012). Hydraulic infrastructure used to do this includes drilling industrial boreholes to access groundwater, building dams to collect flowing surface water, and even the exceptionally energy-intensive activity of desalinating seawater. Many mines are developing technology to reduce the amount of water required for mining (Northey, Mudd, Saarivuori, Wessman-Jääskeläinen, & Haque, 2016). However, as this chapter emphasizes, many of the world's largest and more accessible ore bodies have already been mined and new mining operations are pushed into areas that are more remote – distant from infrastructure, population centres *and* natural resources such as water. For example, miners are now extracting diamonds from the icebound limits of northern Canada and, as discussed above, the arid extremes of the Gobi Desert (Bochove, 2016; Jackson, 2014). On-going mineral demand and constricting supply is also bringing mines in closer, pushing into protected areas and spaces of other significance to nature and society (Durán, Rauch, & Gaston, 2013). These are places where risk of water system change or contamination would have amplified impacts.

During the boom years of 2000 to about 2013, with the rising commodity cycle and high prices for minerals, many corporations had the funds and expected income to afford to invest in and create the means to make mining possible in these difficult locations, both technically and politically. As the commodity supercycle has wound down, many firms no longer have such resources (Humphreys, 2010). Further, many state governments strategically aligned their development plans with the income from major extractive projects and are still expected to deliver these projects (SDSN, 2013). In an often mutual ambition to continue mining and meet development goals, there is now a growing number of shared infrastructure developments where both the state and mine contribute, for example to a water treatment plant, with the expectation that it will serve the community after the life of the mine (ICMM, 2015).

In this way the 'production' of water for mining becomes further entangled in the role of nature and the state in development. Here we see the intersection of the new challenges of governing water and mining relating to three particularly socio-technical aspects of governing mining and development. First, the development of water infrastructure for mining is contingent on the macro-economy because both the commodities cycle and national development strategies affect the interest and ability to build environmental infrastructures related to water and energy. Second, such infrastructure often requires the deployment of new technology to build, manage, and monitor. Third, such situations are afloat in hopes and expectations of imagining the use and benefits of both mineral and water resources in the future. All three spaces call for governance approaches that work with the different expertise these spaces require as well as approaches that make the most of how these spaces overlap.

Changing governance

As we have seen, large-scale mining around the world impacts water resources in both their quality and quantity. Although largely technical problems, we have also seen how the management of these issues and how risk is understood, is socially and politically mediated. This next section provides some further context to situate the different governance approaches that direct this social process of governing technical environmental impacts. Overall, the environment and local communities have long borne the burden of mining's effects with little support. However, resistance to large-scale mining from local communities and international NGOs, increased regulation by some governments, improved technologies by some mining companies, and increased public awareness of the long-term consequences, have more recently re-framed the issues of governing mining, and even more recently governing water and mining. Despite this increased attention to the impacts of mining, solutions are lagging behind the changing impacts as highlighted by the location of new mines and encroaching climate change. The following describes these changes in mining governance and why it is in many ways still insufficient.

The process of governing water and mining has historically been part of the process of governing mining writ large and a significant feature in the transformation of the social and environmental awareness of the industry since the early 1990s (Franks, 2015). This awareness came about due to the confluence of several new ways of making social and environmental impacts public. This included the emergence of large international NGOs, a new focus from academics on the economic development of resource-rich countries, and new approaches to using transparency, reputation, and other forms of 'soft' governance. Some of these efforts worked directly to affect and support governments in their work to regulate mining, but many looked to make mining corporations directly accountable for the impacts of their work.

The full complexity of these governance changes is beyond the scope of this chapter; however; what this section aims to highlight is that the socio-technical mechanisms of mining governance, as is most relevant for governing water and mining, have already seen considerable change. These changes have brought new stakeholders into the governance of the mining sector and, as a result, have built the foundations on which emerging socio-technical governance spaces can stand. The section also highlights why it is now time, yet again, to embrace new approaches to governing the mining industry as new challenges, like its impact on water, become more urgent.

This urgency has come before: the 1990s are perhaps recognized as a time when the mining industry around the world was faced with an eruption of conflict and criticisms (Franks, 2015). Large-scale environmental disasters, displaced communities, and violent conflict were all too often co-occurrences with mine projects (Bice, 2016). In response, the advancing power and reach of civil society organizations including NGOs and environmentalists effectively made public the tremendous environmental harm from numerous large mine sites around the world (Ballard & Banks, 2003).

Simultaneous to the public and media attention on mining, a flurry of academic research on the economic benefits, or lack thereof, of extractive industries to the host country emerged and defined a phenomenon named the 'resource curse' (Auty, 2002, 1995; Mikesell, 1997; Ross, 1999; Sachs & Warner, 2001). The 'resource curse' or economic evidence of weak or slowed development in resource-rich countries compared to countries not dominated by resource economies, undermined the argument that resource extraction drove development and created economic growth for its host countries.

Collaborative initiatives evolved in response to these challenges against the social and environmental harm caused by the mining industry. The first initiatives of scale and impact were the Mining, Minerals, and Sustainable Development Project (MMSD) as well as the Global Mining Initiative (GMI). In different ways both worked to assess the current practices of the mining industry and offer guidance on ways to improve (Franks, 2015).[2] Both of the programmes reached out to civil society collaborators and used an ambition to contribute to sustainable develop as a foundational objective. Around the same time, those financing the mining industry, such as the International Finance Corporation (IFC), the private sector lending arm of the World Bank, were also forced to evaluate their own role in directing the conduct of mining. After a long and iterative process of evaluation and reform, the IFC's environmental and social performance standards were developed and now set a benchmark for best practice in the management of large private infrastructure with development impact (Hunter, 2007). These standards, guidelines, and forums pushed mining and their government regulators to consider the impacts of mining on not only environment and development, but also on human rights, conflict, and transparency (Franks, 2015).

Out of these comprehensive agendas specific initiatives emerged to address individual problems. They included guidelines for specific industry processes such as the International Cyanide Management Programme for the Gold Mining Industry and the Kimberley Process Certification Scheme for Diamonds, which help to control the negative environmental effects of cyanide contamination in water sources and the perpetuation of conflict from the illegal mining and trade of diamonds respectively (Nelson, 2007). These processes require the commitment of government, industry, and civil society to co-manage and monitor their performance. Their success is contingent on these extra-governmental governance processes and their cross-sectorial integration (Ruggie & Kell, 1999). For example, while not addressing the complex resource entanglement as we see it today with food, energy, and water, these initiatives started working within the intersections of environment, human rights, corporations and governments in ways that were innovative in their time (Fransen & Kolk, 2007).

Another emerging tool of the 'soft' governance space was transparency. The success of making information public led to the understanding that public scrutiny has the power to hold organizations accountable. This approach came to the mining industry through the Extractive Industry Transparency Initiative (EITI) as a sub-set of the work of Transparency International's (TI) anti-corruption agenda. It works to promote the accountability of the extractive industries and governments by requiring transparent reporting of revenues paid and received. Civil society is then called upon to review this information (Haufler, 2010). This is a similar approach to that used by Global Reporting Initiative (GRI), which brings corporations to report on further environmental and social metrics of their performance for the benefit of the public and their investors. Helping to coordinate the deployment and advancement of these practices are organizations such as the International Council of Mining and Metals (ICMM) a CEO-led group or the more civil society lead Natural Resource Charter. While there is on-going support for transparency and reputation driven initiatives from the EU, UN, G8, G20 as well as the sponsoring NGOs, it is hard to find specific evidence of their effectiveness. For example, many countries that have adopted the EITI do not show improved performance after they are EITI compliant or compared to non-EITI adopting countries (Sovacool, Walter, Van de Graaf, & Andrews, 2016). As result, there is still a gap between awareness, action, and measurable improvements in the social and environmental impacts of mining.

The growing number and specificity of efforts directed to reform the mining industry and find ways to capture positive development benefits is, however, only one narrow part, some may say tip of the iceberg, of the complex hierarchies and spaces, and multitude of actors involved in shaping and contesting the governance of the mining industry. Although many of the governance approaches discussed above are intended to provide governance in the absence of strong governments, these governance initiatives cannot, and many argue should not, replace the role of laws and governments (Bebbington, 2010; Jordan, Wurzel, & Zito, 2005). Important processes from the mandating of environmental and social impact assessments, to the oversight and participation in environmental monitoring programmes, and the responsibility to protect and backstop the rights of citizens, are essential from governments. Most of the above governance approaches require the participation of the state and local forms of government; however, although now attuned to the urgency of the governance mandate, many of these governments still lack the tools and skills to contribute effectively (Franks, 2015; Ross, 1999).

Currently, new environmental challenges, from the remote locations where further ore deposits are located to the increasing variability of water systems due to climate change, have raised the urgency of addressing the governance of water and mining (Bebbington & Williams, 2008). The effects of changing hydrology and water contamination are amplified when access to clean water is less reliable or available. Competition for produced water resources is at a crucial impasse between communities' needs and industrial development. Clearly, the above governance integration of multiple stakeholders and non-local awareness is still essential in addressing the challenge. But, it is not enough.

In particular there are three challenges that help identify how the current stakeholder lead governance approach is insufficient at addressing, or cannot be appropriately leveraged as is, for the emerging water and mining challenges. First, the current scalar approach to governance is rife with contestations over what containers of space are captured by different hierarchies of governance. For example, a recent effort to govern industrial water effects at the catchment level has produced a sea of constructive criticism (Bebbington & Williams, 2008; Budds & Hinojosa, 2012; ICMM, 2015; Kunz, 2016). Second, the technical challenges of improved water use require expertise and financing that is most concentrated within industry. Mining companies must improve their water use performance dramatically and they must have the support of, if not pressure from, their investors to justify this expenditure. Third, the governance of the water sector is contingent on access to and the production of both energy and food for the mine and for the surrounding communities. However, the governance of these resources is literally 'beyond the fence' of mining operations and, in most firms, organizational expertise to engage in this extended resource co-governance is minimal.

Opportunities to improve mining and water governance

Opportunities to start addressing some of these weaknesses in the governance approach to water and mining under new global conditions are emerging and require innovative re-conceptualization in order to merge them with incumbent governance approaches. The following section illustrates how this could be done for two such opportunities: technology related to networks and spatial imagining; and water–energy linked infrastructure. The presentation of how these technologies are relevant will focus on the ways they engage with the governance challenges of socially mediated technical governance, scale of governance, cost and financial mandate, and expertise.

Technology: networks and spatial imagining

It seems imperative to seek novel governance tools in emerging technologies, as it is new technologies that are making mining possible in remote and precarious places (Li, 2016). From the constraints of working at extremely high altitudes to working in terrain that is nearly always frozen, technological advancements are making this mining possible. It seems they also might facilitate making it better – more accountable for its social and environmental impacts.

Socially, increased access to data networks through mobile phones and community computer centres can increase the connectivity between remote communities affected by mining to both each other and external expertise as we have seen in Chapter 12 on the increased use of ICT for food, energy, and water governance. NGOs pro-environment and critical of mining, such as MiningWatch Canada, have demonstrated the effectiveness of using social media to unite people globally in support of local protesters, as was shown in the above example of the Pascua Lama mine in Chile/Argentina. While access continues to improve, one can expect to see more media and academic reports documenting the ways mining communities are finding far-flung support and accessing foreign technical expertise and advice. Such connections have already been made for the development of other recent large-scale protests such as those during the Arab Spring (Breuer, Landman, & Farquhar, 2015), G20 protests in Toronto (Poell, 2014), and the Occupy movement (Agarwal, Bennett, Johnson, & Walker, 2014). Further, Hecht (2010) documented the difficult process of finding external expertise to evaluate local science and monitoring of uranium mine workers' radiation exposure in Namibia in the 1990s. Facilitating such independent scientific review and expertise can now be greatly advanced by extended social media networks.

Technology again pushes back against the challenges of remote governance through advancements in spatial imaging. Digital imagining technologies like that of Google Earth are becoming widely available and with high resolution. Governments, financial institutions, and NGOs alike can view the location of various mining activities however far away or inaccessible by other means. For governments in particular, improved licensing, and monitoring, of resource extraction, is made much more cost effective with such technologies. When this data can be paired with spatial data of water, biodiversity risks such as protected areas or wildlife corridors there is tremendous potential for improved planning and regulation of mining. The WWF has been pursuing different ways of improving conservation advocacy through this spatial imagery in East Africa through the Africa Land Use Planning and Warning System (ALES) (Tibaldeschi & Schmitt, 2014). Even academic research has been helping to apply combined wildlife data sets and spatial imaging technology to conservation decision making for the oil industry as with the Local Ecological Footprinting Tool (LEFT) designed by zoologists at the University of Oxford with Statoil (Willis et al., 2012). Spatial imaging provides transparent access to mine sites at elastic scales of physical governance. It also presents integrated resource use and impacts, thus advancing the ability to see the effects of mining within complex ecological systems.

Infrastructure: water, energy, climate change, and communities

While technology has made remote mining more feasible and potentially more governable, it has not explicitly addressed the threat of climate change that is making the governance of remote mining areas all the more urgent. For this we turn to technological advancement in the form of infrastructure. To meet their own production needs and environmental standards of performance, mines are increasingly building different forms of water infrastructure. During times when commodity prices are high, or at the height of commodities cycles as we saw in the mid-2000s, the mining industry is more open to new investments and risks such as remote locations and the infrastructure to make those operations feasible (Erten & Ocampo, 2013; Radetzki, Eggert, Lagos, Lima, & Tilton, 2008). Further, to promote the development of mines, international actors on behalf of neoliberal states have advocated for the reduction of rents, or the taxes and royalties, paid to the state for the rights to mine (Bridge, 2004). The justification being that the firms involved take on considerable market and financial risk, or neoliberal risk, to develop the project and can serve development interests with the infrastructure they build (Emel & Huber, 2008).

In some models of infrastructure sharing, the assets are jointly invested in at the start of operations, but, for others, infrastructure used for mining is expected to be handed over at closure as a legacy asset (Admiraal, Sequeira, McHenry, & Doepel, 2017; Toledano & Roorda, 2014). Several law firms have even started to publish short pieces explaining their expertise to help broker such arrangements (McCullough Robertson, 2015). The market for development is no longer only framed in the neoliberal risk or trickle-through contributions. The mine operators with the best reputation *as well as* the best additional development investments will be more likely to gain mining access in competitive situations.

In the context of the effects of climate change on water quantity and quality in remote mining areas investors are also increasingly aware of the risk to their investments and have started to take more seriously corporate reporting related to environmental, social, and governance issues (ESG) (Eccles & Serafeim, 2013). A research programme at the Columbia Water Centre, a part of the Columbia Earth Institute, has been funded by Norges Bank to better understand and model the risks water poses to mining investments. One of the key features of the work is identifying asset-level risk and understanding water risk across the different lifecycle stages of a mining operation (Durucan, Korre, & Munoz-Melendez, 2006). The research is being used to voice support for new US Securities and Exchange Commission (SEC) regulations that would require mining companies to provide technical reports of site water risks and long-term management plans (Condon, 2016).

The combined pressures for new infrastructure, increased risk, and investor oversight partially obscure the opportunity that this infrastructure may offer to make the surrounding communities and regions more resilient to the effects of climate change. In this book in Chapter 9, Alex Money has discussed the ways corporations may be incentivized to make such an investment based on the logics of transaction cost theory. Here we advocate that beyond the incentive to build such infrastructure there is the possibility of designing and governing the use of this infrastructure to support these precarious locations as they become more arid or awash in floods. It could provide integrated resource management on a scale of investment that would otherwise not be accessible to these places. For example, the large copper mining industry in the Atacama Desert of northern Chile struggles to maintain access to enough water for its operations, as water availability and the frequency of drought gets worse year on year. According to Desalination and Water Use news website ("Chile's copper mines", 2015), there are 16 mining-related desalination plants, estimated to cost USD 10 billion, being built or planned while nine are already operating to provide the necessary water. However, the desalination and the process of bringing the water from sea-level to the mines at altitude in the Andes requires a tremendous amount of energy. This was energy that the Chilean state was unable to consistently provide without building new and highly contested hydroelectric dams in Patagonia. So, mines are building their own renewable energy systems (Servert, Cerrajero, & Fuentealba, 2016). These are just the types of infrastructure that will be needed for the future of the cities of the Atacama where, despite its remoteness, over a million people live.

Beyond the market for development that has long been attached to the development of new mining projects, there is now also a market for sustainability. This relationship has been evidenced in the lower cost of capital for corporations across a range of industries (Clark, Feiner, & Viehs, 2015). Investors can now correlate measured and transparent environmental performance with overall sound corporate governance practices and reduced investment risks, potentially increasing the likelihood of investments in environmental infrastructure that could help adapt to a changing climate.

Reconceptualizing integrated resource governance in challenging sectors

The use of digital technologies and new infrastructure helps to address some of the challenges to the existing governance of water and mining. They also pose new risks of their own. As advantages, the use of these technologies requires social mediation by a range of different stakeholders. A government organization alone does not have the expertise to make use of these technologies and the ways in which they are brought to bear on governance challenges will inherently be collaborative. Also, these new tools are elastic in their scale. Governance organizations can use the information they provide at variable scales, and organizations across hierarchies of governance are increasingly gaining equal access as the costs, especially of spatial imaging data, decline. For new infrastructure, the market for sustainability has helped to expand the possibilities for financing. Finally, and perhaps most significant, these tools help us to rethink complex environmental governance as a space of opportunities and not just restraint.

However, the exuberance with which novel and technologically driven optimistic solutions are seized should be tempered. There are already a number of obvious risks that could worsen the power dynamics between communities, states, and mines as a result of the poor implementation of these tools. For example, it is essential that the power and equalizing effects of spatial data are not captured only in the hands of the powerful. If so, the ability to monitor and regulate mining operations will be undermined and communities could once again be subjected to disproportionate burdens of mining impacts and unequal access to the resources to contest this.

Similarly, with the development of water and energy infrastructure, there is a considerable risk of the perpetuation of the neoliberal extractive models of accumulation by dispossession – or the wealth of international businesses deriving from the dispossession of less powerful communities resources (Bridge, 2004; Emel & Huber, 2008). Appropriate participation of the state, NGO watchdogs, and communities are essential in the planning of any new large-scale infrastructure. In some places, partnerships for infrastructure are also attached to larger geo-political relationships related to market access and spheres of influence. For example, the Mongolian government is eager to leverage the growing mining sector to build further bridges for Mongolia to western democracies (Jackson, 2015). Or, in contrast, many African nations are receiving infrastructure through Chinese investments, but are often disappointed when the relationship only goes as far as the infrastructure transaction (Larmer, 2017). Addressing resource scarcity and integrated resource governance through new mine-related infrastructure must occur in ways that are careful not to repeat the social and political risks that have historically been involved in such processes.

In using these new tools to fill gaps in current governance approaches, it is important not to disregard the carefully and painfully made gains in improved governance approaches. As we have seen with research about the governance of the commons, time-tested, local and collaborative approaches already in practice have merit (Ostrom, 2015). New tools should not erase these approaches, but enhance their abilities to operate in a more complex and rapidly changing world. We can perhaps recognize new governance tools and perspectives such as those illustrated in this chapter as a way of augmenting and re-conceptualizing the governance of the commons in a new era of change – not a new approach to resource governance overall. Perhaps then, in time, it may be less surprising to expect long-term environmental benefits and only short-term, and controlled, environmental destruction from industries.

Conclusion

This chapter attempts to illustrate some of the emerging complexities in the environmental governance of mining and water and suggests new tools to address them as well as new ways to conceptualize using these tools. A pessimist may see that there is still a lack of will, coordination, and vastly diffuse and differentiated problems to address. An optimist may see technological advances creating as much opportunity for environmental sustainability as destruction. Since both are in some ways correct, we must work to make this shift in governance as significant for the survival of all parties and the environment as it was in the 1990s for the mining sector. Thinking in terms of the responsibility to govern with the nexus of integrated resource management in mind is a good place to start.

Notes

1 When the tailings dam released, a wave of mud, clays, and other liquids tore downhill killing 19 people, demolishing villages, and damaging wildlife (Phillips, 2016). And now another river still runs red as the result of mining activities while the local communities wait for the Samarco mine to finance the recovery of 5,000 streams along the River Doce, reforest 10,000 hectares of forest and restore another 30,000 hectares of land (Phillips, 2016). The extent to which the accident was preventable if better maintenance and monitoring of the tailings dam occurred is still contested.

2 The Voluntary Principles on Security and Human Rights were developed slightly after the MMSD and GMI, but were lead by the oil and gas industry with the support of Rio Tinto and Freeport McMoran from the mining sector.

References

Admiraal, R., Sequeira, A. R., McHenry, M. P., & Doepel, D. (2017). Maximizing the impact of mining investment in water infrastructure for local communities. *The Extractive Industries and Society*.

Agarwal, S. D., Bennett, W. L., Johnson, C. N., & Walker, S. (2014). A model of crowd enabled organization: Theory and methods for understanding the role of twitter in the occupy protests. *International Journal of Communication*, 8, 27.

Akcil, A., & Koldas, S. (2006). Acid Mine Drainage (AMD): Causes, treatment and case studies. *Journal of Cleaner Production*, *14*(12), 1139–1145.

Auty, R. M. (1995). Industrial policy, sectoral maturation, and postwar economic growth in Brazil: The resource curse thesis. *Economic Geography*, *71*(3), 257–272.

Auty, R. M. (2002). *Sustaining Development in Mineral Economies: The Resource Curse Thesis*. London: Routledge.

Ballard, C., & Banks, G. (2003). Resource wars: The anthropology of mining. *Annual Review of Anthropology*, *32*(1), 287–313.

Bebbington, A. (2010). Extractive industries and stunted states: Conflict, responsibility and institutional change in the Andes. In R. Raman (Ed). *Corporate Social Responsibility: Discourses, Practices and Perspectives* (pp. 97-115) London: Palgrave MacMillan.

Bebbington, A., & Williams, M. (2008). Water and mining conflicts in Peru. *Mountain Research and Development*, 28(3–4), 190–195. doi:10.1659/mrd.1039.

Bice, S. (2016). *Responsible Mining: Key Principles for Industry Integrity*. London: Routledge.

Bochove, D. (2016, July 12). Hunting for diamonds under Canada's frozen Tundra. *Bloomberg News*. Retrieved from www. business.financialpost.com.

Breuer, A., Landman, T., & Farquhar, D. (2015). Social media and protest mobilization: Evidence from the Tunisian revolution. *Democratization*, *22*(4), 764–792.

Bridge, G. (2004). Mapping the bonanza: Geographies of mining investment in an era of neoliberal reform. *The Professional Geographer*, 56(3), 406–421.

Budds, J., & Hinojosa, L. (2012). Restructuring and rescaling water governance in mining contexts: The co-production of waterscapes in Peru. *Water Alternatives*, 5(1), 119.

Chile's Copper mines will turn to desalination for half their water (2015, November 24). Retrieved from www.desalination.biz/news/0/Chiles-copper-mines-will-turn-to-desalination-for-half-their-water/8246/.

Clark, G. L., Feiner, A., & Viehs, M. (2015). From the stockholder to the stakeholder: How sustainability can drive financial outperformance. Retrieved from SSRN https://ssrn.com/abstract=2508281 or http://dx.doi.org/10.2139/ssrn.2508281.

Condon, M. (2016, October 3) Comments to the SEC encourage environmental risk disclosure.

Durán, A. P., Rauch, J., & Gaston, K. J. (2013). Global spatial coincidence between protected areas and metal mining activities. *Biological Conservation*, *160*, 272–278.

Durucan, S., Korre, A., & Munoz-Melendez, G. (2006). Mining life cycle modelling: A cradle-to-gate approach to environmental management in the minerals industry. *Journal of Cleaner Production*, *14*(12), 1057–1070.

Eccles, R. G., & Serafeim, G. (2013). The performance frontier. *Harvard Business Review*, *91*(5), 50–60.

Emel, J., & Huber, M. T. (2008). A risky business: Mining, rent and the neoliberalization of "risk". *Geoforum*, 39(3), 1393–1407.

Erten, B., & Ocampo, J. A. (2013). Super cycles of commodity prices since the mid-nineteenth century. *World Development*, *44*, 14–30.

Franks, D. M. (2015). *Mountain Movers: Mining, Sustainability and the Agents of Change*. London: Routledge.

Fransen, L. W., & Kolk, A. (2007). Global rule-setting for business: A critical analysis of multi-stakeholder standards. *Organization*, *14*(5), 667–684.

Haufler, V. (2010). Disclosure as governance: The extractive industries transparency initiative and resource management in the developing world. *Global Environmental Politics*, *10*(3), 53–73.

Hecht, G. (2010). Hopes for the radiated body: Uranium miners and transnational technopolitics in Namibia. *The Journal of African History*, *51*(2), 213–234.

Humphreys, D. (2010). The great metals boom: A retrospective. *Resources Policy*, *35*(1), 1–13.

Hunter, D. B. (2007). Civil society networks and the development of environmental standards at international financial institutions. *Chicago Journal of International Law*, 8, 437.

ICMM (2015). *A practical guide to catchment-based water management for the mining and metals industry*. Retrieved from www.ICMM.com.

Jackson, S. L. (2014). Building a mineral nation? The oyu tolgoi copper-gold mine and contested infrastructure development in Mongolia.

Jackson, S. L. (2015). Imagining the mineral nation: Contested nation-building in Mongolia. *Nationalities Papers*, *43*(3), 437–456.

Jordan, A., Wurzel, R. K., & Zito, A. (2005). The rise of 'new' policy instruments in comparative perspective: Has governance eclipsed government? *Political Studies*, *53*(3), 477–496.

Kunz, N. C. (2016). Catchment-based water management in the mining industry: Challenges and solutions. *The Extractive Industries and Society*, *3*(4), 972–977.

Larmer, B. (2017, May 2) Is China the world's new colonial power? *The New York Times*. Retrieved from www.nytimes.com/2017/05/02/magazine/is-china-the-worlds-new-colonial-power.html?_r=0.

Li, F. (2016, September 3). The defeat of Pascua Lama. *North American Congress on Latin America* (NACLA). Retrieved from http://nacla.org/news/2016/03/09/defeat-pascua-lama.

Loechel, B. (2013, April 28). Mining companies are underprepared for climate change. *The Conversation*. Retrieved from http://theconversation.com/mining-companies-are-underprepared-for-climate-change-13091.

McCarthy, T. S. (2011). The impact of acid mine drainage in South Africa. *South African Journal of Science*, *107*(5–6), 01–07.

McCullough Robertson (2015, September 16). Sharing mining infrastructure. Retrieved from www.mccullough.com.au/page/Home/Updates/Sharing_mining_infrastructure/.

Mikesell, R. F. (1997). Explaining the resource curse, with special reference to mineral-exporting countries. *Resources Policy*, *23*(4), 191–199.

Morton, K. L., & Van Mekerk, F. A. (1993). A phased approach to mine dewatering. *Mine Water and the Environment*, *12*(1), 27–33.

Nelson, J. (2007). The operation of non-governmental organizations (NGOs) in a world of corporate and other codes of conduct. *Corporate Social Responsibility Initiative*.

Northey, S. A., Mudd, G. M., Saarivuori, E., Wessman-Jääskeläinen, H., & Haque, N. (2016). Water footprinting and mining: Where are the limitations and opportunities? *Journal of Cleaner Production*, *135*, 1098–1116.

Ostrom, E. (2015). *Governing the Commons*. Cambridge, U.K.: Cambridge University Press.

Pahl-Wostl, C., Holtz, G., Kastens, B., & Knieper, C. (2010). Analyzing complex water governance regimes: The management and transition framework. *Environmental Science & Policy*, *13*(7), 571–581.

Phillips, D. (2016, October 15) Samarco dam collapse: One year on from Brazil's worst environmental disaster. *The Guardian*. Retrieved from www.theguardian.com/sustainable-business.

Poell, T. (2014). Social media and the transformation of activist communication: Exploring the social media ecology of the 2010 Toronto G20 protests. *Information, Communication & Society*, *17*(6), 716–731.

Radetzki, M., Eggert, R. G., Lagos, G., Lima, M., & Tilton, J. E. (2008). The boom in mineral markets: How long might it last? *Resources Policy*, *33*(3), 125–128.

Ross, M. L. (1999). The political economy of the resource curse. *World Politics*, *51*(2), 297–322.

Ruggie, J. G., & Kell, G. (1999). *Global Markets and Social Legitimacy: The Case of the 'Global Compact'*. Toronto: York University.

Sachs, J. D., & Warner, A. M. (2001). The curse of natural resources. *European Economic Review*, *45*(4), 827–838.

Schoenberger, E. (2016). Environmentally sustainable mining: The case of tailings storage facilities. *Resources Policy*, *49*, 119–128.

SDSN (2013). Harnessing natural resources for sustainable development: Challenges and solutions. *Sustainable development solutions network, A global initiative for the United Nations*. Retrieved from www.unsdsn.org.

Servert, J., Cerrajero, E., & Fuentealba, E. (2016). Synergies of solar energy use in the desalination of seawater: A case study in northern Chile. *AIP Conference Proceedings*, 1734.

Sovacool, B. K., Walter, G., Van de Graaf, T., & Andrews, N. (2016). Energy governance, transnational rules, and the resource curse: Exploring the effectiveness of the Extractive Industries Transparency Initiative (EITI). *World Development*, *83*, 179–192.

Tibaldeschi, P., & Schmitt, S. (2014). Africa land use planning and early warning system. *WWF Fact Sheet 2014*. Retrieved from http://assets.wwf.org.uk/downloads/ales_factsheet.pdf.

Toledano, P., & Roorda, C. (2014). Leveraging mining investments in water infrastructure for broad economic development: Models, opportunities and challenges. *Columbia Centre on Sustainable Investment Policy Paper*. Retrieved from http://ccsi.columbia.edu/files/2014/05/CCSI-Policy-Paper-Leveraging-Mining-Related-Water-Infrastructure-for-Development-March-20141.pdf.

Urkidi, L. (2010). A glocal environmental movement against gold mining: Pascua: Lama in Chile. *Ecological Economics*, *70*(2), 219–227.

Willis, K. J., Jeffers, E., Smit, M. G. D., Hagemann, R., Collin-Hansen, C., & Weissenberger, J. (2012). LEFT: Local ecological footprint tool. Retrieved from: www.biodiversity.ox.ac.uk/wp-content/uploads/2012/09/Mahamavo_GBIF.pdf.

Weszkalnys, G. (2008). Hope & oil: expectations in São Tomé e Príncipe. *Review of African Political Economy*, *35*(117), 473–482.

14 Conclusion

Laura M. Pereira, Alexandra M. Girard and Caitlin A. McElroy

This book set out to explore governance strategies for the resources of food, energy, and water systems. Each chapter seeks to outline examples of complex system or integrated resource management approaches – nexus-based approaches – to environmental governance in practice. In Chapter 1, we introduced you to the conceptual framework underlying this book – the inverted pyramid of sustainability underpinned by socio-technical innovation, complexity thinking and then, finally, governance – that is a constant thread running through all the chapters in this book. Whilst the introduction provided a rich theoretical background, we intended to present a wide variety of expertise and approaches on how to navigate the complexity of the food–energy–water systems nexus, and, with this conclusion, we aim to provide the reader with some practical take-home messages emerging from each chapter.

The book chapters are divided into two sections: the first focussing more specifically on case study examples from around the world of how FEW challenges have been addressed by a variety of societal actors, governments, and businesses. The second half of this book addresses some broader lessons from natural resource governance that are equally enlightening when a FEW lens is applied to them. We propose looking at the messages emerging from each chapter following the same organisational pattern.

Take-home lessons from case study chapters

Two sets of three case studies in the first half of the book provide examples of the challenges decision makers face when dealing with food, energy, water systems, and in implementing sustainable strategies for the future. Although each case has a dominant focus on one of the three systems, they all reference the complexities of dealing with interdependencies, historically prevailing existing system, structures, and uncertainty.

FEW management in practice: case studies

We start the book in India with examples of how the confluence of the food, energy, and water sectors can complicate the management of any one of these individual resources. Chapter 2 explains why addressing issues with the food–energy–water sectors cannot be done in isolation of each other. The four cases studies of a water–energy policy scheme ("*Jyotigram Yojana*"), the National Solar Mission, food pricing, and trade laws to promote crop diversification, and foreign direct investments to build sufficient cold-storage infrastructure, illustrate the multi-dimensional and interdependent aspects of the food–energy–water nexus in India's new governance landscape. The chapter acknowledges that well-intended government policies that are created within strong sectorial boundaries can result in complex unintended socio-economic challenges that are difficult to untangle. In particular, securing food, energy, and water resources is difficult when governance landscapes fail to recognise the interdependencies existing between these sectors. It concludes that growing demands in the food–energy–water sectors will require more partnerships between the public and private sectors, as well as further changes in domestic and global trade laws and regulations. In this new governance landscape, it is hoped that improved market access and coordination between public and private institutions can better position India to target the interdependent challenges in food, energy and water sectors.

Moving to another continent, Chapter 3 takes an in-depth look at the politicised history of corn in Mexico and provides an example of how a social movement, Slow Food International (SFI), is enabling a local revitalisation of blue corn as a heritage food. The chapter shows that social movements are limited in facilitating local small-scale farmers and cooperatives' access to global markets because of limited funds, lack of selling expertise, and the lack of being able to coordinate leadership amongst producers. However, they do provide some marketing expertise that can support the link between local farmers and niche markets by commoditising local and traditional products in Mexico. To move them from substance agriculture to export agriculture necessitates the coordination of various actors globally as well as mobilising funds unaccessible to farmers. This entails that investment is usually necessary for enterprising endeavours to operate successfully. The chapter concludes that international social movements, such as Slow Food International, have a role in fostering transnational ties and supporting transnational entrepreneurship; however, this narrative can be problematic as it simultaneously promotes commodification of heritage for a global elite of "foodies". It underscores the messiness of creating cross-scale governance institutions that bridge diverse knowledge systems, and urges reflexivity in establishing partnerships to ensure social and environmental sustainability.

Chapter 4's discussion on the energy system in South Africa concludes the initial triumvirate of food–energy–water case studies. Complexity thinking is essential in avoiding trade-offs and recognising synergies between interventions across FEW systems; transitioning out of infrastructure intensive systems is particularly challenging as these cost-intensive investments create lock-ins and path dependencies that are difficult to destabilise. Chapter 4's sustainability challenge is a locked-in fossil-fuel based energy system that is among the most polluting in the world and that does little to reduce the economic and social inequalities of South Africa. Using an historical approach to discuss the beginnings of South

African electrification, the analysis makes it clear that the deep connections between the energy and the mineral sectors have always been present in South Africa; the mineral–energy complex. The chapter shows that the lessons learnt from history can help South Africa to transform its energy sector. From the point of view of developments in renewable energy, the past investment patterns seem to indicate that the best solutions for South African energy needs have actually come from distributed and sustainable infrastructures rather than from the development of national monopolies. Whilst the mineral–energy complex still operates, its failure to deliver to citizens in recent years has resulted in a push towards more innovative solutions, mainly based in decentralised renewable energy, that have the potential to create a more sustainable energy system if fostered. Innovative government policies such as the Renewable Energy Procurement Program (REI4P) offer hope of a transition away from a fossil fuel-intensive energy system to one that is both fairer and cleaner.

The second set of case study chapters begins with energy governance in China and its implications for the water system. Unlike most of the other chapters in this section, Chapter 5 provides a very technical, and quantitative, reasoning towards understanding the complexity of the FEW governance. As explained in the chapter, issues around natural resources governance are sometimes down to simple questions, such as the use of inefficient technology or the mismatch between where natural resources are located and where they are needed. It provides an engineering angle to the challenges of water as an indispensable input in almost every life-cycle stage of the energy sector. The chapter details how water is predominantly used as a cooling medium for power production. Power production's water intensity differs significantly by energy source and cooling technology used. Due to significant regional variations and fragmented institutions, potential water–energy conflict is an especially pronounced issue in China. Uncoordinated or "one-size-fits-all" energy polices may result in violations of other policy domains and thus raise governance challenges. Furthermore, there are possible conflicts between China's water demands from its energy development because its target for national water use is likely to be exceeded under a variety of scenarios. On top of this, some regions are projected to experience water variability under climate change. Overcoming these socio-technical challenges will require institutional transformation and inter-sectorial and cross-scale (between national and regional governing bodies) policy coherence.

Chapter 6 analyses the implications of Peru's asparagus industry on the water system. In this chapter, as in the previous one, climate change is incorporated as an additional uncertainty factor to the complexity of governing FEW systems. The chapter notes that changing climate regimes, coupled with Peru's desire to sustain its agricultural production of asparagus, are drivers of change that have numerous short- and long-term implications upon the country's water resources. It suggests some key interventions that the country must make in order to get onto a more sustainable trajectory and to avoid negative trade-offs that could increase Peru's exposure and sensitivity to climatic risks in the longer term. This would compromise the country's economic security, which is highly dependent on agricultural exports. The first finding is that it is important to incorporate both climate information about asparagus and other water-intensive agro-exports into water governance decision making. The second involves the need to recognise, across all levels of governance, the vulnerability of the agriculture and the water resources sectors to changing climate regimes. Ultimately, the chapter calls for a drastic change in the way agricultural production and export management are evaluated by incorporating the long-term implications of climate change. It provides evidence for the always politically difficult argument that it is necessary for countries to invest in the long term rather than focus on building economic successes now that will compromise the ability of future generations to prosper.

The final chapter in the first section of the book, Chapter 7, turns to the Australian Water Market, which emerged out of clear political motivations and environmental concerns. In this water market system, the economic value for water is primarily a supply–demand proposition and weather is the dominant short-term driver of water prices. With its progressive adaptation to market issues and reducing trade impediments, increasing accessibility to credible market information, and widespread familiarity with the water-trading process, participation in the market keeps increasing. The Australian Water Market is hence considered a successful market-based solution to shifting agricultural practices in response to increased variability in water availability. It is thus the perfect example through which to observe the dynamic making and remaking of the complex interactions between food, energy and water. A key take-home message from Chapter 7 is that, despite the best intention for it to do so, the Australian Water Market does not function according to rational market-based decision making. Beneath the discourse of economic rationality lie limitations linked to questions of materiality, access to capital and technical expertise, farmers' diverging agencies linked to socio-cultural structures, and government interference. A more accurate picture emerges by situating water markets within wider social and ecological structures, in which individual water traders fit into evolved arrangements characterised by complex interactivity with different actors with different agency. The chapter concludes by saying that the water market in Australia is not a one-model-fits-all solution. Different contexts mean that different actors will have varying abilities to benefit from the market. Policy makers interested in implementing a water-market based on the Australian model should therefore map out water-trading behaviour as a nuanced and multifaceted system and evaluate the actors less likely to benefit from the system. This will allow policy makers to adapt the market's framework to protect those "weaker" shareholders to benefit rural communities and the agricultural sector.

Broader lessons from natural resource governance

The second half of the book focuses less on specific systems and more on broader lessons from natural resource governance that help provide a more reflective understanding of how to make the most effective decisions under situations of institutional/situational/organisational complexity. The first four chapters look more specifically at natural resource governance through the broader, more resource-inclusive and sometimes less geographically specific lens, of the food–energy–water systems, whilst the remaining two chapters take a look at more technological innovations that have the potential to aid FEW governance, but that must be implemented with due caution.

Chapter 8 uses a very context-specific case of sugar cane governance in Kwa-Zulu Natal, South Africa, to bring insights into the role of human agency, social learning, and collaboration in enabling the emergence of novel governance arrangements to support transitions to sustainable farming practices. Through the lenses of institutional bricolage and expansive social learning, the study demonstrates the importance of collaboration across institutions and scales, along with the building of novel institutional platforms, to address the complex sustainability challenges facing agricultural industries. The management and decision-making tool called SUSFARMS®, and an enabling collaborative governance platform called the SUSFARMS® 2018 Collaboration, provide a good example of new governance mechanisms that have arisen out of a combination of formal, and informal institutions across governance levels – referred to as institutional bricolage. The potential for this innovation to enable more diverse knowledge systems to be included as a way of measuring sustainability in agriculture could be an important extension. The chapter clearly finds that collaboration and institution building are enabled through facilitation of a trusted intermediary and thorough local actors who are ready to make the most of "windows of opportunity" along with social cohesion, commitment, and strong leadership at the local level.

Chapter 9 is a theoretical exercise to understand how corporate water risk is understood: it explores if a proposed theoretical framework based on transaction costs theory can also help logically explain or justify investments for corporate water return. The theoretical framework offers a five-factor (5FFs) corporate decision-making guide that it illustrates through a fictional-case of a food and beverage corporation. By using this framework, the chapter explores the ways the logics of traditional corporate governance theory can be extended to support decisions that could support corporate actions for environmental sustainability – especially for shared natural resources like water and energy. This work is important for reconceptualising nexus governance because it show how business incentives can be used to reconsider the roles and relationships between corporations, states, and communities as users of and investors in natural resources. It also illustrates a more forward thinking and agenda setting aspect to governance than some of our other chapters. It helps to illustrate how the ways in which future environmental governance is structured is an accessible and open field for innovations today.

Chapter 10 deviates slightly from a specific focus on FEW systems, but turns instead to discuss natural resource management strategies, focussing on marine protected areas. The links to managing food systems are evident, but, more importantly, the chapter provides insights into how interventions need to be designed, implemented, and managed to encourage the long-term support of affected communities if they are to be successful. Natural resource management strategies have a wide variety of costs and benefits (including non-monetary impacts) on different individuals within communities and these need to be assessed to ensure their support and success in the long term. There is an imperative for those implementing interventions in these systems to assess, communicate, and to be honest about the repercussions of interventions for different individuals and groups across short- and long-term time scales. The cost-reduction benefits that NRM strategies can provide in the longer term need to be emphasised rather than focussing on the potential specific individual benefits that can be harnessed by certain individuals from an intervention. More importantly, this chapter raises the question of the perspective of the management of FEW by asking whether the rationale for better managing natural resources should be purely based in economics or more simply characterised as a moral duty towards future generations.

The following chapter takes us through an exploration of the relevance of indigenous knowledge systems for meeting sustainability challenges. Using an example from the Sahtu region of the Northwest Territories in the Canadian Arctic, Chapter 11 once again explores the role of climate change in creating urgency for good environmental governance, but this time by exploring different worldviews and perceptions. Indigenous Arctic communities are facing increasing geopolitical challenges linked to the commodification of their traditional lands. This chapter warns that the choice to extract natural resources from a society's familiar territory or traditional lands can only be sustainable if it is made in accordance with the local ontologies, i.e how community members' understand and experience the world, what they perceive as being reality, and how they make themselves a part of it. Simply put, 'natural resources' are not experienced in the same way by someone who believes that they are alive and inhabited by spirits and someone who only sees a utility or exchange value in them. It is therefore imperative to recognise these differences when implementing sustainable resource governance strategies and to understand how these strategies are mediated into people's everyday lived realities. The chapter concludes that Sustainable Social Environmental Enterprise (SSEE) could offer a route that is ontologically suitable for energy and commodity-intensive development strategies to be successful in Indigenous communities. Through SSEEs, these strategies could not only preserve the integrity of these communities' critical natural capital, but also have a geopolitically emancipatory function by serving as decolonial means of empowerment.

Chapter 12 begins the shift into more technological discussions by looking at the role of Information and Communication Technologies (ICTs) and FEW governance. It finds that the expansion of ICTs in the fields of development and environmental sustainability has created new opportunities for governing through "small infrastructures" that support information gathering and exchange, which in turn are changing relationships between different governance actors (namely, civil society, state, and business actors) in FEW-related sectors. New ICTs' capabilities for dealing with large datasets can play a central role in helping governance actors become more responsive and better able to navigate changing demands and cross-sectoral complexities demanded in FEW governance. There is a potential role of citizen-orientated involvement with ICTs as part of data collection and dissemination, which can make a difference in terms of strengthening governance rescaling and for enabling connections across scales to become more visible. It is also important not to limit the governance capabilities to ICTs that are typically associated with 'big data'-gathering interventions, but instead to explore how the capabilities may form a basis for smarter forms of public dialogue around ICTs that involve stronger engagement of concerned citizens, which can open up new directions for achieving broader societal goals (such as the Sustainble Development Goals) while empowering actors that are not always part of the policy debate. The integration of ICTs with governance programs requires new principles, policies and practices with regards to data privacy and protection of individuals in order to ensure trust between users and decision makers.

Finally, the book ends with Chapter 13 and an insight into the complexities, urgencies, and opportunities for improving the governance of the mining sector and especially the relationship between mining and water. This chapter begins by detailing the specificity and technicality of many of the interactions between the mining sector and water. It also shows how, through this interaction with water, the mining industry has further impacts on the use and production of energy and food. While the governance and assessment of these risks are still socially mediated, since the 1990s a number of cross-scale and multi-stakeholder governance mechanisms have been established. However, as we have seen in several other chapters these approaches are no longer sufficient for environmental governance of challenges facing rapid changes from climate change and economic forces. The chapter offers two potential new tools that could be adopted to augment the environmental governance of the mining sector as it is being pushed to operate in more remote and precarious places. These are first, digital technologies that allow for greater connection between affected communities and spatial imagining that creates opportunities for monitoring and oversight. Second, it highlights the potential for the environmental infrastructures that mines must build to make such locations operable to also serve as means of assisting adaptation to climate change in areas unlikely to otherwise source such investments. In doing so, this chapter taps into the themes of ICT potential discussed in Chapter 12 and of corporate infrastructure investment theorised in Chapter 9. It concludes optimistically, but with a warning, that reflects many of the lessons throughout this collection of case studies. The chapter emphasises that although the resource challenges are urgent and have commonalities in their governance demands, such as the appropriate scale of governance and difficulty in maintaining equitable and participatory governance, there are many current successes that need to be preserved and enhanced by new approaches to resource governance. In essence, moving forward with emerging resource governance strategies is, as this book illustrates, a process of also taking stock of the past, present, and future of resource governance.

Some cross-cutting themes

Before drawing to a close we thought it would be helpful to discuss some of the cross-cutting themes in this book. To start we often saw the transition into a more sustainable system requires the destabilisation of a dominant system or way of doing things. The case of South Africa's energy system in Chapter 4 requires a careful combination of innovation and preparation and institutional enabling. Chapter 8's reference to institutional bricolage and social learning provides some insight into how to navigate this in practice. Although by no means perfect, learning from these more local examples has broader implications. The same kind of institutional bricolage to shift a dominant system is required in the Chinese energy sector if it is to become more sustainable, especially in terms of its water usage (Chapter 5). The same case goes for the Peruvian export agriculture sector in its use of water that is becoming a limiting factor (Chapter 6). Lessons from successful governance innovations in India in Chapter 2 show that it is possible – across levels of governance – to come up with solution to complex challenges if one is cognisant of the systemic and interconnected nature of food, energy, and water.

From an academic perspective, what is interesting is that diverse disciplinary lenses can provide different insight into these complex problems. Chapter 9 provides a hypothetical, yet important, innovation around corporate water return that is also trying to navigate the challenge of the business-as-usual path dependency. Chapter 4 makes use of a historical perspective to point out where governance choices in the past have led with the hopes that this understanding allows for better decisions in the future. Further, Chapter 3 and Chapter 11 used socio-cultural perspectives that remind us of the significance of culture, traditional values, and place-based ways of life that are integral to the diversity of our societies and future abilities to solve problems creatively. And yet, these socio-cultural-led accounts perhaps also stick with us so that in reading about the technical challenges of governing water and energy in China, we are attuned to how even these decisions are mediated by culture through the politics and institutions of a country. Remarkably, through these chapters, scientists, anthropologists, economists, geographers, and historians present a common understanding of governance obstacles and refreshingly different takes on ways to think about solutions.

Climate change is also a definite cross-cutting theme in the book and indeed it is often included in nexus discussions because of its close associations with the food, energy, and water systems (see Chapter 1). Climate change is an important discussion point, not only because of its pressing nature, but also because it requires of FEW governance systems both mitigation and adaptation. It brings with it all the complex problems associated with sustainability – of inter and intra-generational equity, fairness, and justice that is imperative for FEW governance to tackle if it is to be even remotely successful. Chapters 5, 6, 11, 12, and 13 all refer explicitly to climate change, but it could easily have appeared as an important consideration in each and every single chapter.

The important role of local communities and a recognition of their ways of being and understanding is also a common theme that runs through the book, especially highlighted in Chapters 3,10, 11, and tangentially in Chapter 8. Moving away from the rational agent that is critiqued in Chapter 7, recognising the need for understanding different worldviews, and incorporating the important contributions that these make to governance, is imperative in overcoming the natural resource issues that we face today.

Following-up on the governance trends highlighted in the introduction we think it has been perhaps all too obvious how essential questions over the appropriate and effective scale of governance is for nearly all our cases. Chapter 6 and 12 explicitly discuss the complications and possibilities of multi-scalar and cross-sector governance, but we challenge you to find a chapter where it is not relevant! Further, we have tried to present a diversity of approaches to the relationship between the public and private sector in environmental governance, from the corporate-leading perspective of Chapter 7 to state-led approaches of Chapters 4 and 5, and we show that business itself is rarely easily reconciled as a public or private actor. In Chapter 13 we illustrate the imperative to engage and find opportunities for positive governance even with industries like mining that have long been in cross-hairs for their impacts on the environment.

Finally, the role of technology is always an important one and is something that needs to be carefully considered in governance systems. Chapters 12 and 13 lay out a series of pros and cons around the use of technologies for improving FEW governance. Whilst there will never be a silver bullet, adapting technology to meet the needs of local contexts (bearing in mind the ethical implications of working with local, often vulnerable, communities), is critical and will play an increasingly important role in future governance systems.

Final thoughts

We have sought to explore the question of the governance of FEW by providing examples not only of various approaches, but also of various locations, and various times. Each country, land, territory have natural resources issues specific to their landscape, culture, history, economy, level of development, and politics, highlighting the vast diversity of means to manage FEW across the planet. Moreover, policies and actions aimed at managing FEW are highly fluctuating, constantly being shaped and reshaped by external forces linked to changing international and national politics, economic growth, or available technology. These varied examples have shown us that the management of FEW is oftentimes a question of reinventing, adapting, and building upon existing practices embedded in specific and unique socio-cultural and political contexts as well as "hard" structures (such as pre-existing roads, plants, etc).

We have not sought to enter into deeper discussions in this conclusion, but rather to provide a simple summary of lessons that can be learnt from each chapter. As we have highlighted before (and it has been stressed in multiple chapters), there are a *multitude* of solutions and best practices to the question of the governance of FEW and we therefore do not intend to – nor are capable of – providing a passe-partout for practitioners. By looking back at each chapter and at what key learnings we can get from them, we seek to provide the reader with the necessary varied knowledge tools for their own exploration of a solution to their specific FEW governance problems and location, with the knowledge that the best solution must often be flexible and adaptable to future changes, or bricolage. We hope that, among this collection of cases, you find, as we have as editors, perspectives that you found refreshing, interesting, and at times challenging. Most of all we leave this work positive about the global network of ideas and committed individuals who are will continue to work to improve natural resource governance and all of our sustainable futures.

Index

For Product Safety Concerns and Information please contact our EU representative GPSR@taylorandfrancis.com Taylor & Francis Verlag GmbH, Kaufingerstraße 24, 80331 München, Germany